Macromolecular Symposia

Symposium Editors: K. Horie, A. Abe

Editor: I. Meisel
Senior Associate Editor: S. Spiegel
Associate Editor: A. Carrick
Assistant Editor: M. Staffilani

Executive Advisory Board: M. Antonietti, M. Ballauff,
S. Kobayashi, K. Kremer, T.P. Lodge,
H.E.H. Meijer, R. Mülhaupt,
A.D. Schlüter, H.W. Spiess, G. Wegner

201

pp. 1–326

September 2003

Macromolecular Symposia publishes lectures given at international symposia and is issued irregularly, with normally 14 volumes published per year. For each symposium volume, an Editor is appointed. The articles are peer-reviewed. The journal is produced by photo-offset lithography directly from the authors' typescripts.
Further information for authors can be obtained from:
Editorial office "Macromolecular Symposia"
Wiley-VCH Verlag GmbH & Co. KGaA,
Boschstrasse 12, 69469 Weinheim,
Germany
Tel. +49 (0) 62 01/6 06-2 38 or -5 81; Fax +49 (0) 62 01/6 06-3 09 or 5 10;
E-mail: macro-symp@wiley-vch.de
http://www.ms-journal.de
Suggestions or proposals for conferences or symposia to be covered in this series should also be sent to the Editorial office at the address above.

Macromolecular Symposia:
Annual subscription rates 2003 (print only or online only)*
Germany, Austria € 1318; Switzerland SFr 2168; other Europe € 1318; outside Europe US $ 1568.
Macromolecular Package, including Macromolecular Chemistry & Physics (18 issues), Macromolecular Bioscience (12 issues), Macromolecular Rapid Communications (18 issues), Macromolecular Theory & Simulations (9 issues) is also available. Details on request.
* For a 5 % premium in addition to **Print Only** or **Online Only**, Institutions can also choose both print and online access.
Packages including Macromolecular Symposia and Macromolecular Materials & Engineering are also available. Details on request.
Single issues and back copies are available. Please inquire for prices.

Orders may be placed through your bookseller or directly at the publishers:
WILEY-VCH Verlag GmbH & Co. KGaA, P. O. Box 10 11 61, 69451 Weinheim, Germany, Tel. +49 (0) 62 01/6 06-400, Fax +49 (0) 62 01/60 61 84. E-mail: service@wiley-vch.de

Macromolecular Symposia (ISSN 1022-1360) is published with 14 volumes per year by WILEY-VCH Verlag GmbH & Co. KGaA, P. O. Box 10 11 61, 69451 Weinheim, Germany. Air freight and mailing in the USA by Publications Expediting Inc., 200 Meacham Ave., Elmont, NY 11003, USA. Application to mail at Periodicals Postage rate is paid at Jamaica, NY 11431, USA. US POSTMASTER please send address changes to: Macromolecular Symposia, c/o Wiley-VCH, III River Street, Hoboken, NJ 07030, USA.

International Union of Pure and Applied Chemistry

Macromolecular Division

Invited lectures from the
IUPAC Polymer Conference on the
Mission and Challenges of Polymer Science and Technology (IUPAC PC2002)

held in
Kyoto, Japan
December 2–5, 2002

organized by
Society of Polymer Science, Japan, The Science Council of Japan

Symposium Editors
Kazuyuki Horie, Tokyo University of A & T
horiek@cc.tuat.ac.jp

Akihiro Abe, Tokyo Institute of Polytechnics
aabe@chem.t-kougei.ac.jp

International Advisory Committee
Akihiro Abe (Japan); Masuo Aizawa (Japan); P. de Gennes (France);
J. M. J. Fréchet (USA); R. G. Gilbert (UK); J. -I. Jin (Korea);
Toshiro Masuda (Japan); Ryoji Noyori (Japan); Hideki Shirakawa (Japan);
R. F. T. Stepto (UK); Akiyoshi Wada (Japan); Fosong Wang (China);
G. Wegner (Germany); W. J. Work (USA)

Organizing Committee
Seiichi Nakahama (Chairman); Kazuyuki Horie (Vice chairman); Akihiro Abe;
Takuzo Aida; Isao Ando; Takeshi Endo; Akira Hasegawa; Koichi Hatada;
Shohei Inoue; Yasuto Iwai; Tisato Kajiyama; Mureo Kaku; Masahiro Kakugo;
Mikiharu Kamachi; Toyoki Kunitake; Toshiro Masuda; Makoto Misonou;
Katsuhiko Nakamae; Toshio Nishi; Norikazu Nishino; Takuhei Nose;
Yoshio Okahata; Yoshio Okamoto; Hitoshi Ohtaki; Hideki Sakurai; Kohei Sanui;
Shigetoshi Seta; Ken-ichi Shida; Yoichi Shimokawa; Naohiro Soga; Tamio Ueno;
Akiyoshi Wada; Akio Yamamoto; Hitoshi Yamaoka

Contents of Macromolecular Symposia 201

IUPAC PC2002
Kyoto (Japan), 2002

Preface
K. Horie, A. Abe

Session A: Polymer Concepts in Chemistry, Physics, and Biology

Session B: Frontiers of Polymer Science

Session C: Advanced and Emerging Polymer Technologies

* Asterisks indicate the name(s) of the author(s) to whom inquiries should be
addressed.

Author Index

Preface

The IUPAC Polymer Conference on the Mission and Challenge of Polymer Science and Technology (IUPAC PC2002) was held as the first strategic symposium of the IUPAC Macromolecular Division (MMD) on December 2–5, 2002 at the Kyoto International Conference Hall, in order to determine the present status of polymer science and technology and clarify their mission and challenges in the future. The conference was planned as a part of the activities celebrating the 50th anniversary of the Society of Polymer Science, Japan, and cosponsored by IUPAC Macromolecular Division (MMD) and the Science Council of Japan.

The scientific program consisted of plenary lectures, six scientific oral and poster sessions, and a panel discussion summarizing the conference and discussing the role and activities of IUPAC MMD in the world polymer community. Six themes arranged from a strategic viewpoint for the conference were:
Polymer Concepts in Chemistry, Physics, and Biology
Frontiers of Polymer Science
Advanced and Emerging Polymer Technologies
State of the Art in "Bio-Polymers"
Polymers and the Environment
Commodity Polymers and the World Economy

The conference was attended by 762 participants from 29 countries; 4 plenary and 41 invited lectures were presented. This volume contains invited lectures, which are published in the same order as they were presented in the original program of the conference. Plenary lectures and a keynote article of the conference will be published in Pure and Applied Chemistry, IUPAC.

K. Horie
A. Abe

Macromol. Symp. **2003**, *201*, 1—9

Shape Persistence as a Concept in the Design of Macromolecular Architectures

Gerhard Wegner

Max-Planck-Institute for Polymer Research, 55128 Mainz, Germany
Email: wegner@mpip-mainz.mpg.de

Summary: Shape persistent macromolecules are objects defined by a time independent overall shape. In most cases such macromolecular objects are characterized by an exterior surface structure and an internal architecture giving coherence to the shape and topology. Shape persistent macromolecules are essential to the development of a polymer based nanotechnology and serve as constitutive units of larger scale architectures either via self-organization or via processes in which they are assembled to give constructs of nanoscale defined patterns.

Keywords: nanocomposites, nanotechnology, self-organization, stiffness, supramolecular structures

Introduction

Most of the industrially important polymer materials are composed of flexible macromolecules. Their solutions, melts, glassy and even semi-crystalline forms are characterized by the random flight nature of the chain trajectory. Important properties like rubber elasticity and the viscoelastic behavior of their melts find explanation in the most probable random distribution of the bond vectors along the chain trajectory and the time/temperature dependent changes and fluctuations of that distribution when external forces are exerted.

Contrary to that biomacromolecules derive their biological function from shape persistence that is a non-fluctuating positioning of chain segments within the volume filled by segments of the individual macromolecules; chain segments far apart from each other along the trajectory have a fixed spatial relationship ("topochemistry"). Although long recognized as an important principle in the world of biomacromolecules, "shape persistence" was only relatively recently accepted and put forward as a concept in the design of synthetic macromolecules and as an objective towards large scale architectures beyond the individual macromolecules.[1-5]

It needs to be mentioned, however, that H. Staudinger already in the late 20^s of the last century has put forward the idea that rod-like and, therefore, shape-persistent macromolecules are at the origin of the strength of certain fibres, in particular poly (oxymethylene) and

DOI: 10.1002/masy.200351101

cellulose.[6] This opinion was not shared by many of his contemporaries and was in fact not based on good experimental evidence but bears visionary potential.

Definition of Shape-persistence

Shape-persistent macromolecules are defined as objects characterized by a time independent overall shape. They have an exterior – that is surface – structure and an internal architecture which gives coherence and stability to the shape but also contributes to the function of the macromolecular object. Restricted fluctuations of the internal or external (surface) structure may be possible and necessary to provide functional properties as long as the overall shape persists.

Figure 1 describes the salient features by which linear flexible and rod-like chains decorated by a dense population of flexible side-groups – so called hairy-rod macromolecules (HRM) – differ in terms of physical concepts.[2, 7, 8]

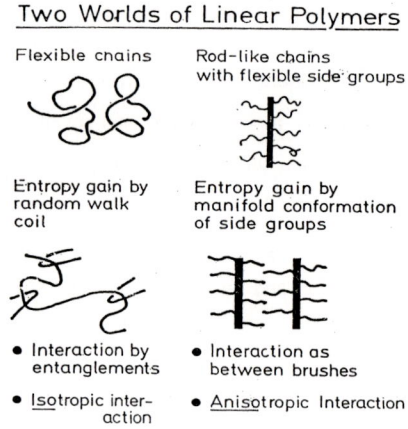

Fig. 1. Comparison between flexible and "hairy-rod macromolecules" (HRM). The latter are prototypes of shape-persistent linear polymers.

The meaning of persistence is exemplified by Figure 2 comparing a (flexible) Kuhn-chain with a worm-like Porod-chain.

The salient magnitude, the persistence length P, is a statistical description of the average shape of the worm-like chain. P is related to the ratio of the work needed to achieve a certain bending of the trajectory and the average thermal energy acting on the chain as given by T. *Odijk*:[9] $P = \pi d^4 E/(64 k_B T)$ and $K = \pi d^2 E/4$ where d is the diameter of the rod-like chain having a Youngs-modulus per unit length of E at temperature T, K is the elastic module of the whole worm-like chain.

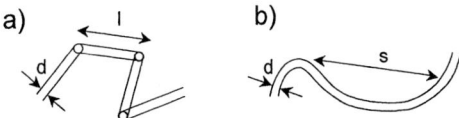

Fig. 2. A flexible Kuhn-chain (a) characterized by random fluctuation of the directors of the statistical elements of the chain of length l and thickness d; A worm-like (Porod-Kratky) chain (b) characterized by continuous variation of the director along the contour.

We note in passing that *Odijk's* formulation tells us how to improve the stiffness of the individual chain by understanding that the diameter of the segment is the most important parameter. This may be an important consideration for further work aiming for chemistry and physics with individual macromolecules at the level of nanotechnology.

Shape persistence is achieved by direct synthesis of a bond pattern of the constitutive elements of the chain which prevents randomisation of the bond directors. Hairy-rod macromolecules (HRM) are a good example. Here, flexible side-chains act like a polymer bound solvent shell, disguise the backbone elements and help to achieve processability.

Figures 1 and 6 summarize what has been learned in the past 15 years on the relation between chemical structure and persistence in linear HRM. Substituted poly (p-phenylene)s[10] and cellulose derivates[11] are key examples where persistence is achieved by fixing the conformational angle between constitutive units.

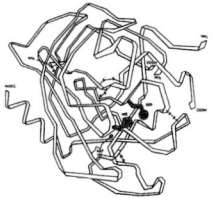

Fig. 3. Example of a shape persistent fold-structure of a globular protein (chymotrypsin) exhibiting fixed spatial relationships between segments far apart on the chain trajectory.

Phthalocyaninatopoly(siloxane)s have attracted interest as examples in which the very dense packing of the constitutive units prevents random fluctuations of the chain trajectory. P is therefore as large as one micrometer.[12] Similarly, helical polypeptides, e.g.

copolyglutamates[13] have very large persistence length. Shape persistence envisages – in fact – any defined shape of an object. The synthesis can be directed such that severe steric hindrance between the constitutive units enforces a particular shape: this is the case for dendrimers of high generation number, dendronized linear macromolecules and the like. A further approach is folding of linear flexible macromolecules into a secondary or tertiary structure then characterized by rigidity. The best known example is the folding of proteins, that is linear biogenic polypeptides into three-dimensionally well defined globular structures for which Figure 4 gives a schematic outline.

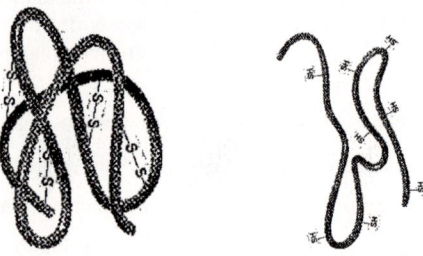

Fig. 4. Folding and unfolding of linear (bio) macromolecules: How to achieve a precise and non-random fold pattern in synthetic macromolecules is problem and challenge.

But also in the field of DNAs and RNAs we find many examples of regularly folded structures. The processes of folding and unfolding remain essentially a puzzle. So far no synthetic analogue exists in which a linear (co-)polymer has been demonstrated to undergo a regular random-coil to specific globule transition as indicated by Figure 4.

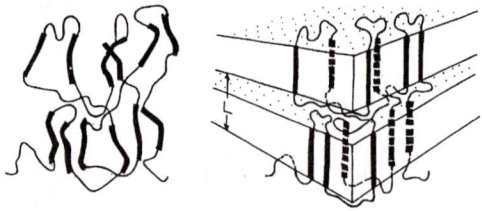

Fig. 5. The solidification model of polymer crystallization as an example how structure in the globular macromolecule relates to supramolecular texture in semicrystaline polymers.[4]

It remains a major challenge to macromolecular chemistry to design polymers which would undergo such transitions and would yield identical fold structures every time the transition is made. The underlying theory on inter- and intramolecular interactions and the molecular dynamics controlling the trajectory of folding are not sufficiently developed to predict progress in the near future. More theory is clearly needed. And yet, it needs to be reminded, that folding mechanisms and consequenses of folding processes are of outmost importance to the field of the industrially important polymers.

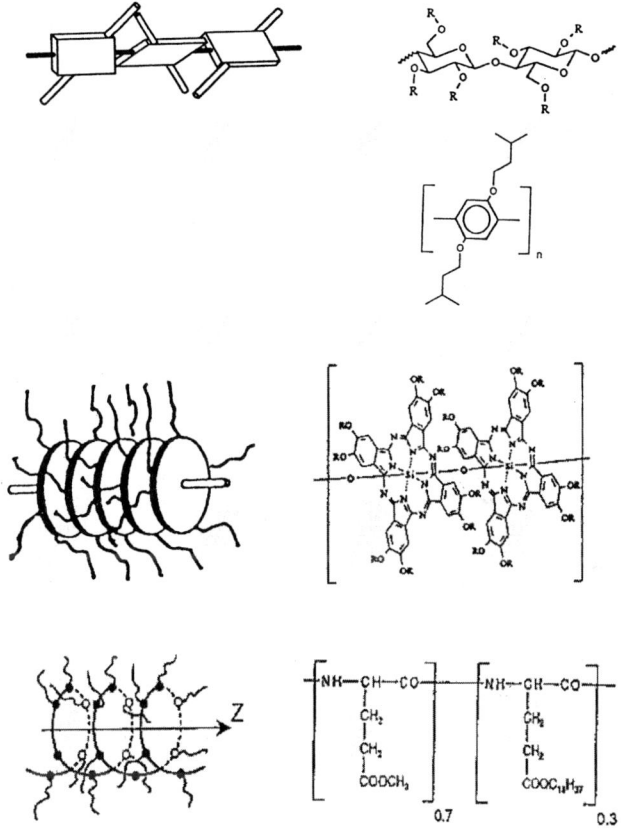

Fig. 6. Structure of HRMs obtained by suppressing bond fluctuations; from top to bottom: Fixation of the connective bond conformational angle; steric hindrance of bond fluctuations along director axis; stabilization of a secondary (helical) structure by weak bonds (H-bridges) along trajectory.[16]

Even here, there remain many open questions as to what precisely determines the fold-length and – thereby – the degree of crystallinity in ordinary semicrystalline polymers and fibres manufactured from such polymers. The "solidification-model"[14] as sketched in Figure 5 has been widely accepted, but its chemical consequences in terms of fixing crystallinity by copolymer structure has not been properly exploited.

Assemblies and Constructs

Although it is very important to understand the chemistry and physics of individual macromolecular objects, the final goal is, nevertheless, to use them as building blocks for large scale "molecular devices" or materials. It is important to differentiate between "macromolecular architecture" and "supramolecular assembly". The former refers to the topology of the individual macromolecule and comprises all stable fold-structures where as the latter refers to the interaction and topology of two or many of the individual objects. Depending on methods and conditions macromolecular objects will undergo mutual interactions and reactions leading to new structures composed of the constituent objects in a process sometimes called "self-organization". Examples for supramolecular assemblies or – what is used as a synonym – supramolecular architectures of HRM are shown in Figure 7.

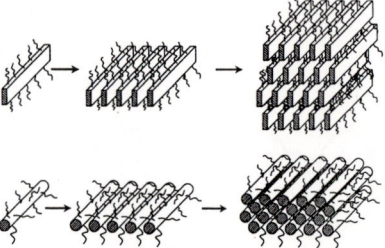

Fig. 7. Forms of supramolecular architectures of HRM.

The relation between the shape of the constituent HRM and the symmetry of packing is self-evident. In this context the word "self-organization" needs to be used with caution. In most cases structures described in literature are merely one out of many phases which can be obtained from the same constitutive units depending on the phase diagram and the parameters described therein. In other words, it remains to be clarified what the precise set of parameters (temperature, solvent, pressure, further components) are under which a specific supramolecular structure can be obtained. The hope[15] that one can make "programmed" (macro)molecules to achieve "programmed" functional architectures remains somewhat elusive on closer inspection.

More interesting are so-called constructs, that is super-structures to be obtained by processing (macro) molecules on the nanoscale level. The methods used for this purpose could be called "synthesis of supra-molecular architectures". The Langmuir-Blodgett assembly of HRM serves as a model case.[11, 16]

Figure 8 shows the principle of the process by which layered assemblies of HRM can be constructed with a deliberately chosen sequence of layers differing in both chemistry and the relative orientation.[17]

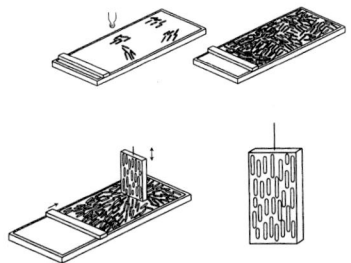

Fig. 8. Processing of HRM by the Langmuir-Blodgett Method to obtain layered assemblies.[2,16-18]

An actual example of a model construct which has been "synthesized" accordingly from two different cellulose derivates, namely isopentylcellulose and pentenylcellulose, is shown in Figure 9 together with the X-ray evidence (grazing incidence diffractograms) which proves the success in achieving a specific pattern of the layer sequence.[11]

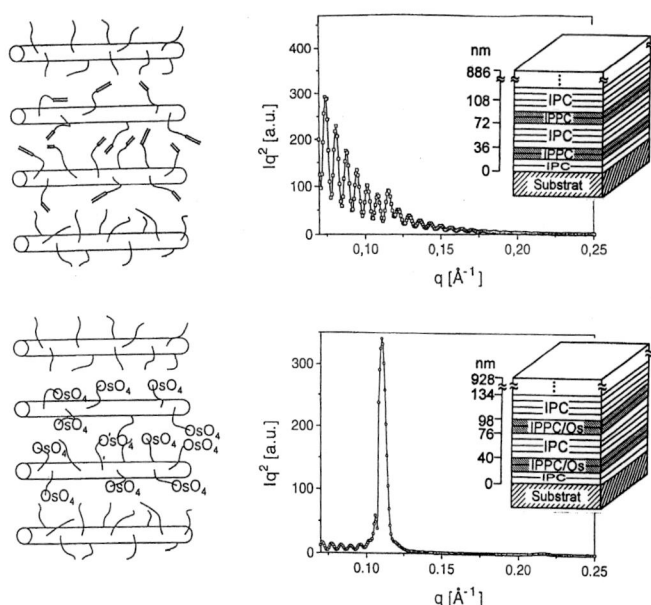

Fig. 9. Example of a construct obtained by processing cellulose based HRM via the LB-Method (above) and subsequent topochemical modification.[11]

The construct obtained was treated with osmium tetroxide which reacted topochemically with only those layers containing double bonds. As a consequence a new architecture was obtained exhibiting the expected diffractogram. Many interesting architectures can be obtained by this combination of molecular design (Hairy-Rod-Macromolecule type) and processing on the nanoscale making use of the LB-Process.

Examples are a construct having selective membrane properties because of defined spaces length between the backbones of the HRM[11, 13] and a device the function of which is entirely based on the oriented deposition of a precise number of monolayers between two electrodes making use of the LB-technique. This device works as a light emitting diode, the light being polarized along the molecular axis of the constitutive HRM which is a poly (p-phenylene) derivate.[18]

Concluding Remarks

It is worthwile noting that "order", "precision" and "perfection" need to be defined with regard to length and time scales. Again, HRM may serve as an example. The side chains attached to the backbone elements are liquid-like and form a disordered shell around the main chain. Nevertheless, a rather perfect layered assembly is constructed from such objects where "perfection" relates to the sequence of and the distance between indivual layers. However, within a given layer, the correltion between adjacent HRM is of nematic Type[12] and there exists in-plane disorder characterized by disclinations on the length scale of 100-1000 nm.

This is not unusual and is seen more or less well expressed in all forms of supramolecular architectures and even globular proteins can only function in their native state because of segmental dynamics which allows transport of substrate, water and salt to and from the reactive sites.

Thus, the understanding and "design" of molecular dynamics as an important feature in supramolecules will also be the key element in the design of artificial "motors", that is functional nano-scale entities which could serve as transporters for molecules in synthetic molecular scale factories. These few hints may suffice to show how shape persistent macromolecules are essential objects in the context of nanotechnology. They have considerable scope in such fields as microelectronics, cosmetics, pharmaceutical technology and biomedicine. The line of thought leading from the concept of design and engineering in macromolecular dimensions to application for advanced technologies combines synthesis of macromolecular objects with practical concepts of processes which allow to handle and organize macromolecular objects on the nano-scale. The technologies addressed comprise among others sensors and actuators, catalysis, membrane separation, controlled transport and energy conversion.[19]

[1] J. M. Rodriguez-Parada, R. Duran, G. Wegner, *Macromolecules* **1989**, *22*, 2507.

[2] G. Wegner, *Ber. Bunsenges, Phys.Chem.* **1991**, *95*,1326.

[3] *Interdisciplinary Macromolecular Science and Engineering*, A Workshop Co-Sponsored by NSF and DoE, S. I. Stupp, Ed., U. of Illinois Printing Services **1998**.

[4] V. Percek, W.-J. Cho, G. Ungar, D. J. P. Yeardley, *J. Amer. Chem. Soc.* **2001**, *123*, 1302.

[5] B. J. de Gans, S. Wiegand, E. Zubarev, S. I. Stupp, *J. Phys. Chem. B.* **2002**, *106*, 9730.

[6] H. Staudinger, *"Die Hochmolekularen Organischen Verbindungen"* Neudruck, Springer Verl. Berlin 1960, p.241.

[7] S. Vanhee, R. Rulkhens, U. Lehmann, C. Rosenauer, M. Schulze, W. Kohler, G. Wegner, *Macromolecules* **1996**, *29*, 5136.

[8] M. Mierzwa, G. Floudas, M. Neidhöfer, R. Graf, H. W. Spiess, W. H. Meyer, G. Wegner, *J. Chem. Phys.* **2002**, *117*, 6289.

[9] T. Odijk, *Macromolecules* **1995**, *28*, 7016.

[10] A. D. Schlüter, G. Wegner, *Acta Polymer* **1993**, *44*, 59.

[11] M. Schulze, M. Seufert, C. Fakirov, H. Tebbe, V. Buchholz, G. Wegner, ACS-Symp. Series 688 **1996** Eds. T J. Heinze and W. G. Glasser, p. 306f.

[12] J. Wu, G. Lieser, G. Wegner, *Adv. Mater.* **1996**, *8*, 151.

[13] S. Iida, M. Schaub, M. Schulze, G. Wegner, *Adv. Mater.* **1993**, *5*, 564.

[14] M. Stamm, E. W. Fischer, M. Dettenmaier, P. Convert, *Faraday Discuss. Chem. Soc.* **1979**, *68*, 263

[15] J. M. Lehn, Angewandte Chem. **1988**, *100*, 92.

[16] G. Wegner, Mol. Cryst. Liqu. Cryst. **1993**, *235*, 1.

[17] R. Silerova (Back), L. Kalvoda, D. Neher, A. Ferencz, J. Wu, G. Wegner, Chem. Mater. **1998**, *10*, 8.

[18] V. Cimrova, M. Remmers, D. Neher, G. Wegner, Adv. Mater. **1996**, *8*, 146

[19] G. Wegner, Acta mater, **2000**, *48*, 253

Dendritic Macromolecules at the Interface of Nanoscience and Nanotechnology

Jean M.J. Fréchet

Department of Chemistry, University of California, Berkeley, CA 94720-1460
and Division of Materials Science, Lawrence Berkeley National Laboratory, USA
E-mail: www.frechet.com

Summary: As a result of their unique architecture and structural as well as functional versatility, dendrimers have generated considerable interest in numerous areas of the physical sciences, engineering, as well as the biological sciences. Both their size - in the 1-10 nm range – and their globular shape resemble those of many proteins suggesting a host of biomimetic and nanotechnological applications. This brief highlight describes some of our recent work with nascent applications of dendrimers as unimolecular nanoreactors, as nanoscale antennae for energy harvesting and transduction, and as nanosized carriers for diagnostic or therapeutic applications. While implementation of some of these applications may still be distant, the impatient critic might remember that new markets are not created overnight as demonstrated by the slow commercial acceptance of many promising molecules and technologies with development frequently extending decades after their initial discovery.

Keywords: catalysis, dendrimers, light-harvesting, nanotechnology, therapeutics

1 Shape, Flexibility and Molecular Ordering of Dendrimers

Dendrimers are size monodisperse, globular macromolecules in which all bonds emerge radially from a central focal point or core with a regular branching pattern and with repeat units that each constitute a branch point.[1-3] Numerous macromolecules possess some of the features of dendrimers, including high degree of branching and multiplicity of chain-ends and reactive sites. These include hyperbranched polymers – both natural (e.g. polysaccharides) and synthetic – as well as hybrid dendritic-linear polymers, dendronized polymers, comb-burst polymers, etc.[2,3] Despite unsupported claims to the contrary, none of these dendritic macromolecules match the ultimate properties of dendrimers. Even with true dendrimers, properties of the dendritic state,[3] such as core encapsulation[4] and unusual intrinsic viscosity behavior in solution,[5] are only reached when globularity is achieved at a certain generation or size threshold. The flexibility of dendrimers depend greatly on the generation or number of layers of repeat units, the choice of monomer repeat unit, including the number of branches, the number and type of bonds between branches, as well as the degrees of freedom available to interbranch bonds. In general, low generation dendrimers are quite flexible while high generation dendrimers may become rigid.[7]

A controversy concerning the shape of dendrimers and the placement of their chain ends either at the "periphery" of the globular macromolecule or back folded within its building blocks had arisen in recent years. This was fueled by a variety of calculations and measurements suggesting either back folding of the chain ends or supporting their peripheral arrangement[6]. In reality, free-energy rules and dendrimers react to their environment (i.e. adjust their shape and the placement of their functional groups) in order to minimize their free energy.[7] Thus a dendrimer might adopt a fully extended conformation and reach a volumetric maximum and an almost spherical shape in a good solvent while collapsing to a more compact volume in another or in the absence of a solvent with its final shape, chain-end placement, and other structural features determined by its intrinsic flexibility, the interactions of its various components (core, internal repeat units, chain-ends) and the interactions with its near neighbors or a surface. Similarly, the location of chain-ends (peripheral or back folded) in all but the most rigid structures is dictated by free energy.[7] If the chains ends possess favorable interactions, such as H-bonding or π-stacking, with the inner building blocks, back folding may be expected to occur, a phenomenon that may be exacerbated or mitigated by solvent. If the absence of favorable enthalpic contributions, entropy considerations usually disfavor the mixing of chain ends with dissimilar inner building blocks.[8] The existence of a "cavity" within dendrimers has been another topic of controversy. The structure of a dendrimer bears some analogy to that of a micelle although the dendrimer is a static unimolecular covalent assembly while the micelle is a dynamic supramolecular assembly. Like micelles, larger dendrimers possess a volume and structural features that enable them to accommodate guest molecules,[9-10] particularly when enlarged by solvation with a good solvent. However, with the possible exception of very specialized structures such as shell-crosslinked dendrimers, dendrimers do not possess a permanent and rigid cavity. Small guests that can penetrate the volume of a dendrimers as a result of favorable enthalpic interactions may remain encapsulated following collapse of a solvated structure.

Encapsulation may become permanent as in Meijer's "dendritic box"[11] if the peripheral density of the dendritic structure is increased to rigidify the macromolecule while guest molecules are located within the extended volume of a dendrimers.

2 Designing Free-energy Driven Unimolecular Nanoreactors

Catalysis with highly branched macromolecules, primarily dendrimers, has been well documented in the literature.[12] Depending on the location of the catalytic site(s), dendritic catalysts with either multiple active sites located at chain-ends or constituting a catalytically active core have been described.[12] The first approach offers the advantage of relatively high catalyst loading, however, the dendritic backbone merely serves as a support and its favorable structural and functional features cannot be completely utilized. In contrast, the second approach enables encapsulation of the active site,[4] thereby offering protection from unfavorable deactivation mechanisms. Moreover, the placement of catalytic moieties at the core or the interior of a globular dendritic structure allows for the fine-tuning of overall molecular properties and catalytic activity by modification of the periphery and the interior environment, respectively. Early work by Moore and Suslick[13] explored dendritic Mn-porphyrins that exhibited improved stability in solution compared to the free metalloporphyrin as a result of encapsulation by the surrounding polymer. The dendritic catalyst also appeared to provide increased regioselectivity in simple epoxidation reactions using various dienes as substrates. The greater significance of this early study, however, was a clear confirmation of previous findings that small molecules could penetrate the congested exterior of the dendritic scaffold for reaction[4a] at the core. Since then, several groups have investigated the effect of the dendritic superstructure on the catalytic activity of encapsulated sites. In most cases few advantages could be associated with the dendritic character of the new catalytic moieties.[14,15] A critical reading of some recently published work suggests that mass transport within the dendritic shell may well be a significant factor and that product inhibition could be responsible for the relatively limited performance of several systems.[15]

In view of our reading of these previous findings, our approach to dendrimers catalysis has included a global design of not only the catalytic site itself, but also its environment to incorporate appropriate mass transport "molecular machinery". Thus the dendrimers should not only provide a reaction center, but it should also function, much like an enzyme, as a unimolecular nanometer size "reactor" transporting and concentrating substrate to the environment that promotes reaction, and removing product from the catalytic site as soon as it is produced. In recent work[16,17] we have demonstrated this concept and shown that by utilizing the chemical nature and structural features of the dendritic building blocks, tailored microenvironments can be generated that assist in transporting substrate and product, as well as in stabilizing crucial intermediates and transition states. The key concept makes use of contrasting polarity between the dendritic inner and outer environments coupled with a polarity difference occurring during the course of the catalyzed chemical transformation. The amphiphilic design leads to preferential accumulation of substrates and, in some instances, stabilization of transition states or intermediates in the interior, while the product is simultaneously expelled into the external medium thereby preventing inhibition. An added feature of this free energy-driven mass transport is the fast kinetics that result from the high local concentration of substrate that can be achieved at the site of catalysis within the dendrimer.

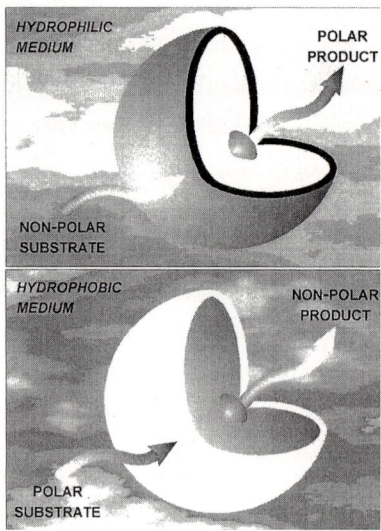

Fig. 1. Design of free-energy based catalytic nanoreactor.

With this key concept in mind, it is possible to delineate criteria pertinent to the general design of efficient polymeric catalysts. Irrespective of the catalytic site(s) to be incorporated into the polymer, the primary consideration for molecular design depends on the intrinsic polarity change associated with the specific transformation. Depending on the relative polarity of the product compared to the substrate, two types of catalysts can be envisioned (Figure 1). In the case of a polarity increase from substrate to product, the reaction will be performed in a more hydrophilic medium utilizing a regular micelle-like but unimolecular, shape-persistent dendrimer. This design is particularly attractive since many catalytic transformations, such as oxidations, involve polarity increases and catalysis in water is of great importance for the development of environmentally friendly chemical processes. In a similar manner, if polarity decreases during the course of the reaction, one would employ a reverse unimolecular micelle in a hydrophobic solvent.

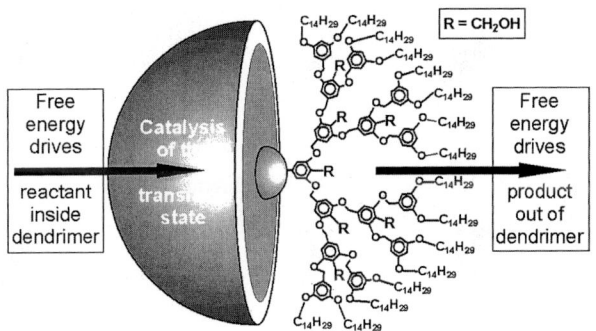

Fig. 2. Transition state catalyst for unimolecular elimination.

As proof of concept, a rather unsophisticated, reverse micelle-like dendrimer (Figure 2) was designed to catalyze an E1-type elimination process.[16] The catalyst consists of a hydrophilic interior that favors a low energy transition state for the carbenium intermediate while also providing for preferential entry of the substrate. Peripheral alkyl chains provide a hydrophobic exterior that offers solubility in non-polar solvents, such as hexanes. In E1-type elimination reactions involving tertiary alkyl halides, high turnover numbers (17400) were observed leading to almost quantitative conversions with very low catalyst loading (<0.01 mol %). Presumably, the slightly polar alkyl halide is drawn and concentrated into the polar core, where formation of the cationic intermediate and subsequent elimination occur to yield the non-polar alkene that is driven from the core to the non-polar corona and then to surrounding solvent. Solid NaHCO$_3$ present in the hexane phase serves as an acid acceptor. In this manner, the free energy of the system can minimized at each event and high efficiencies can be obtained.

The generality of the amphiphilic design concept has also been demonstrated and applied by Hecht[17] to excited state catalysis. Using an amphiphilic dendritic photocatalyst (Figure 3), Hecht was able to affect a [4+2]cycloaddition between singlet oxygen ($^1O_2$) and cyclopentadiene (CP). The cycloadduct is further reduced in situ to the allylic diol in the presence of thiourea. The large polarity increase associated with the overall transformation dictated our choice for the relative polarities of the inner and outer compartments. By encapsulating a $^1O_2$-sensitizing benzophenone core into a globular dendrimer having a hydrophobic interior and hydrophilic surface, it was possible to demonstrate the effect of dendrimer size on the performance of this nanoscale photoreactor. Since singlet oxygen lifetimes are greatest in nonpolar environments, increasing the generation of the dendrimer, led to the expected increase in CP conversion. A noteworthy feature of this system is that the bimolecular reaction benefits from both the high internal substrate concentration and the enhanced lifetime of singlet oxygen in the hydrophobic core environment.

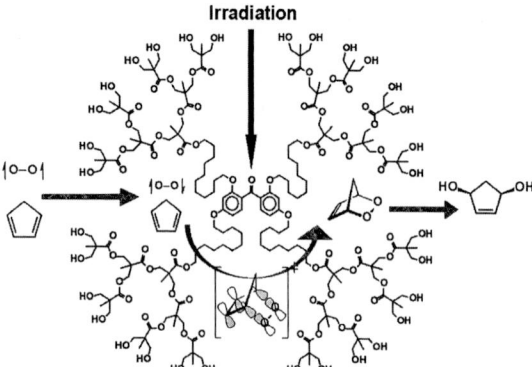

Fig. 3. Photocatalyst for bimolecular cycloaddition.

Although this area of research is still in its infancy, these initial findings are encouraging and demonstrate the importance of the amphiphilic motif for rational catalyst design. We are currently extending this theme exploring a variety of reactions with different polymer architectures to gain further insight into important aspects regarding the possible roles of the backbone and of the cooperativity of multiple catalytic sites. Furthermore, a strategy that combines light harvesting with photocatalysis is also under active investigation as a step towards more evolved systems that begin to mimic the complexity and combination of function found in photosynthesis.

3 Dendrimers as Nanoscale Light Harvesting Antennae

The very structure of dendrimers, with a radially emanating layered arrangement of functionalities around a central core, makes them uniquely suited for applications involving energy harvesting and conversion.[18] Thus, a molecule may be designed that incorporates multiple peripheral dyes used to harvest energy and capable of electronic communication with a functional core where the concentrated energy is "reprocessed". For example, harvested broadband radiation may be up- or down-converted into mono-chromatic light, or transformed into electrical or chemical energy. Thus, dipole dipole interactions between chromophore may be used to effect Förster energy transfer between an array of terminal donor chromophores and a core acceptor dye, which may also be excited independently of the periphery. Since emission is observed from the core only, the system serves as a spatial and spectral energy concentrator or "molecular lens" (Figure 4).

Essentially, this mimics the primary events in photosynthesis, where the light-harvesting complex funnels its excitation energy to the special pair leading to subsequent charge separation. Two types of systems, which either use the dendritic architecture solely as a scaffold,[19,20] or involve the dendrimer backbone in the energy transfer event,[21-23] have been explored. Our group has shown that an amplification of the core acceptor emission may be achieved in high generation dendrons labeled with multiple peripheral donor chromophores.[20] The amplification effect, misunderstood by some, has its origin in the enhanced donor absorption cross-section and the extremely fast rate of through-space energy transfer from the peripheral chromophores to the core, therefore giving rise to efficient light harvesting.[20] As for many other common forms of "amplification", this amplification is simply a "reprocessing" of energy that, in this case, enables the core chromophore to emit more light (including also energy transferred from the multiple peripheral dyes) that it possibly could solely via its direct excitation. In a key finding, Moore et al. demonstrated that a significant acceleration of energy transfer could be achieved within dendrimers having an internal energy gradient, resulting from a stepwise decrease of the HOMO-LUMO gaps of the branching units when progressing toward the acceptor.[22] Balzani et al. have constructed bipyridine-based polynuclear metal complexes capable of controlling the direction of energy transfer via alteration of the excited state energies by introducing appropriate metals.[23] This strategy impressively demonstrated how supramolecular chemistry could be used to assemble multiple chromophores while controlling their relative orientation. The performance benefit of the dendritic architecture over that of a linear polymer has also been demonstrated recently.[24]

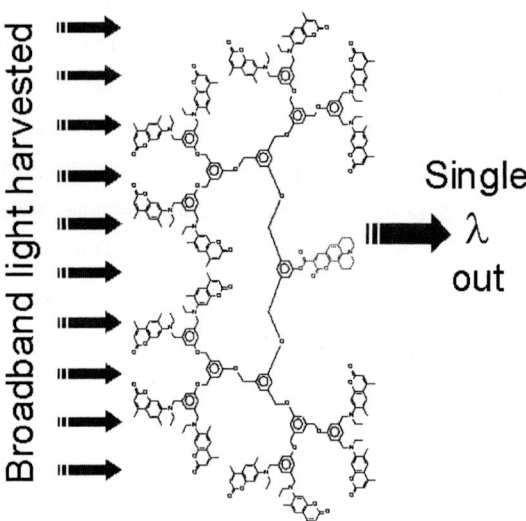

Fig. 4. Dendritic light harvesting antenna.

Recent work from Müllen et al., and from our own laboratory, has shown that cascade energy transfer involving a multiplicity of chromophores is possible.[25] Similarly, we have used two photon absorbing chromophores to effect up-conversion of energy using dendritic antennae.[26]

4 The Potential of Dendrimers in Macromolecular Therapeutics

The efficient and targeted delivery of therapeutic agents is one of the great challenges of today's medicine. In particular, macromolecules may be used to modify the bioavailability and pharmacokinetics of known drugs as exemplified by Schering-Plough's PEG-Intron™, in which a short (MW = 12,000 D) strand of monomethoxy poly(ethylene glycol) [PEG] is attached to Interferon α-2b, a water-soluble protein (MW = 19,271D) produced by recombinant DNA techniques. The conjugate, a potent drug for the treatment of hepatitis C, vastly outperforms the free Interferon α-2b also used to treat the same disease. The conjugation of drugs to PEG ("PEGylation") is now studied extensively as a means of improving the performance of known as well as new drugs. However, with low molecular weight drugs, the conjugation of one or even two molecule of drug to one molecule of monomethoxy-PEG or PEG to form a prodrug leads to issues of low molar concentrations of active or high viscosity of the conjugate solution.[27]

Linear polymers may be used for the conjugation of multiple copies of a drug onto a single polymer chain as exemplified by Kopečeck and Duncan's polymeric drug conjugates based on poly(N-2-hydroxypropyl-methacrylamide) [HPMA].[28] While linear polymers such as HPMA show great promise, dendritic structures offer several advantages including lower polydispersity, better accessibility of reactive sites, better control over their number, better defined nanometer size, more compact shape, lower solution viscosity, and better ability to interact with receptor sites in multivalent fashion, etc.[29] Our program in the development of delivery vehicles for therapeutic agents has included both micro- and nanoparticulates based on pH-degradable polymers[30] for the delivery of vaccines and genes, and dendritic macromolecules for the conjugation and delivery of drugs.[31] General design concepts include structural features affording low toxicity water-soluble carriers with high drug loading capacities as exemplified by our linear-dendritic hybrids based on aliphatic polyester dendrons.[31] In early studies, targeting to tumor cells was based on Maeda's Enhanced Permeation and Retention[32] effect, which mandates relatively long plasma residence times, and thus requires optimization of the size of the dendritic-drug conjugate. Release of the free drug from the multivalent conjugates is based on the low pH that prevails in tumor tissue and the use of acid-labile linkages for attachment of the drug molecules to the dendritic carrier. In early work,[31,33] we have evaluated a number of different dendritic architectures based on aliphatic polyester dendrons for their suitability as drug carriers both *in vitro* and *in vivo*. These water soluble and non-toxic systems, can be used to conjugate potent anticancer drugs such as doxorubicin via acid-labile hydrazone linkages (Figure 5). Attachment of the drug to the dendritic carrier reduced its cytotoxicity and biodistribution experiments showed little accumulation of the DOX-polymer conjugate in vital organs while the serum half-life of the doxorubicin conjugate was significantly higher than that of the free drug. Thus, this new nanoscale drug carrier system exhibits promising characteristics for the development of new polymeric drugs.[31]

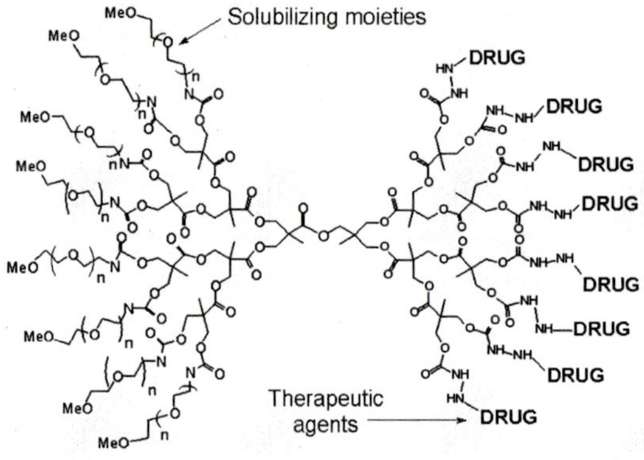

Fig. 5. Dendritic drug delivery system (from ref. 33).

5 Conclusion

It is clear that dendrimers have much to offer in numerous areas of the broad fields of nanoscience and nanotechnology. With a size typically ranging from 2-10nm, quite comparable to that of many proteins, dendrimers are large enough to allow significant tailoring of their structural features, physical properties, as well as surface and inner chemistry, yet they are small enough to allow their easy characterization and their use as mimics of many important biological macromolecules. There is little doubt that molecules possessing dendritic features and polyvalent character will find use in high added value applications ranging from nanotechnology to medicine. Whether these will be true dendrimers, dendritic hybrid, or even hyperbranched structures, really does not matter but the body of knowledge acquired through the fundamental study of structure-property-function relationships of dendrimers with precise and well-characterized structures will surely be of great value in the development of numerous classes of highly branched functional macromolecules with designs optimized for specific applications.

Acknowledgements

Financial support of this research by the AFOSR, NIH, NSF, and DOE (BES) is acknowledged with thanks. The author also thanks his many gifted Berkeley coworkers whose names appear in the references cited, as well as Dr. Craig Hawker (IBM Almaden).

1. Tomalia, D. A.; Baker, H.; Dewald, J.; Hall, J. M.; Kallos, G.; Martin, R.; Ryder, J. *Polym. J.*, **1985**, *17*, 117.
2. Newkome, G. R.; Moorefield, C. N.; Vögtle, F. *"Dendritic Molecules: Concepts, Syntheses, Perspectives"*, VCH: Weinheim 1996.
3. Fréchet, J.M.J., Tomalia, D.A. (Eds)*"Dendrimers and other dendritic polymers"*, Wiley Series in Polymer Science, John Wiley and Sons, Chichester, UK. 2001.
4. 4a. Hawker, C.J.; Wooley, K.L.; Fréchet, J.M.J., *J. Am. Chem. Soc.*, **1993**, *115*, 4375; 4b. Hecht, S., Fréchet, J.M.J., *Angew. Chem. Int. Ed*, **2001**, *40*, 74.
5. Mourey, T.H., Turner, S.R., Rubinstein, M., Fréchet, J.M.J, Hawker, C.J., Wooley, K.L. *Macromolecules*, **1992**, *25*, 2401.
6. Bauer, B.J.; Amis, E.J. Chapter 11 in reference 3, pages 271-4, 2001.
7. Fréchet, J.M.J., *Proc. Nat. Acad. Sci.* **2002**, 4782. Fréchet, J.M.J. *Science*, **1994**, *263*, 1710.
8. Tully, D.C., Fréchet, J.M.J, *Chem. Commun.* **2001**, 1229. Schenning A.P.H.J.; Elissen-Roman, C.; Weener J.W.; Baars, M.W.P.L.; van der Gaast, S.J.; Meijer, E.W. *J. Am. Chem. Soc.* **1998**, *120,* 8199. Wiener, J.W.; Baars, M.W.P.L.; Meijer, E.W. in ref. 3, chapter 11, pp. 387-424, 2001
9. Hawker, C.J.; Wooley, K.L.; Fréchet, J.M.J. *J. Chem. Soc. Perkin. Trans. I,* **1993**, 1287. Newkome, G.R.; Moorefield, .N.; Baker, G.R.; Saunders, M.J.;Grossman, S.H. (1991) *Angew. Chem. Int. Ed. Engl.,* **30**, 1178-80. Mattei, S.; Seiler, P.; Diederich, F.; Gramlich, V. *Helv. Chim. Acta*, **1995**, *78*, 1904. Stevelmans, S.; van Hest, J.C.M.; Jansen, J.F.G.A.; van Boxtel, D.A.F.J.; de Brabander-van den Berg, E.M.M.; Meijer, E.W., *J. Am. Chem. Soc.,* **1996**, *118*, 7398.
10. Liu, M.; Kono, K.; Fréchet, J.M.J., *J. Controlled Release*, **2000**, *65*, 121.
11. Jansen, J.F.G.A.; Meijer, E.W.; de Brabander-van den Berg, E.M.M. *J. Am. Chem. Soc.*, **1995**, *117*, 4417.
12. Astruc, D.; Chardac, F., *Chem. Rev.* 2001, *101,* 2991. Oosterom, G. E.; Reek, J. N. H.; Kamer, P. C. J.; van Leeuwen, P. W. N. M., *Angew. Chem., Int. Ed.* 2001, *40*, 1828. Van Heerbeek, R.; Kamer, P. C. J.; Van Leeuwen, P. W. N. M.; Reek, J. N. H., *Chem. Rev.* 2002, *102,* 3717. Twyman, L.J.; King, A.S.H.; Martin, I.K. Chem. Soc. Rev. **2002**, *31,* 69.
13. Bhyrappa, P.; Young, J. K.; Moore, J. S.; Suslick, K., *J. Am. Chem. Soc.* 1996, *118*, 5708.
14. (a) Mattei, S.; Seiler, P.; Diederich, F.; Gramlich, V., *Helv. Chim. Acta* 1995, *78(8),* 1904. Smith, D. K.; Zingg, A.; Diederich, F., *Helv. Chim. Acta* 1999, *82,* 1225.
15. Rheiner, P. B.; Seebach, D., *Chem. Eur. J.* 1999, *5*, 3221. Habicher, T.; Diederich, F.; Gramlich, V. *Helv. Chim. Acta* 1999, *82,* 1066.
16. Piotti, M.E.; Rivera, F.; Bond, R.; Hawker, C.J.; Fréchet, J.M.J., *J. Am. Chem. Soc.,* **1999**, *121*, 9471.
17. Hecht, S.; Fréchet J.M.J., *J. Am. Chem. Soc.* **2001**, *123*, 6959.
18. Adronov, A.; Fréchet, J.M.J.,*Chem. Commun.* **2000**, 1701.
19. Stewart, G.M., Fox, M.A., *J. Am. Chem. Soc.* 1996, *118*, 4354. Plevoets, M., Vögtle, F., De Cola, L., Balzani, V., *New J. Chem.* **1999**, 63. V. Vicinelli, P. Ceroni, M. Maestri, V. Balzani, M. Gorka, and F. Vögtle, *J. Am. Chem. Soc.*, 2002, *124*, 6461.

22

20. Gilat, S.L., Adronov, A., Fréchet, J.M.J., *Angew. Chem. Int. Ed.* **1999**, *38*, 1422. Adronov, A., Gilat, S.L., Fréchet, J.M.J., Ohta, K., Neuwahl, F.V.R., Fleming, G.R. *J. Am. Chem. Soc.* **2000**, *122*, 1175. Adronov, A., Malenfant, P.R.L., Fréchet, J.M.J., *Chem. Mater.* **2000**, *12*, 1463. Neuwahl, F. V. R.; Righini, R.; Adronov, A., Malenfant, P.R.L., Fréchet, J.M.J.; *J. Phys. Chem. B* **2001**, *105*, 1307.

21. Jiang, D.L., Aida, T., *J. Am. Chem. Soc.* **1998**, *120*, 10895. Ng, A.C.H., Li, X.Y., Ng, D.K.P. *Macromolecules* **1999**, *32*, 5292. Kawa, M., Fréchet, J.M.J., *Chem. Mater.* **1998**, *10*, 286. Li, F., Yang, S.I., Ciringh, Y., Seth, J., Martin, C.H., Singh, D.L., Kim, D., Birge, R.R., Bocian, D.F., Holten, D., Lindsey, J.S., *J. Am. Chem. Soc.* **1998**, *120*, 10001.

22. Devadoss, C., Bharathi, P., Moore, J.S., *J. Am. Chem. Soc.* **1996**, *118*, 9635;

23. Serroni, S., Juris, A., Venturi, M., Campagna, S., Resino, I.R., Denti, G., Credi, A., Balzani, V., *J. Mater. Chem.* **1997**, *7*, 1227; Balzani, V. Campagna, S., Denti, G., Juris, A., Serroni, S., Venturi, M., *Acc. Chem. Res.* **1998**, *31*, 26.

24. Adronov, A., Malenfant, P.R.L., Robello, D. R.; Fréchet, J.M.J.; *J. Polym. Sci. A*, **2001**, *39*, 1366.

25. T. Weil, E. Reuther, and K. Müllen, *Angew. Chem., Int. Ed.*, **2002**, *41*, 1900. Serin, J.M.; Brousmiche D.W.; Fréchet, J.M.J. *Chem. Commun.* **2002**, 2605.

26. Adronov, A.; Fréchet, J.M.J.; He, G.S.; Kim, K.S.; Chung, S.J.; Swiatkiewicz, J; Prasad, P.N. *Chem Mater.* **2000**, *12*, 2838. Brousmiche, D.; Serin, J.M.; Fréchet, J.M.J; He, G.S.; Lin, T.C.; Chung, S.J.; Prasad, P.N.; *J. Am. Chem. Soc.* **2003**, *125*, in press.

27. Greenwald, R.B., Conover, C.D., Choe, Y.H. *Crit. Rev. Ther. Drug*, **2000**, *17*, 101. Choe, Y.H., Conover, C.D., Wu, D., Royzen, M., Gervacio, Y., Borowski, V., Mehlig, M., Greenwald, R.B., *J. Controlled Rel.* **2002**, *79*, 55.

28. Duncan, R. *Anticancer Drugs*, **1992**, *3*, 175. Omelyanenko, V., Kopeckova, P., Gentry, C; Kopeček, J. *J. Control. Release* **1998**, *53*, 25-37. Kopeček, J., Kopečekova, P., Minko, T., Lu, Z-R., *Eur. J. Pharmaceut. Biopharmaceut.* **2000**, *50*, 61.

29. Haensler, J., Szoka, F. C. Jr. *Bioconjugate Chem.* **1993**, *4*, 372. Liu, M.; Fréchet, J.M.J.; *Pharmaceut. Sci. Technol. Today.* **1999**, *2*, 393. Malik, N.; Wiwattanapatapee, R.; Klopsch, R.; Lorenz, K.; Frey, H.; Weener, J. W.; Meijer, E. W.; Paulus, W.; Duncan, R. *J. Control. Release*, **2000**, *65*, 133.

30. Murthy, N.; Thng, Y.X.; Schuck, S., Xu, M.C.; Fréchet, J.M.J. *J. Am. Chem. Soc.* **2002**, *124*, 12398.

31. Ihre, H.R.; Padilla De Jesús, O.; Szoka, F.C., Jr.; Fréchet, J.M.J. *Bioconj. Chem.* **2002**, *13*, 443. Ihre, H., Padilla de Jesús, O.L., Fréchet, J.M.J. *J. Am. Chem. Soc.* **2001**, 123, 5908. Padilla De Jesús, O.; Ihre, H.R.; Gagne, L.; Fréchet, J.M.J.; Szoka, F.C., Jr.; *Bioconj. Chem* **2002**, *13*, 453.

32. Maeda, H., Seymour, L. W., Miyamoto, Y. *Bioconjugate Chem.* **1992**, *3*, 351-362. Maeda, H., Wu, J., Sawa, T., Matsumura, Y., Hori, K. *J. Control. Release*, **2000**, *65*, 271.

33. Gillies, E.; Fréchet, J.M. J., *J. Am. Chem. Soc.* **2002**, *124*, 14137.

Macromol. Symp. **2003**, *201*, 23—28   23

On/Off Switching on Polymer Conformation

Kenichi Yoshikawa

Department of Physics, Kyoto University & CREST, Kyoto 606-8502, Japan
E-mail: yoshikaw@scphys.kyoto-u.ac.jp

Summary: The manner of folding transition from elongated coil to compact globule of single polymer chain is discussed. Based on theoretical consideration, it is argued the semi-flexible polymer chain exhibits large discrete transition on the level of individual single chains, whereas the transition looks continuous, or cooperative, on the ensemble of chains. As the experimental verification, in the present article thermodynamic and kinetic aspects of folding transition of single giant DNA molecules are described. It is shown that rich variety of nano-ordered structures are obtained from single DNA molecules through suitable setup of the experimental conditions. The stability of such nano-structures generated from single polymer chain is discussed in relation to the ordered compact structure with large number of chains in semi-dilute and concentrated polymer solutions.

Keyword: coil-globule transition, DNA condensation, nano-ordered structure, polyelectrolyte, single chain observation

Introduction

According to the standard description on texbooks of polymer science, single polymer chain undergoes coil-globule trantion through the so-called θ-state.[1,2] This means that the transition has been regarded as continuous, or cooperative, in general. In contrary to this, recently it has become clear the individual giant DNA molcules exhibit large discrete transition accompanied with the change of the segement density of the order of 10^4-10^5. The common characteristics of the folding transition of single giant DNA molecules are summarized as follows:[3,4]

1) Individual DNA molecules exhibit an all-or-none transition between an elongated coil and a compact globule, irrespective to the chemical nature of condensing chemicals, such as polyvalent cation, hydrophilic polymer, cationic surfactant, etc.

2) There is rather wide parameter area on the coexistence between the unfolded and folded states. The folding transition appears to be steep but continuous for the ensemble average of DNA molecules.

3) The free energy of a single DNA is interpreted with the profile of double minima. The transition is classified as first-order phase-transition under the criterion of Landau.

          DOI: 10.1002/masy.200351103

4) The negative charge on DNA disappears with the foding transition, except for the surface on the compact state. This means that individual globules, or compact DNAs, behave as soluble colloid.

In the present article, we will argue that the discrete nature of the folding transition is a general property of semi-flexible polymer chains.

ON/OFF Switching of Single Semi-flexible Chain[3-6]

Let us discuss the effect of chain stiffness on the manner of folding transition in a polymer molecule. We consider a polymer chain with the contour length L and Kuhn length λ, corresponding to the length of a segment. Thus, the number of segments is given as $N = L/\lambda$. When a chain is dissolved in a good solvent, the characteristic one-dimensional size R_c (such as end-to-end distance, radius of gyration, hydrodynamic radius) of an elongated polymer chain is represented as in the following relationship.

$$R_c \cong \lambda N^{3/5} \qquad (1)$$

On the other hand, the size of a compact folded chain is given as eq. (2), where s is the cross sectional area of the chain.

$$R_g \cong (\lambda s N)^{1/3} \qquad (2)$$

If the cross section can be represented by a circle of diameter D, the area is represented as $s = \pi D^2/4$, i.e., $D^2 \cong s$. Here, we introduce a parameter on chain stiffness as $\eta = \lambda/D$. When the stiffness paramer is much larger than unity, $\eta \gg 1$, the change in R becomes significant. In the case of DNA, it is known that $s = 2nm$ and $\lambda = 100nm$. The contour lenght in natural DNA is rather large; bacterial DNA is on the order of mm and manmalian DNA is above cm. As an example, let us calculate the change of the size on a DNA chain with 30 kilo base pairs, which has the contour lengh of ca. 10 μm and the number of the Kuhn segments $N = 100$. From eqs.(1) and (2), the diffrence of the density between the globule and coil is obtained as $\rho_g/\rho_c = (R_c/R_g)^3 \cong 10^5$. For comparison, it is to be noted that the change of the density on the transition between liquid water and vapor is only ca. 2000. Now, it has become evident that the density

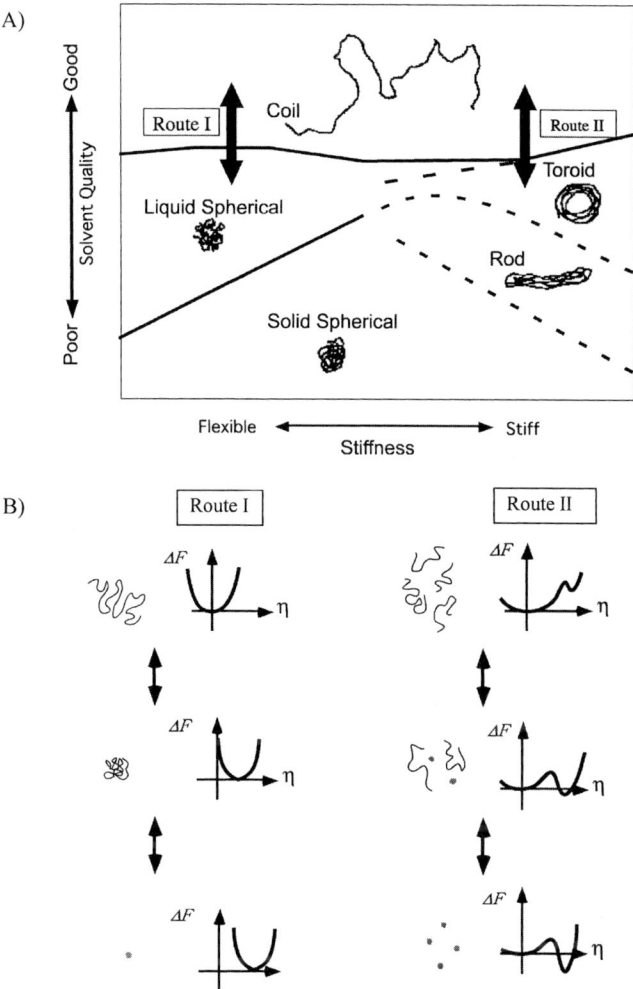

Fig. 1. A) Phase diagram on the conformation of a single polymer chain (modified from the original figure in reference[7]). B) The manner of coil-globule transition is markedly different between the two Routes (see Fig.1 A). The conformational change is schematically represented, together with the corresponding profile of free energy on individual chains. η: density of segments.

difference before and after the folding transition of giant DNA is very significant. Actually, from detailed theoretical considerations it has been established that long stiff plymer, including giant DNA, exhibits large discrete transition, i.e., first-order phase-transition.[7,8]

Phase Diagram in a Single Chain

Figure 1 shows the phase diagram of a single chain with the change of chain stiffness, deduced from a theoretical calculation of Multi-canonical Monte Carlo method.[7] When the stiffness is large enough, the transition from elongated coil into compact state is discrete, i.e., first-order phase-transition. It is noted that depending on the change in the quality of the solvent, or in the pair interaction between the segments, different solid-like states are generated as the most stable conformation, such as toroid, rod and spherical globule. On the other hand, the transition is diffuse or continuous for the molecular chain with lower stiffness, where liquid-like spherical state is generated after the transition. Such continuous nature concides with the current picture of the coil-globule transition in standard textbooks.[1,2] In between the discrete and continuous transitions, a region of the intermediate nature of transition exists. When a polyelectrolyte chain exhibits such intermediate stiffness, intrachain segregated state, or pearling structure, is generated.[9,10]

Nano-ordered Structures from a Single Chain

Figure 2 shows the electron micrograph on variety of structures made from single T4 DNA (165kbp, contour length 57 μm).[3,4] With cationic condensing agents, such as spermidine(3+), spermine(4+), cobalt ion(3+), tightly packed toroids with the diameter of 60-80nm are formed after the coil-globule transition. With the excess of the cationic reagents, the toroids swell into larger size as a result of the weaker attractive interaction between the segments.[11] With a cationic surfactant having multi-cationic head group, liposhermine, spool-like structure is generated.[12] With the addition of PEG-A, amino-pendant polyethelene glycol, partially compact state with the mini-globule(s) is induced.[9] Similar partially segregated structure is generated with polycations having laregе number of cationic groups.[13] When DNA is complexed with dicationic surfactant, pearling chain is generated where plural number of mini-toroids are found along a single chain (A. Zinchenko, et. al., unpublished result).

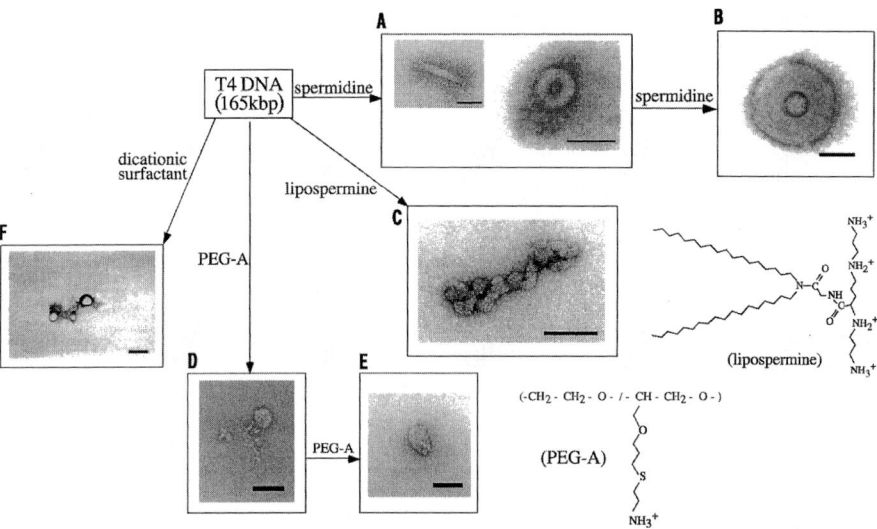

Fig. 2. Electron microscopic images of compact states generated from single T4 DNA molecules. The bars indicate 0.1µm. (The original references are A:[14], B:[11], C:[12], D and E:[9], F: Zinchenko, et. al., unpublished).

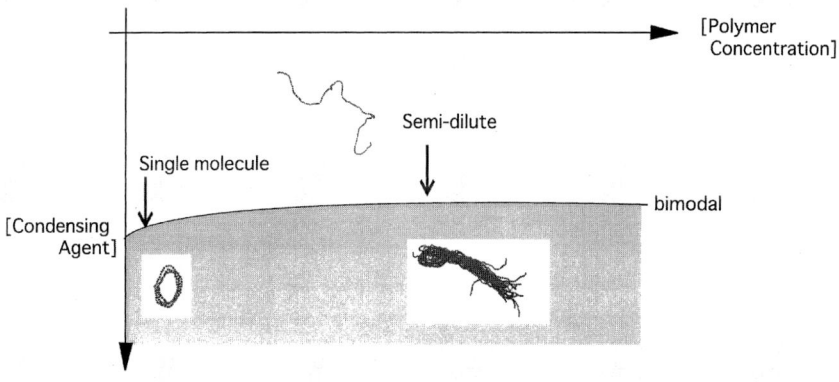

Fig. 3. Diagram of transition of stiff polymer. With the increase of polymer concentration, bundle with multiple chains is generated. (T. Sakaue, et. al., unpublished).

Competition Between the Events of Single Chain and Multiple Chains

With the increase of the polymer concentration in stiff chains, assembly of multiple chains becomes to be generated as is schematically shown in Figure 3. Interesting to say, it has been found, form the experiments on giant DNA, that the necessary concentration to cause the compaction of single chain at low polymer concentrations and also the condensation of multiple chains at high polymer concentrations are essentially the same (T. Iwataki, et. al., unpublished result). It is of scientific value to extend further the experimental and theoretical studies on the conformational change of polymers by comparing single and multiple-chain processes.

[1] P. J. Flory, "Statistical mechanics of chain molecules", Interscience, New York, **1969**.

[2] P. G. De Gennes, "Scaling concepts in poymer physics", Cornell University Press, Ithaca, **1979**.

[3] K. Yoshikawa, *Adv. Dru. Del. Rev.*, **2001**, 52, 235.

[4] K. Yoshikawa and Y. Yoshikawa, "Compaction and condensation of DNA", in "Pharmaceutical perspectives of nucleic acid-based therapeutics", eds., R. I. Mahato, et.al, Taylor & Francis, **2002**, 137.

[5] A. Yu. Grosberg and A. R. Khokhlov, "Statistical physics of macromolecules", American Institute of Physics, New York, **1994**.

[6] K. Yoshikawa, *J. Biol. Phys.*, **2002**, 28, 701.

[7] H. Noguchi and K. Yoshikawa, *J. Chem. Phys.*, **1998**, 109, 5070.

[8] T. Sakaue and K. Yoshikawa, *J. Chem. Phys.*, **2002**, 117, 6323.

[9] K. Yoshikawa, Y. Yoshikawa and T. Kanbe, *J. Am. Chem. Soc.,* **1997**, 119, 6473.

[10] S. Takagi, K. Tsumoto and K. Yoshikawa, *J. Chem. Phys.*, **2001**, 114, 6942.

[11] Y. Yoshikawa, K. Yoshikawa and T. Kanbe, *Langmuir*, **1999**, 15, 4085.

[12] Y. Yoshikawa, N. Emi, T. Kanbe, K. Yoshikawa and H. Saito, *FEBS Lett.*, **1996**, 396, 71.

[13] Y. Yoshikawa, Y. S. Velichko, Y. Ichiba and K. Yoshikawa, *Eur. J. Biochem.*, **2001**, 268, 2593.

[14] H. Noguchi, S. Saito, S. Kidoaki and K. Yoshikawa, *Chem. Phys. Lett.*, **1996**, 261, 527.

Macromol. Symp. **2003**, *201*, 29—45

The Influence of Hydrogen Bonds on the Globular Structure of HP-copolymers

Yury Kriksin,*[1,4] Pavel Khalatur,[2,4] Alexei Khokhlov*[3,4]

[1]Institute for Mathematical Modelling of RAS, Miusskaya pl. 4a, 125047 Moscow, Russia
Fax: +7 (095) 972-0723; E-mail: kriksin@imamod.ru, hq@imamod.ru
[2]Department of Physical Chemistry, Tver State University, Sadovy per. 35, 170002 Tver, Russia
[3]Physics Department, Moscow State University, 117234 Moscow, Russia
[4]University of Ulm, Department of Polymer Science, Albert-Einstein-Allee 11, Ulm, D-89069, Germany

Summary: We present the results of computer modeling of coil-globule transition of HP-copolymer which can form hydrogen bonds between some specific monomer units (hydrogen bond units). Langevin dynamics approach is used for the simulation of coil-globule transition. We study the influence of the number and distribution of hydrogen bond units along primary sequence on the formation of globular conformations.

Keywords: computer modeling, copolymers, hydrogen bonds, primary sequence

Introduction

The properties of homopolymer globule formed by saturating bonds were studied theoretically in the paper.[1] The purpose of this work is to carry out computer experiments which simulate the dynamics of copolymers with saturating (hydrogen) bonds in order to understand the influence of copolymer character of the chain on the formation of a globule induced by hydrogen bonds. It is well-known that hydrogen bonds play very important role in conformational properties of biopolymers,[2-3] and that they have the property of saturation. In our model only specific pairs of units (hydrogen bond units, HBU) can form a hydrogen bond. In the previous paper[4] we have studied HP-copolymers with hydrophobic (H) and hydrophilic (P) monomer units. In particular, we have considered the so-called protein-like conformations with hydrophobic monomer units in the core of the globule and hydrophilic units in the envelope of this core. Such protein-like copolymers have various interesting physical and

 DOI: 10.1002/masy.200351104

chemical properties,[5-12] in particular, special primary sequence obeying the so-called Levy-flight statistics.[11] In this paper the main emphasis is the general problem of hydrogen bonds influence on the stability of copolymer globular conformations, rather than the discussion of specific features of coil-globule transition for different primary sequences.

The Model of Hydrogen Bonds in Computer Experiment

To simulate model copolymers, we use Langevin dynamics.[13] The generalization of our previous simulations[4] is connected with the introduction of intramolecular potential taking into account hydrogen bonds.

We consider model freely-jointed chain, with fixed bond lengths l, consisting of N monomer units of the mass m. Let us define the vector r_n as the position of n-th monomer unit in the space and introduce the notation N_α ($\alpha = $ H, P) for the number of monomer units of the type α in the chain ($N_H + N_P = N$).

We will use a modified form of Lennard-Jones potential slightly different from that applied in the previous paper[4]

$$u(r) = \begin{cases} \varepsilon_0[(r/\sigma)^{-12} - 2(r/\sigma)^{-6}] + \varepsilon_0 - \varepsilon, & r \in (0, \sigma], \\ \varepsilon[(r/\sigma)^{-12} - 2(r/\sigma)^{-6}] & r \in (\sigma, +\infty), \end{cases} \tag{1}$$

where

$$r = |r_n - r_k|, \quad |n - k| > 1 \tag{2}$$

The potential (1) includes three parameters (σ, the characteristic size, ε, attractive energy parameter and, ε_0, excluded volume energy parameter).

The additional factor included in the mathematical model is connected with hydrogen bonds (HB). In our model HB is constituted of only two monomer units. Monomer units capable of forming of HB are called *hydrogen bond units*, HBU. HBU can take part only in one HB (saturation effect).

Here HB potential has the form

$$
u_H(r) = \begin{cases}
\varepsilon_{0H}[x^{-12} - 2x^{-6}] + \varepsilon_{0H} - \varepsilon_H, & x \in (0,1], \\
\varepsilon_H[\mu^{-2}(x-1)^2 - 1], & x \in (1, 1 + \mu^2(\mu + \lambda)^{-1}], \\
-\varepsilon_H \lambda^{-1}(2\mu + \lambda)^{-1}(x - 1 - \mu - \lambda)^2, & x \in (1 + \mu^2(\mu + \lambda)^{-1}, 1 + \mu + \lambda], \\
0, & x \in (1 + \mu + \lambda, +\infty),
\end{cases}
\tag{3}
$$

where

$$
x = r / \sigma_H, \qquad r = |\, r_n - r_k \,|, \qquad |\, n - k \,| > 1. \tag{4}
$$

The parameters σ_H, ε_H and ε_{0H} have the similar meaning to the parameters σ, ε and ε_0 in equation (1). Dimensionless parameters μ and λ define the shape of potential curve. Only some monomer units (HBU) can be involved in HB. If two HBUs form HB, they can not participate in other HB. When two free HBUs are nearer than critical distance, i.e. $r < (1 + \lambda + \mu)\sigma_H$, they form one HB. On the contrary if the distance between two units participating in HB becomes more than critical one (i.e. $r > (1 + \lambda + \mu)\sigma_H$), the HB is destroyed (HB potential vanishes).

We use protein-like HP-copolymer[4] as suitable model for copolymer sequence and we consider the case of uniformly distributed HBU in the primary sequence. We consider three kinds of HBU distributions along the primary sequence: (a) distribution related to protein-like sequence, (b) random distribution, and (c) periodical distribution.

Let us describe the algorithm of the arrangement of hydrogen bonds related to protein-like sequence (see Fig. 1). First, we form dense globular protein-like "parent" conformation. Then we choose some points belonging to the body of parent globule and encircle them by spheres of radius R. In order to avoid the ambiguity, we put these points in the nodes of hexagonal grid with distance d between neighbors. Finally, within each sphere of radius R inside the body of the globule we find nearest pairs of monomer units and consider them as HB participants (HBUs).

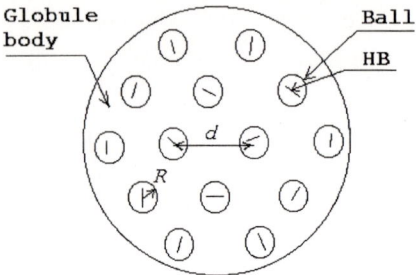

Fig. 1. The arrangement of hydrogen bonds related to protein-like sequence.

We obtain random arrangement of HBUs as a result of random choice of monomer units in the primary sequence.

In the case of periodical arrangement, the distance between two nearest HBUs along the chain is constant.

Freely-jointed chain has $2N+1$ degrees of freedom. Constraint dynamics of model polymer chain is described by the corresponding system of equations of motion.[14-17] Langevin forces are considered as linear combination of random noise and linear dissipation, taking into account heat action.[13] The numerical integration of equations of motion is carried out using the standard difference scheme.[18]

We use the same method of temperature estimation and the same cooling procedure as in our previous paper.[4]

The new way to estimate the temperature of coil-globule transition of HP-copolymer is proposed. Let us consider the finite chain of homopolymer. The temperature of coil-globule transition is usually defined as the inflection point of the dependence $R_g^2(T)$ of the square of gyration radius on temperature. This point corresponds to $\partial^2[R_g^2(T)]/\partial\tau^2 = 0$. From mathematical viewpoint the numerical differentiation is an ill-posed problem.[19] Thus we need higher accuracy of the values of $R_g^2(T)$ to estimate the temperature of coil-globule transition with necessary reliability. In order to achieve this level of accuracy we have to carry out long-time computations. When we consider copolymers the problem of evaluation becomes more complicated in comparison with homopolymers because different types of monomer units have their own temperature of coil-globule transition.

To overcome this difficulty, we use the dependence of the total energy $E(T)$

$$E(T) =<U_{LJ}>+<U_{HB}>+(2N-5)k_B T,\tag{5}$$

on temperature, where $<U_{LJ}>$ and $<U_{HB}>$ are time-averaged potential energies of van der Waals and hydrogen bonds interactions, respectively (the energy of covalent bonds between neighboring monomer units is excluded). The value of $(2N-5)$ in Eq. 5 is the number of internal degrees of freedom of freely-jointed chain. We consider $E(T)$ as an analogue of full mechanical energy. Negative values of full energy correspond to constrained motion of particles. Positive ones correspond to infinite motion. The curve $E(T)$ is characterized by monotonous growth. From mechanical viewpoint, the zero point of $E(T)$ is the boundary between constrained and infinite motion. We suppose that the zero point T_z of $E(T)$ is directly proportional to the temperature of coil-globule transition, T_{cg}, for a chain of the same length, N. In order to estimate the corresponding factor η_{cg}, we find the ratio T_{cg}/T_z for homopolymer and assume that $\eta_{cg} = T_{cg}/T_z$. Then we assume that the temperature of coil-globule transition of HP-copolymer can be obtained from the equation $T_{cg} = \eta_{cg}T_z$, where T_z is the zero point of $E(T)$ for HP-copolymer.

The purpose of our study is to understand how hydrogen bonds influence the structure of a globule. This question subdivides into the following items:

a) How does HBU arrangement in the primary sequence influence the protein-like structure (hydrophobic in the core, hydrophilic units in the envelope)?

b) How does the number of HBUs in the primary sequence influence the protein-like structure?

It is necessary to have a criterion for quantitative estimation of the quality of globular protein-like structure. We use the method of convex polyhedron described in our previous paper.[4] We consider time-averaged values of criterion function Λ[4] and study how they depend on temperature T at various arrangements and numbers of HBUs.

Results of Computer Simulations

In this study we assume that every monomer unit has unit mass ($m = 1$), and covalent bond between nearest neighbors has a fixed length ($l = 1$), while the number of monomer units in the chain is equal to $N = 256$.

Let us define other model parameters. The parameters of modified Lennard-Jones potential (1) are the following:

$$\varepsilon_0 = 1, \qquad \sigma = 1. \tag{6}$$

The energy parameters ε for HP-copolymer will be considered as a variable depending on the kind of interacting monomer units ($\varepsilon_{HH} = 2.0$, $\varepsilon_{PP} = 0.2$, $\varepsilon_{HP} = \sqrt{\varepsilon_{HH}\varepsilon_{PP}} = 0.6325$) The values of $\varepsilon_{\alpha\beta}$ ($\alpha, \beta = H, P$) correspond to optimum recovery of parent globule structure.[4] The primary sequence is protein-like. The parameters of HB potential have the following values:

$$\varepsilon_{0H} = 5, \qquad \varepsilon_H = 10, \qquad \sigma_H = 1, \qquad \lambda = 0.3, \qquad \mu = 0.2. \tag{7}$$

The parameters of HBU related to protein-like sequence (see Fig. 2) have the following values:

$$d = 2, \qquad R = 0.5. \tag{8}$$

As a first step, we define the coil-globule transition temperature for homopolymer chain composed of monomer units of H type ($\varepsilon = \varepsilon_{HH}$) without HB. Fig. 2 shows the temperature dependence of the mean square radius of gyration

$$R_g^2 = \frac{1}{N} \sum_{n=1}^{N} |r_n - R_c|^2, \tag{9}$$

where R_c is the center-of-mass position vector.

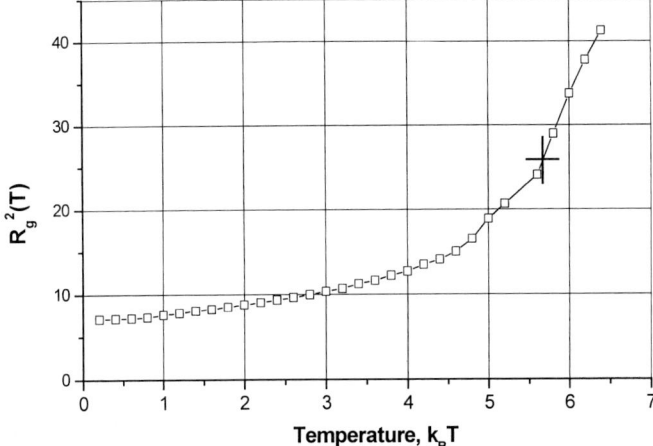

Fig. 2. The value of R_g^2 as a function of temperature for 256-unit homopolymer chain. The point of coil-globule transition is marked by the cross.

In Fig. 2, the inflection point corresponding to the temperature of coil-globule transition, T_{cg}, is marked by the cross ($\varepsilon = \varepsilon_{HH} = 2.0$). As we have mentioned above, it is more simple to find the zero point T_z of intramolecular energy (5) as a function of temperature T. Fig. 3 shows the dependence of intramolecular energy (per one degree of freedom) $e(T) = E(T)/(2N-5)$ on temperature T. Monotonous behavior of intramolecular energy gives unique root of the equation $e(T) = 0$.

We can see from Table 1 that the constancy of the factor $\eta_{cg} = T_{cg}/T_z$ for various types of polymer chains is observed with high accuracy (in the limits of confidence interval).

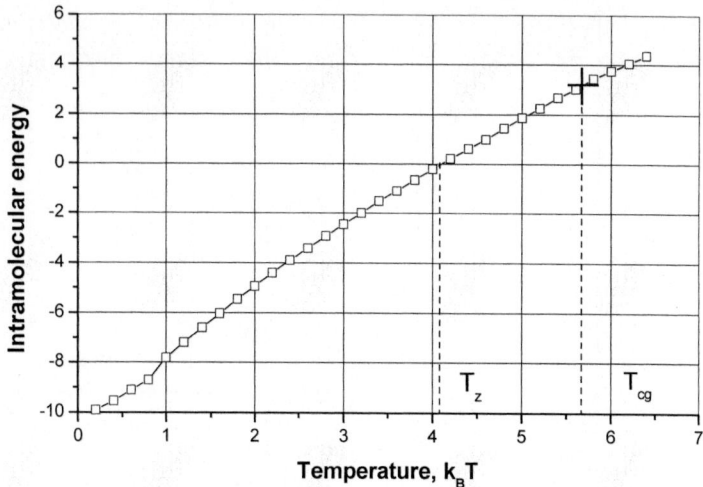

Fig. 3. The dependence of intramolecular energy $e(T) = E(T)/(2N-5)$ on temperature T. The temperature of coil-globule transition, T_{cg}, is marked by the cross (see also Fig. 2).

Table 1. The coefficient η_{cg} for different types of HP-copolymers ($N = 256$).

The type of HP-copolymer	η_{cg}	Error
Homopolymer	1.39	0.02
Protein-like HP-copolymer without HBU	1.38	0.02
Protein-like HP-copolymer with 20 HBU	1.37	0.03
Protein-like HP-copolymer with 256 HBU	1.44	0.05

In order to determine the temperature of coil-globule transition, T_{cg}, we define the factor $\eta_{cg} = T_{cg}/T_z$ and then use the equation

$$T_{cg} = \eta_{cg} T_z \tag{10}$$

Thus, the equation (10) can be considered as a useful empiric rule.

The recovery of protein-like parent globular structure takes place within the temperature interval $[0, T_z]$. Monomer units of both types (H and P) are intermixed in the globule if the temperature $T > T_z$. We have found that the recovery of protein-like parent globular structure occurs in the presence of HB as well as in the absence of them. Let us first estimate the influence of an arrangement of HBU along the chain. We have considered three different distributions of 20 HBUs: (a) related to protein-like sequence, (b) random distribution, and (c) periodical distribution (see above). Three appropriate time-averaged curves of convex polyhedron criterion[4] $< \Lambda(T) >$ are shown in Fig. 4. As we can see, these three curves lie in the nearest vicinity to each other. Therefore the arrangement of HBUs along the chain have the secondary significance in comparison with the number of HBUs. Also, we can see another confirmation of the secondary significance of the arrangement of HBUs in the fact that the values of coil-globule transition temperature, $T_{cg} = 3.6/k_B$, found for three types of arrangement (a)-(c) practically coincide.

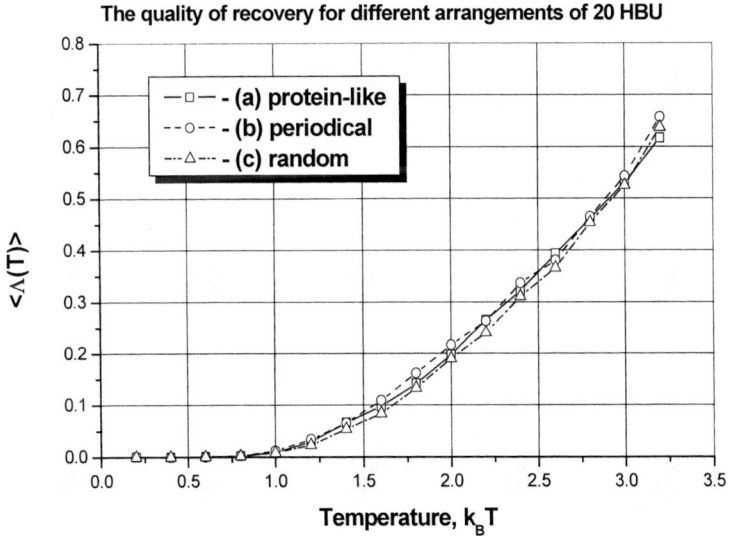

Fig. 4. Time-averaged convex polyhedron criterion $< \Lambda(T) >$ for the following arrangements of HBUs: (a) related to protein-like sequence, (b) random sequence, and (c) periodical sequence.

The conclusion that the HBU distribution has the secondary significance in the formation of parent globular structure allows us to consider only periodical HBU arrangement in the further investigation. It is interesting to study how the quality of the recovery depends on the number of HBU in the chain. The criterion functions $< \Lambda(T) >$ for different numbers of periodically arranged HBU in the chain are shown in Fig. 5.

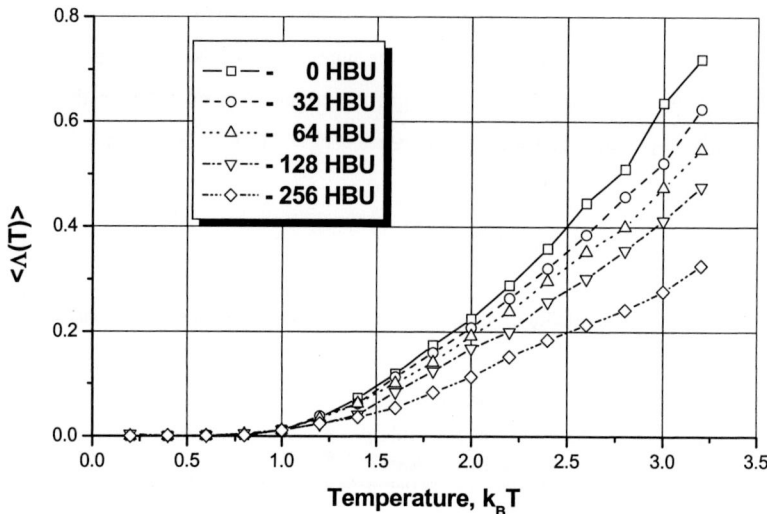

Fig. 5. Time-averaged criterion functions $< \Lambda(T) >$ for different numbers of periodically arranged HBU.

We have included minimum and maximum possible numbers of HBU, that is 0 and 256. The curves, corresponding to the numbers of HBU equal to 0 and 256, form the boundaries for the region in which other curves lay. The conclusion is evident: an increase in the number of HBU stabilizes protein-like parent globular structure.

It is interesting to consider the dependence of time-averaged quality of recovery on the number of HBU at a fixed temperature. The corresponding data are shown in Fig. 6 for three values of temperature T. These dependencies are monotonically decreasing functions. If the temperature is fixed, an increase in the number of HBU results in a decrease in the values of criterion function $< \Lambda(T) >$. In other words, the quality of recovery improves with increasing HBU.

We can see also that the dependence of the coil-globule transition temperature T_{cg} on the number of HBU is monotonically increasing and convex (see Fig. 7).

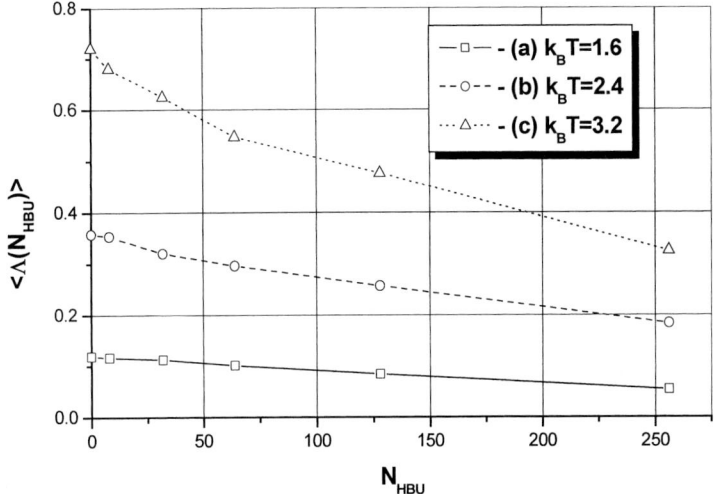

Fig. 6. The dependencies of time-averaged recovery quality $<\Lambda>$ on the number of HBU N_{HBU} for fixed temperatures T: (a) $T = 1.6/k_B$, (b) $T = 2.4/k_B$, and (c) $T = 3.2/k_B$.

It is also interesting to discuss the possibility of reconstruction of unique parent conformation. Our numerical experiments show that the parent globular conformation is never reproduced. To understand the role of HB in the possible conservation of unique parent conformation, the parent and recovered globular conformations were compared with the help of RMSD (root mean square distance)

$$RMSD = [\min_W \sum_{i=1}^{N} |W(r_i^{(2)} - R_c^{(2)}) - r_i^{(1)} + R_c^{(1)}|^2]^{1/2}, \qquad (11)$$

where W is the matrix of rotations, $r_i^{(k)}$ is the position of i-th monomer in the k-th chain ($k = 1,2$), and $R_c^{(k)}$ denotes the center of mass of the k-th chain.

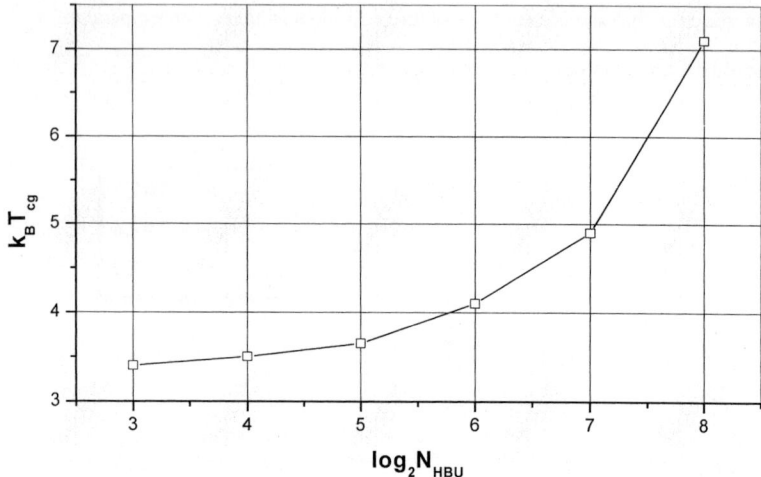

Fig. 7. The dependence of the temperature of coil-globule transition T_{cg} on the number of periodically arranged HBU N_{HBU}.

RMSD vanishes if two conformations of the same polymer chain coincide and it has the order of the size of the globule if they are different. In order to reveal how hydrogen bonds influence the process of reconstruction, we have considered three variants of HP-copolymer chain: (a) without HB, (b) with standard HB ($\varepsilon_H = 10$), and (c) with strong HB ($\varepsilon_H = 50$). In the cases (b) and (c), we used the HBU arrangement related to protein-like sequence.

The main feature of strong HBs is their inability to be destroyed in the temperature range $0 \leq T \leq T_{cg}^0 = 3.3/k_B$, where $T_{cg}^0 = 3.3/k_B$ is the temperature of coil-globule transition for HP-copolymer chain without HB. Standard HB may be destroyed within this temperature interval.

We see that RMSD decreases as the intensity of HB increases. But the order of magnitude of RMSD is in all the cases of order of the globule size ($R_g \approx 2.6$). The conclusion is evident: HBs do not retain the information enabling the reconstruction of unique parent globular conformation.

Table 2. The RMS distances (in the units of σ from eq.(6)) between parent and recovered conformations for different types of HB.

The type of HP-copolymer	Distance, σ
With 20 strong HBU	2.45
With 20 standard HBU	3.13
Without HBU	3.51

Another point characterizing the uniqueness of the recovery of parent globule is the structure of HBs itself. We have carried out one hundred runs on "cooling" of protein-like HP-copolymer chain with 20 HBUs in order to discover the recovery of HBs related to protein-like sequence. It was found that the formation of HBs is completely random. The formation of HBs is more probable for nearest HBUs. An increase in distance between HBUs leads to a decrease in the probability of HB formation. We should stress that at the same time, the recovery of the parent globular structure takes place in all 100 cases with high quality (the values of criterion function are $\Lambda = 0$ in 95% of cases and $\Lambda < 10^{-3}$ in remaining 5% of cases for the temperature $T = 0.2 / k_B$).

We can suppose that the distribution of monomer units in the globule and the structure of HBs are random, while the parent globular structure with H-units in the core and P-units on the surface is reproduced with high accuracy. This situation can be explained by the degeneracy of energetic levels corresponding to the conformations providing for parent protein-like globular structure. The degeneracy of energetic levels is the fundamental property of the chain consisting of monomer units interacting via central potential forces. For example, we can not distinguish right and left (mirror reflected) conformations from energetic point of view. We believe that only more complicated structure of monomer unit could enable the reconstruction of unique parent globular conformation.

We also consider the stability of HBs. Two main characteristics of the stability have been investigated: (a) average HB lifetime and (b) the number of breaks of HBs per time unit (i.e., the frequency of HB breaks). We have considered protein-like HP-copolymer with maximum number of HBUs ($N = 256$). HBs are stable for the temperatures in the range $0 < T \leq 1.0 / k_B$ and the breaks of HBs are very rare. It is necessary to carry out the enormous volume of numerical calculations to get the statistical data on HBs in this temperature interval. We have performed appropriate computer experiments for the temperature $T = 3.1 / k_B$. To determine relative average lifetime of HBs composed of two monomer units i and j ($|i - j| > 1$), we used the ratio

$$\tau_{ij} = (\Delta t)^{-1} \sum_{n=1}^{L_{ij}} t_{ij}^{(n)} ,$$ (12)

where $t_{ij}^{(n)}$ is the lifetime of uninterrupted existence of a given HB before n-th break ($n = 1,..., L_{ij}$) for time interval $\Delta t = 20000$ ($8 \cdot 10^6$ iterations).

In addition, we have determined the frequency of breaks of this HB

$$v_{ij} = (\Delta t)^{-1} L_{ij}$$ (13)

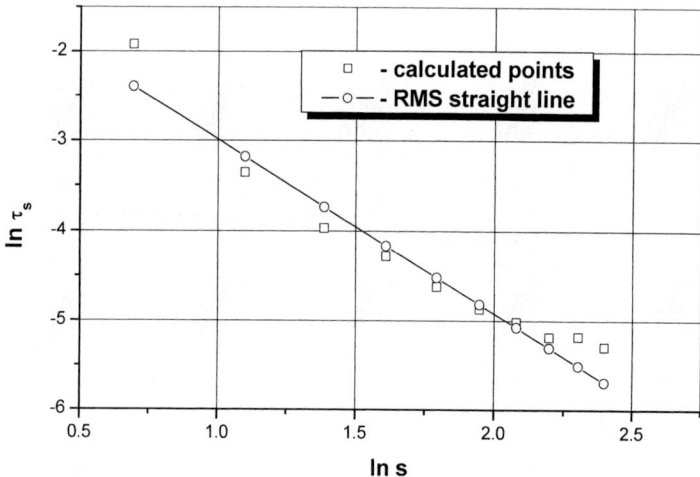

Fig. 8. The dependence of average relative lifetime of HBs τ_s and LSM straight line $\tau = a \ln s + b$ ($a = -1.93$, $b = -1.05$) on the logarithm of the chain distance s for maximum number of HBUs, $N_{HBU} = N = 256$.

As have been mentioned above, the formation of HB with given numbers i and j of monomer units is more or less random event. Therefore we need some parameter as an independent variable to describe statistical properties of relative lifetime and the frequency of HB breaks. To this end, we have chosen the distance $s=|i\text{-}j|$ between HBUs along the chain and defined average relative lifetime τ_s and average frequency of breaks v_s

$$\tau_s = (N-s)^{-1} \sum_{|i-j|=s} \tau_{ij}, \qquad v_s = (N-s)^{-1} \sum_{|i-j|=s} v_{ij}, \; s = 2,...,N-1. \qquad (14)$$

These quantities are shown in Fig. 8 (τ_s) and Fig. 9 (v_s) in double logarithmic scale as functions of chain distance s, at $s = 2,3,4,5$. Most of HB has the value of s in this region. Straight lines presented in Figs. 8 and 9 are calculated by LSM (least-squares method). We can suppose power laws for τ_s and v_s and estimate the corresponding exponents (i.e., the parameter a in the equations $\tau = a \ln s + b$ and $v = a \ln s + b$). It is well known that the probability of the formation of loops in Gaussian chain scales as $\sim s^a$ with $a = -1.5$. The values of a found for the model under consideration are close to that observed for Gaussian chain. Weak distinctions can be explained by excluded volume effects.

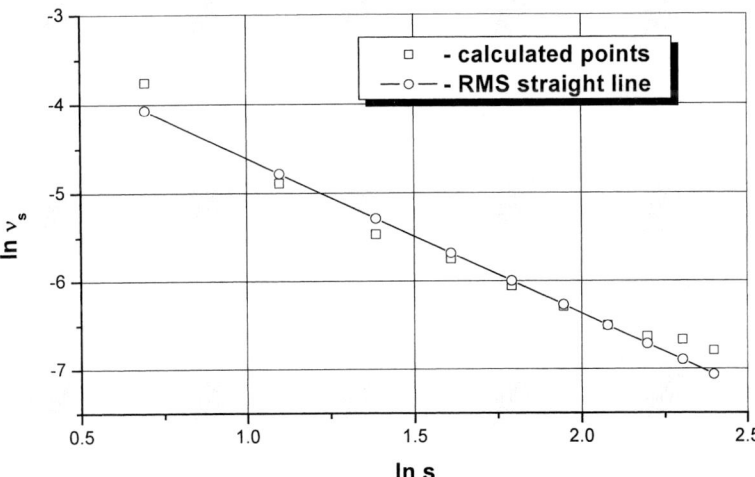

Fig. 9. The dependence of HB breaks frequency v_s and LSM straight line $v = a \ln s + b$ ($a = -1.76$, $b = -2.85$) on the logarithm of the chain distance s for maximum number of HBUs, $N_{HBU} = N = 256$.

Conclusions

1) We proposed the model of HP-copolymer with hydrogen bonds (HB) for computer simulation of the formation of globular structure.

2) The new way for estimation of the temperature of coil-globule transition for HP-copolymers is proposed.

3) The primary structure of parent globule and the HB primary structure do not recover during slow "cooling"[4] of protein-like HP-copolymer.

4) HBs make protein-like globular structure more stable at higher temperatures. The more is the number of hydrogen bond units (HBUs) in the chain the more stable is the protein-like globular structure. The arrangement of HBUs along the chain has secondary significance.

5) The probability of the formation of HBs decreases with increasing the distance s between appropriate HBUs along the chain. We assume power dependence of this probability on s.

Acknowledgements

We acknowledge the financial support from Alexander-von-Humboldt Foundation, Program for Investment in the Future (ZIP), INTAS (project 01-0607), and Russian Foundation for Basic Research (grant 01-01-00474a).

[1] I.M.Lifshitz, A.Yu.Grosberg, A.R. Khokhlov *Zh. Eksp. Teor. Fiz (Soviet Physics - JETP)*, **1976**, *71*, 1634.
[2] F. Cordier, S. Grzesiek, *J. Amer. Chem. Soc.* **1999**, *121*, 1601.
[3] G. Cornilescu, J.-S. Hu, A. Bax, *J. Amer. Chem. Soc.* **1999**, *121*, 2949.
[4] Y.A. Kriksin, P.G. Khalatur, A.R. Khokhlov, *Macromol. Theory and Simul.* **2002**, *11*, 213.
[5] A.R. Khokhlov, P.G.Khalatur, *Physica A* **1998**, *249*, 253.
[6] P.G. Khalatur, V.I. Ivanov, N.P. Shusharina, A.R. Khokhlov, *Rus. Chem. Bull.* **1998**, *47*, 855.
[7] A.R. Khokhlov, V.I. Ivanov, N.P. Shusharina, P.G. Khalatur, "Engineering of synthetic copolymers: Protein-like copolymers", in: *The Physics of Complex Liquids*. F. Yonezawa, K.Tsuji, K. Kaij, M. Doi, T. Fujiwara Eds., World Scientific, Singapore 1998, p. 155.
[8] A.R. Khokhlov, P.G.Khalatur, *Phys. Rev. Lett.* **1999**, *82*, 3456.
[9] V.A. Ivanov, A.V. Chertovich, A.A. Lazutin, N.P. Shusharina, P.G. Khalatur, A.R. Khokhlov, *Macromol. Symp.* **1999**, *146*, 259.
[10] A.V. Chertovich, V.A. Ivanov, A.A. Lazutin, A.R. Khokhlov, *Macromol. Symp.* **2000**, *160*, 41.
[11] E.N. Govorun, V.A. Ivanov, A.R. Khokhlov, P.G. Khalatur, A.L. Borovinsky, A.Y. Grosberg . *Phys. Rev. E*, 2001, *64*, 040903.
[12] A.R. Khokhlov, A.Y. Grosberg, P.G. Khalatur, V.A. Ivanov, E.N. Govorun, A.V. Chertovich, A.A. Lazutin, "Conformation-Dependent Sequence Design of Protein-Like AB-Copolymers", in: *Proceedings of the International School of Physics "Enrico Fermi"*, IOS Press, 2001, p. 313.

[13] N.G. van Kampen, *"Stochastic Processes in Physics and Chemistry"*, North-Holland publ. co., Amsterdam 1981.

[14] M.P. Allen, D.J. Tildesley, *"Computer Simulation of Liquids"*, Clarendon Press., Oxford 1990.

[15] T.C. Bradbury, *"Theoretical Mechanics"*, Wiley, New York 1968.

[16] J. Orba, J.P. Ryckaert, "Methods in Molecular Dynamics". Rapport d'activite Scientifique du CECAM, 1974.

[17] J.P. Ryckaert, G. Ciccotti, H. Berendsen, *J. Comput. Phys.* **1977**, *23*, 327.

[18] V.V. Kislov, Y.A. Kriksin, I.V. Taranov, "Mathematical models of surface monolayers of organic molecules", in: *Mathematical methods in contemporary chemistry*, S.I.Kuchanov, Ed., Gordon and Breach Sc. Pub., Amsterdam, 1996. P.413-441.

[19] A.N. Tihonov, V.J. Arsenin, *"Methods of solving ill-posed problems"*, Winston, Washington, D.C.; Wiley, New York, 1977.

Polymer Modeling: Where Has it Been and Where is it Going?

David Rigby, *¹ B.E. Eichinger²

1. Accelrys, Inc., 9685 Scranton Road, San Diego, CA 92121-3752, USA
2. Department of Chemistry, University of Washington, Seattle, WA 98195, USA
Email: david@accelrys.com , eichinge@chem.washington.edu

Summary: The development of the area of polymer modeling often referred to as molecular modeling has been reviewed from its early beginnings to the present day. Key forces influencing the development include computational power, algorithmic advances and access to computational resources. The desire to apply modeling techniques to predict the properties of increasingly complex polymer-containing systems, taken in conjunction with a number of current limitations discussed in this brief review, is expected to define in part some essential future developments.

Keywords: equation of state, force field, molecular modeling, polymer conformation, simulations

Introduction

The term *Polymer Modeling* has been used to describe a number of techniques whose aim is to describe or predict experimentally measurable properties of polymers in a wide variety of situations including dilute and concentrated solutions, bulk systems, crosslinked materials, chains adsorbed at interfaces or restricted to confined geometries, and polymers in liquid crystalline or other types of mesophase. Typical methods used for modeling these systems range from quantum mechanical and force field based simulations, through various types of coarse-grained approaches, to semi-empirical correlation-based methods and pure thermodynamic methods of varying degrees of sophistication. For the purpose of the present review, based on work presented at the IUPAC-PC2002 conference on the Mission and Challenges of Polymer Science and Technology held in Kyoto Japan, we restrict our focus mostly to the area of polymer modeling often referred to as molecular modeling, involving construction, usually using a computer, of atomic-scale models representing the geometry, spatial configuration, topology and energetics of systems containing chain molecules. The closely related topic of mesoscale modeling, which directly or indirectly probes behavior associated with longer length scale aspects of structure, includes studies of morphology and

DOI: 10.1002/masy.200351105

evolution of morphology in systems undergoing microphase separation, and studies of the various topological features found in elastomeric network forming systems and their consequences. Mesoscale modeling is discussed at length elsewhere in this volume and accordingly is not covered further here.

Early Polymer Models

The use of molecular models in polymer science emerged naturally from the early 1920s work of Staudinger and others aimed at establishing the validity of the macromolecular hypothesis, which is described in some detail in the historical review by Morawetz.[1] Following the general acceptance that high molecular compounds such as caoutchouc were indeed composed of long sequences of covalently bonded repeating units, the stage was set for a whole era of scientific enquiry into the spatial configurations of these flexible macromolecules. Thus, for example, we begin to see the development of expressions for describing characteristic spatial properties such as the moments of the end-to-end distance distribution and the radius of gyration, treated initially within the framework of the statistics of the random walk, which had been discussed much earlier by Pearson[2] and treated mathematically by Lord Rayleigh.[3] In an application to rubber elasticity, Treloar derived the exact expression for the probability distribution for the end-to-end distance of walks of any length,[4] which rapidly approaches the Gaussian form, and later constructed some of the first physical models of random walk chains using wooden sticks, wire and glue.[5]

Contemporaneous studies of chain molecules of different types, including natural polymers starch, cellulose and proteins together with a range of commercial synthetic polymers contributed to rapid increases in sophistication of what initially were mostly mathematical models. Theoretical studies of these models quickly established that any flexible chain molecule in which the range of interactions is local – as represented by chemical bond lengths, valence angles, and chain backbone torsional angles (possibly restricted by near-neighbour steric interactions along the backbone) – would exhibit universal behavior, for which the mean-squared end-to-end distance would always be given by the expression $<r^2_o> = C_n n l^2$, where n denotes the number of backbone bonds of length l, and where the characteristic ratio C_n becomes a constant for chains of sufficient length. Although developed based on the treatment of isolated chain molecules, it was proposed by Flory that the same overall statistics would be followed by multichain systems in two situations, namely polymers in dilute solutions of certain solvents (*theta* solvents, for which the second virial coefficient vanishes) and chains in the melt, a postulate which has subsequently been confirmed by experiment on many occasions. In other situations, chain molecules experience net non-local interactions (sometimes ambiguously termed long-range interactions) leading to mean-squared dimensions scaling with a power of n greater than unity (approximately 1.18 in good solvents). A description of the development of theories and related experimental work aimed at quantifying behavior of perturbed chains can be found in Yamakawa's comprehensive 1971 monograph.[6]

Further development of actual physical models continued into the 1950s with noteworthy contributions from Corey and Pauling[7] in their development of space filling models, made from hardwood and plastic, for visualizing conformations of proteins. In the early 1960s, these models evolved into the well-known CPK molecular models following the redesign work of Koltun initiated at the U.S. National Institutes of Health Biophysical Laboratory. While such models, and their modern-day computer-generated counterparts, continue to be useful for gaining insight into various aspects of polymer conformation, especially in the biological field, the overwhelming number of conformations available to flexible chains inevitably necessitates the introduction of the machinery of statistical mechanics. Thus, given some model of a polymer chain – with specified bond lengths, valence angles, etc. – combined with a prescription for calculating energies of different conformers and for quantifying nonbonded interactions between remote segments of a chain, or perhaps with another chain or small molecule, it is straightforward to write expressions for the partition function and associated averages of properties over phase space. However until the advent of electronic computers, evaluation of such averages was not routinely feasible.

Beginnings of Computational Polymer Modeling

As already implied, much of the framework on which polymer modeling would eventually be based was established prior to the development of the first electronic computers, though numerical applications had yet to emerge. This situation began to change during the 1950s as a result of a number of important developments. First, increasing access to centralized computational facilities, coupled with standardization of programming languages, for the first time made it possible to undertake tasks such as Monte Carlo sampling or complete enumeration of all possible conformations accessible to simple self-avoiding lattice chain models[8] (of relevance to the so-called excluded volume problem), and to contemplate evaluation of the integrals over phase space required to compute the properties of systems containing polymers. Secondly, this era witnessed the emergence of a number of simple and efficient algorithms destined to find widespread use in simulations, together with a comprehensive and rigorous formalism for treating the configurational statistics of unperturbed chains, as described briefly below.

One of the most significant algorithmic contributions to studies of equilibrium behavior is found in the so-called Metropolis Monte Carlo algorithm,[9] which for the first time made it possible to sample configurations of a model system in an efficient but unbiased manner, such that the probabilities of the resulting configurations would conform to canonical ensemble statistics. This algorithm was widely applied in polymer modeling, both in studying basic polymer models (e.g. lattice-based chains), and also in simulations of chemically more realistic models of chains either in isolation or in condensed phases.

The complementary area of molecular dynamics simulation, in which Newton's laws are solved iteratively for the atoms (or other subunits) comprising model polymers yielding information on both equilibrium and dynamical behavior, also benefited from key early algorithmic developments. Some of the most important include the simple integration algorithm introduced by Verlet,[10] and various prescriptions for controlling the temperature and pressure in simulations of condensed phases, as proposed by Berendsen,[11] Nose,[12] Hoover[13] and Parrinello and Rahman.[14] A summary of these algorithms may be found in the now-classical text of Allen and Tildesley.[15]

Another significant development of this early era was the establishment of the rotational isomeric state (RIS) scheme, which replaces the integration over all chain backbone dihedrals contained in the configurational partition function by a discrete sum over preferred, and optionally energetically-weighted states. The basic formalism for computing the second moment of the end-to-end distance distribution of a polymer chain was treated by a number of workers at the end of the 1950s, and summarized in the works of Volkenstein[16] and of Birshtein and Ptitsyn.[17] Shortly thereafter the application of the technique was greatly expanded by Flory and coworkers, beginning with the treatment of polymethylene chains subject to (realistic) interdependent torsional potentials by Abe et al.,[18] with subsequent extension to a wide variety of polymers and extensions to other properties as described in a number of volumes including Flory's 1969 monograph[19] and subsequent work of Mattice and Suter.[20]

Other important polymer-specific algorithms which emerged during these early years include pivot algorithms as employed by Lal[21] and by Stellman and Gans[22] for implementing Monte Carlo moves aimed at rapid sampling of conformation space, as used subsequently in the RIS Metropolis Monte Carlo method.[23] Also at this time, we see the emergence of conformational sampling methods which change the coordinates of sequences of 3 or 4 bonds, leaving the rest of the chain fixed,[24] which were used for later modeling of condensed phases.

The Modern Era

The past twenty five years or so has seen a rapid acceleration in the amount of research activity involving molecular modeling of polymers, and as previously this development is linked to advances in the power and availability of computational resources. Thus at the beginning of this period, widespread access to powerful supercomputers from vendors such as CDC, Cray, Fujitsu and IBM led to comprehensive studies of hitherto inaccessible phenomena, increasingly focusing on dense polymer systems in contrast to the isolated chain models studied in the earlier work. A few examples taken from the 1991 book *Computer Simulations of Polymers*[25] include: PVT relations of melts and glassy polymers, dynamical motions of long chains in dense melts, diffusion studies of light gases in dense systems, polymer melting and crystallization, and mechanical behavior of amorphous polymers. By the middle of this period – around 1988-1992 – powerful UNIX workstations affordable by most laboratories began to shoulder much of the computational burden formerly offloaded to the supercomputers, in addition to providing a platform for high quality graphical systems, often empasized as part of commercial modeling software systems. During the past few years, fast microprocessors have found their way into desktop and laptop computers, and into high end massively parallel computers with phenomenal processing power.

The supercomputer and workstation era also saw the development of further ingenious Monte Carlo move and condensed phase model building algorithms. A number of noteworthy examples include the introduction of the slithering snake (sometimes called reptation) algorithm,[26] as well as a number of end and internal bridging algorithms,[27] and chain regrowth algorithms.[28] Other methods well suited to rapid equilibration of dense systems

52

include applications of the bond fluctuation method,[29] and use of high coordination lattices.[30] It should be remarked that these methods have led to an impressive increase in the range of systems and problems amenable to modeling, though there are frequently limitations in application to arbitrary chain topology and chemistry.

Force Field Based Simulations

From a survey of recent advances in polymer modeling,[31] as well as in a subsequent comprehensive review of the field,[32] it is apparent that while progress continues to be made in the area of algorithms and understanding the physics of high polymer systems, there is also an increasing emphasis on modeling the behavior of chemically diverse polymer systems and complex situations encountered in modern materials (mixtures, interfaces, confined systems, etc.). One of the essential requirements for such studies to be ultimately successful is the availability of accurate force fields. In this regard, the last two decades have witnessed

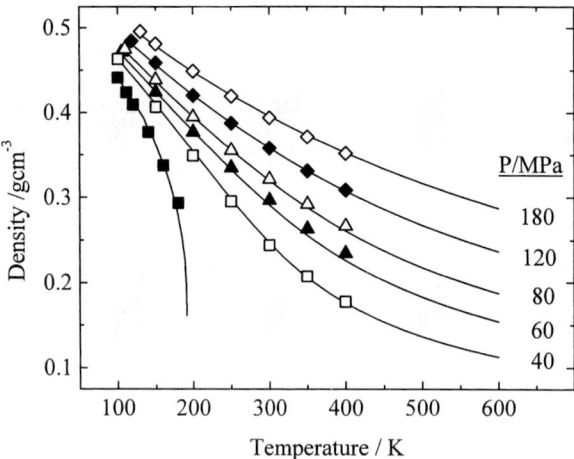

Fig. 1. Density of methane calculated from constant pressure molecular dynamics simulations using the COMPASS force field.[33,34] 100ps dynamics simulations used for each point. Solid curves denote experimental data.

significant advances in the sophistication and accuracy of force fields for modeling organic material, originating mostly from simulations of macromolecules conducted in the life sciences. In recent years, the methodologies developed for deriving force field parameters for these biologically-oriented force fields, usually involving high level ab initio quantum mechanical calculations, have been applied to polymers and small molecule organics (with careful attention also given to parameterizing intermolecular nonbonded interactions using non quantum based methods). An example of what can now be achieved with small molecules is shown in Figure 1.

Here it is observed that PVT bevaviour can be accurately predicted over broad ranges of temperature and pressure. In some of our earlier work, we have also performed similar calculations on melts of systems containing oligomers and high polymer, where similarly good agreement with experiment can be obtained (with extrapolation of oligomer data to infinite molecular weight when studying high molecular weight polymer). A typical example is shown in Figure 2.

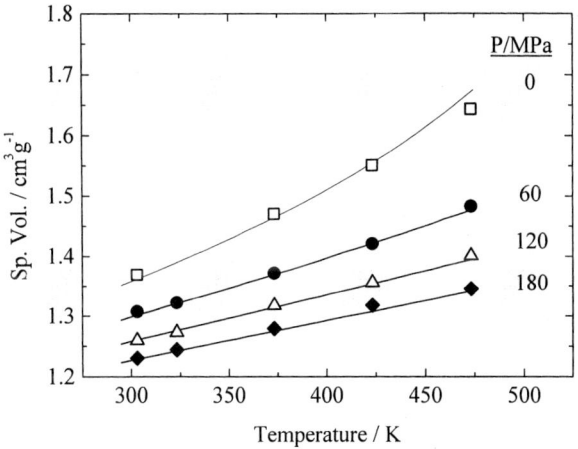

Fig. 2. Specific volume-temperature curves for $C_{11}H_{24}$ obtained as in Figure 1.

Other behavior which must be accurately predictable if modeling is ever to be used to make quantitative calculations of *excess* mixture properties such as the enthalpy of mixing includes cohesive properties. Here also, it is found that heats of vaporization or, alternatively, solubility parameters, of organic solvents can be accurately predicted when a carefully parameterized force field is used. Moreover, although the range of experimentally-reported solubility parameters for polymers is often quite broad, there are good indications that molecular modeling can also be used to make reliable and accurate predictions of this important quantity. As an example, Figure 3 illustrates the method of estimating the solubility parameter at 298K for high molar mass poly(ethylene oxide), which was shown in a previous publication to be in excellent agreement with experiment.[35]

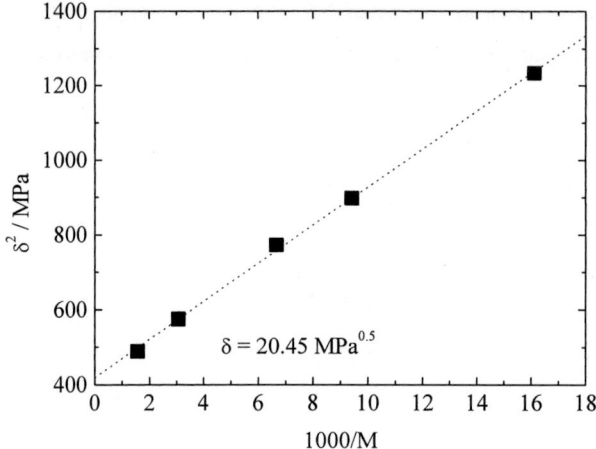

Fig. 3. Calculation of solubility parameter δ of high molar mass poly(ethylene oxide) by extrapolation of oligomer data vs. the reciprocal of the molar mass M.

Concluding Remarks

Polymer modeling has evolved dramatically since the advent of the first electronic computers, with early applications using simple, often lattice-based, models of isolated chains to investigate the factors which govern the spatial configuration of chain molecules upon which many important phenomena such as solution viscosity depend. Widespread access to supercomputers later effected a shift in focus of polymer modeling activity to a broader range of properties, mostly pertaining to bulk melt, glassy and crystalline systems. This trend has continued through the era of the laboratory graphical workstation and most recently into a new era of desktop computers and large scale parallel systems, while the breadth of applications continues to expand, ranging from studies of fundamentals of high polymer physics, through property prediction for bulk polymer and solvent, to studies of chemical phenomena (such as degradation), and increasingly involving studies of complex systems encountered in materials research.[36] Parallel developments in the field involving longer length scale simulations have fueled an interest in generating the molecular level properties required as input to these mesoscale simulations using data obtained purely from atomistic simulation.

Although it is difficult to speculate on the future direction of polymer modeling, it is nonetheless possible to identify some current limitations which in principle should drive future developments.

In the little-explored area of polymer chemistry there are a few needs:

1. Improved understanding of polymer degradation, perhaps involving reactive force fields.

2. More extensive, quantum mechanics based, studies of polymerization behavior, including ring formation, transesterification, etc.

In the area of force field based simulations, some limitations are as follows:

1. The accuracy of current force fields is impressive. However work is still required to establish whether the current level of accuracy is adequate for direct prediction of excess thermodynamic properties of mixtures, for example.

2. Alternative (e.g. Gibbs-ensemble) methods for studying phase behavior[37], while successful, are hampered by limited coverage of the united atom force fields currently used, while usage of widely available and more extensively parameterized all atom force fields is limited by low efficiency of the associated simulation algorithms.

3. Force field-based methods for accurately and routinely predicting properties of glassy polymers (e.g. volume-temperature behavior below and around T_g) are still in need of further development.

4. Although promising work involving use of atomistic methods to study aspects of viscoelastic behavior has recently been reported[38], this has so far been rather limited.

5. Methods for prediction of diffusivities and solubility of molecules of moderate size (e.g. of the size of drugs with 30-50 heavy atoms) in amorphous polymer have yet to be perfected.

In the area of long length scale simulations, perhaps the weakest link currently is the difficulty in obtaining accurate values of the necessary input parameters, especially thermodynamic interaction parameters, directly from simulations. In this regard, investigations of the sensitivity of the characteristic morphologies and their rate of development to variation/errors in input parameters would also be of interest.

56

[1] H. Morawetz, *Polymers. The Origins and Growth of a Science*, J. Wiley & Sons, New York 1985, Ch 10.

[2] K. Pearson, *Nature* **1905**, 77, 294.

[3] Lord Rayleigh, *Phil. Mag.* **1919**, 37, 321.

[4] L. R. G. Treloar, *Trans. Faraday Soc.* **1946**, 42, 77.

[5] see L. R. G. Treloar, *The Physics of Rubber Elasticity*, 3rd ed., Oxford Univ. Press 1975.

[6] H. Yamakawa, *Modern Theory of Polymer Solutions*, Harper & Row, New York 1971.

[7] R. B. Corey, L. Pauling, *Rev. Sci. Instr.* **1953**, 24, 621.

[8] see, for example, F. T. Wall, J. J. Erpenbeck, *J. Chem. Phys.* **1959**, 30, 634; M. E. Fisher, B. J. Hiley, *J. Chem. Phys.* **1961**, 34, 1253.

[9] N. Metropolis, A. W. Rosenbluth, M. N. Rosenbluth, A. H. Teller, E. Teller, *J. Chem. Phys.* **1953**, 21, 1087.

[10] L. Verlet, *Phys. Rev.* **1967**, 159, 98.

[11] H. J. C. Berendsen, J. P. M. Postma, W. F. van Gunsteren, A. DiNola, J. R. Haak, *J. Chem. Phys.* **1984**, 81, 3684.

[12] S. Nose, *J. Chem. Phys.* **1984**, 81, 511.

[13] W. G. Hoover, *Phys. Rev. A.* **1985**, 31, 1695.

[14] M. Parrinello, A. Rahman, *J. Appl. Phys.* **1981**, 52, 7182.

[15] M. P. Allen, D. J. Tildesley, *Computer Simulation of Liquids*. Oxford Univ. Press 1987.

[16] M. V. Volkenstein, *Configurational Statistics of Polymeric Chains*, Interscience, New York 1963.

[17] T. M. Birshtein, O. B. Ptitsyn, *Conformations of Macromolecules*, transl. S. N. Timasheff and M. J. Timasheff, Interscience, New York 1966.

[18] A. Abe, R. L. Jernigan, P. J. Flory, *J. Am. Chem. Soc.* **1966**, 88, 631.

[19] P. J. Flory, *Statistical Mechanics of Chain Molecules*, Interscience, New York 1969.

[20] W. L. Mattice, U. W. Suter, *Conformational Theory of Large Molecules*, Interscience, New York 1994.

[21] M. Lal, *Mol. Phys.*, **1969**, 17, 57.

[22] S. D. Stellman, P. J. Gans, *Macromolecules* **1972**, 5, 516.

[23] J. D. Honeycutt, *Comp. Theor. Polym. Sci.* **1998**, 8, 1.

[24] A. T. Clark , M. Lal, *British Polymer Journal* **1977**, 9, 92.

[25] R. J. Roe (Ed.), *Computer Simulations of Polymers*, Prentice-Hall, New Jersey 1991.

[26] M. Vacatello, G. Avitabile, P. Corradini, A. Tuzi, *J. Chem. Phys.* **1980**, 73, 548.

[27] P. V. K. Pant, D. N. Theodorou, *Macromolecules* **1995**, 28, 7224.

[28] J. J. de Pablo, M. Laso, U. W. Suter, *J. Chem. Phys.* **1992**, 96, 2395.

[29] I. Carmesin, K. Kremer, *Macromolecules* **1988**, 21, 2819.

[30] P. Doruker, W. L. Mattice, *Macromol. Theor. Sim.* **1999**, 8, 463.

[31] B. E. Eichinger, D. Rigby, *Current Opinion in Solid State and Materials Science* **2001**, 5, 445.

[32] B. E. Eichinger, R. Khare, *Molecular Modeling* in: *Encyclopedia of Polymer Science and Technology*, Interscience, New York 2003.

[33] H. Sun, *J. Phys. Chem. B* **1998**, 102, 7338.

[34] D. Rigby, revised COMPASS parameters for methane, unpublished.

[35] D. Rigby, H. Sun, B. E. Eichinger, *Polymer International* **1997**, 44, 311.

[36] see, for example, R. Khare, J. J. de Pablo, A. Yethiraj, *Macromolecules* **1996**, 29, 7910.

[37] S. K. Nath, F. A. Escobedo, J. J. de Pablo, *J. Chem. Phys.* **1998**, 108, 9905.

[38] O. Byutner, G. D. Smith, *Macromolecules* **2002**, 35, 3769.

Polymer Concepts in Microscopically-Viewed Phase Transition Behavior of Crystalline Polymers

Kohji Tashiro,* Yayoi Yoshioka, Hisakatsu Hama, Akiko Yoshioka

Department of Macromolecular Science, Graduate School of Science, Osaka University, Toyonaka, Osaka 560-0043, Japan
Email: ktashiro@chem.sci.osaka-u.ac.jp

Summary: Structural evolution process in crystalline phase transition, isothermal crystallization from melt and solvent-induced crystallization has been investigated for crystalline polymers on the basis of the time and/or temperature dependence of small-angle X-ray scattering (SAXS), wide-angle X-ray scattering (WAXS) and infrared/Raman spectra. As examples of isothermal crystallization study, polyethylene and polyoxymethylene were investigated. As for polyoxymethylene, the time-resolved infrared spectral measurement clarified the appearance of infrared bands characteristic of folded chain crystal (FCC) in the early stage of crystallization process, followed by the detection of infrared bands characteristic of extended chain crystal (ECC) in a later stage. From the time-resolved SAXS measurement, the stacked lamellar structure of ca. 14 nm long period was detected at first and it was followed by an appearance of the new lamellae in the amorphous region sandwiched between the original lamellae, resulting in the tightly stacked lamellar structure of 7 nm long period. The timing of detecting the SAXS 14 nm peak was almost the same with the observation of infrared FCC bands and the SAXS 7 nm peak with the infrared ECC band. This correspondency allowed us to speculate that some chains bridging several lamellae became fully extended when the new lamellae were formed in the amorphous region sandwiched between the original lamellae. As a result the ECC-like structural parts were formed, giving the ECC infrared bands. As other examples, the Brill transition of aliphatic nylons and the solvent-induced crystallization of syndiotactic polystyrene have been reviewed.

Keywords: Brill transition, isothermal crystallization, polyoxymethylene, solvent-induced crystallization, syndiotactic polystyrene

Introduction

One of the most characteristic structural features of a flexible polymer is a sensitive variation of its molecular conformation depending on the environmental condition. Typical example is seen in the phenomenon of polymorphism or the existence of various kinds of crystal

© 2003 WILEY-VCH Verlag GmbH & KGaA, Weinheim   DOI: 10.1002/masy.200351106

modification. Besides, by changing such an external condition as temperature, stress, etc. the polymer chains experience phase transitions among these crystalline modifications. In the crystallization phenomenon also the polymer chains change their conformation remarkably from the random coil to the regular form. Such a drastic change of chain conformation results in remarkable change in the aggregation structure of the chains and also the change in higher order structure.

In order to understand the essential features of the phase transition and crystallization phenomenon, we need to obtain the information on the structural formation process as well as the structure itself. In other words we need to perform the time-resolved measurements of X-ray diffraction, infrared and Raman spectra, and so on. For the X-ray diffraction measurement, we can make a rapid collection of the one- and two-dimensional data by using a combination of quite powerful synchrotron radiation and highly-sensitive detector such as PSPC (position sensitive proportional counter) or CCD camera. Fourier-transform vibrational spectrometer allows us to make a rapid-scanning measurement of infrared/Raman spectra.

In the present paper we will describe the experimental results about the structural changes in the isothermal crystallization processes of POM and PE, in the solvent-induced crystallization of syndiotactic polystyrene (sPS) and in the Brill transition of nylons, which were obtained by utilizing the above-mentioned modern techniques.

Isothermal Crystallizations of POM and PE

POM is known to show characteristic infrared spectra, which are quite sensitive to the morphology of crystalline phase: folded chain crystal (FCC) and extended chain crystal (ECC).[1] These two kinds of POM sample show the infrared bands at the positions different by 100 cm^{-1}! By measuring the infrared spectra during the isothermal crystallization, therefore, we may trace the morphological change concretely. By combining this infrared spectral technique with the WAXS and SAXS methods, the structural change in the isothermal crystallization process of POM was investigated from the various points of view.

When the isothermal crystallization was performed at $130^{\circ}C$, the 1139 cm^{-1} infrared band of FCC morphology started to appear immediately after the temperature jump from the melt and increased in intensity with time. The 902 cm^{-1} band of ECC morphology was observed to appear around 150 sec later and increased in intensity. The time-resolved measurement of SAXS profile showed similar two-stages change. In the time region of detecting the FCC infrared bands, the SAXS peak (L_1) with the long period of ca. 14 nm started to appear. After that, the peak (L_2) of ca. 7 nm long period increased in intensity just when the infrared bands intrinsic of ECC morphology were detected, and the L_1 peak decreased in intensity in parallel. The generation of stacked lamellar structure of L_1 in the early stage of crystallization and the change to the lamellar structure with the long period almost a half of L_1 can be observed also in the crystallization of polyethylene.[2] These SAXS data could be interpreted reasonably in such a way that the amorphous region sandwiched between the original lamellae changes into a new crystalline lamella at the secondary crystallization stage (lamella insertion model). The SAS data were quantitatively analyzed on the basis of 1-dimensional lamellar stacking structure model with the second kind of paracrystalline disorder,[3] from which the

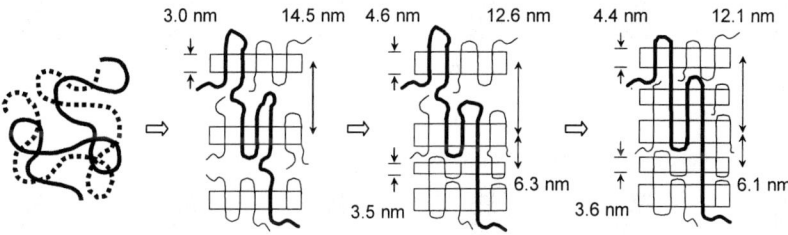

Fig. 1. A model of structural evolution process of POM in the isothermal crystallization at 130°C.

structural information was obtained. By combining this information with the good correspondence between the X-ray and IR data (L_1 and FCC, and L_2 and ECC), we may derive the structural change shown in Figure 1 as a possibility. Immediately after an occurrence of crystallization, stacked lamellar structure with ca. 14 nm long period is generated. The molecular chains are folded and give the infrared bands of FCC. As the time passes furthermore, new lamellae are generated from the amorphous part sandwiched between the original lamellae. Some of the amorphous chain segments have a probability to pass through the neighboring two lamellae. When this amorphous region regularizes into a new lamella, the amorphous chain segment changes into straight stem and the total length of the extended segment becomes longer to form a taut tie chain.

Solvent-induced Crystallization of sPS

Amorphous sPS sample was exposed in an atmosphere of such organic solvent vapor as toluene.[4] The time dependence of infrared spectra was measured during this exposure. In parallel to the increment of toluene band intensity, the amorphous bands decreased in intensity and the crystalline bands characteristic of T_2G_2 helical chain conformation increased in intensity. Some of crystalline bands were found to appear earlier than some other crystalline bands. This phenomenon could be detected also in the Raman spectra. This detection time difference among the various bands is considered to come from a difference in sensitivity of an infrared (and Raman) band to the effective helical chain length or a difference in critical sequence length: a crystalline band can be detected for the first time when the length of a regular helical segment is beyond a critical value intrinsic of this band.[5] That is to say, the bands with relatively short critical sequence length were detected at first after injection of toluene, and the bands having longer critical sequence length were detected much later than the former bands, indicating a growth of helix. In parallel to this spectroscopic study, the time-resolved measurement of X-ray diffraction was made during exposure of the sample into a toluene gas. The timing to detect the X-ray diffraction intensity was almost the same with the timing to observe the Raman bands with long critical sequence length. Therefore, we may speculate that the random coils in the amorphous phase change at first into regular and short helical segments, which grow into longer helical sequences and gather together to form a crystalline lattice as being observed in the X-ray diffraction.

This solvent–induced crystallization of sPS was observed at room temperature, much lower than the original glass transition point (ca. 100°C). It may be easily speculated that the molecular motion in the amorphous phase is activated by a plasticizer effect of absorbed toluene molecules. In order to detect such a molecular motion experimentally, we measured the half-width of infrared amorphous band because the band width is inversely proportional to the relaxation time in general.[6] In fact, the half-width of the amorphous band became wider after injection of solvent, indicating a start of molecular motion. After that, for the first time, the crystalline bands with short critical sequence length began to appear. The half-width of the amorphous band became narrower again when the crystalline lattice was formed, indicating that the amorphous region was sandwiched between the crystalline lamellae and its motion was confined more or less due to some geometrical constraint.

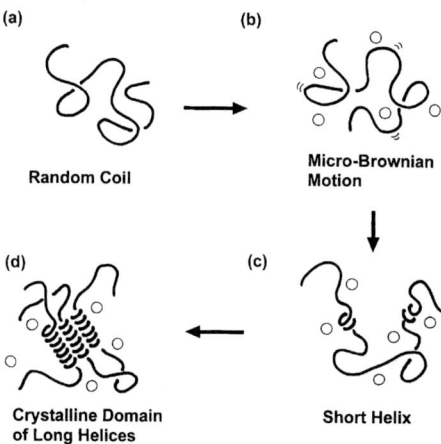

Fig. 2. An illustrated model of solvent-induced crystallization of sPS. Random coil starts to make a micro-Brownian motion by absorbing solvent (open circle). Then short helices are generated, grow longer and form a crystal lattice.

By combining all the experimental data mentioned above, a concrete image could be obtained for the solvent-induced crystallization phenomenon of sPS as shown in Figure 2. By absorbing toluene, the amorphous chains become mobile and change into regular short helical segments, which grow to longer helical sequences and form a crystalline lattice.

Brill Transition of Aliphatic Nylons

Many aliphatic nylons exhibit the so-called Brill transition phenomenon between the α form of triclinic structure and the pseudo-hexagonal form,[7, 8] but the details have not yet been clarified enough well. The conformational change in the methylene sequences is one of the most important problems to be solved about this transition. It was found to be possible to trace this conformational change by investigating the temperature change in a series of progression bands observed in the infrared spectra of aliphatic nylons.

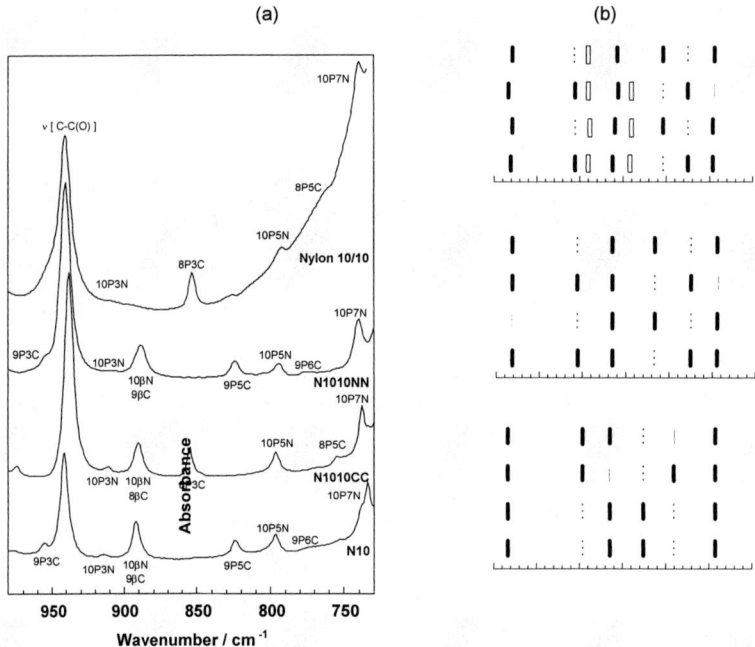

Fig. 3. (a) Observed infrared spectra of nylon 10/10 and its model compounds (refer to the text) in the CH_2 rocking region. The mark 8P5C, for example, indicates that this band is equivalent to the CH_2 rocking mode (P) with $k = 5$ of n-C_8H_{18}.(b) Comparison of the observed data with the predictions made by assuming the effective number of oscillators in different ways. The numbers [p, q] indicate the carbon numbers of equivalent n-alkane molecules including methyl end groups.

The progression bands are observed for a methylene segment of finite length as well known for n-alkanes and so on and can be interpreted on the basis of a simply-coupled oscillator model.[9] The vibrational frequency is a function of phase difference δ between the neighboring methylene units: $\delta = k\pi/(m + 1)$ where $k = 1, 2, \ldots, m$ and m is a total number of zigzag methylene units included in the stationary wave of the vibration. We have found that the progression bands observed for many nylon samples can be assigned reasonably by separating a methylene unit adjacent to the amide group from the other methylene sequence. For example, for nylon 10/10 -[-NH(CH₂)₁₀NHCO-(CH₂)₈-CO-]-, the observed progression bands can be assigned to the vibrational modes of methylene sequences of $(CH_2)_8$ (NH side) and $(CH_2)_6$ (CO side), not $(CH_2)_{10}$ and $(CH_2)_8$. Figure 3 shows this situation clearly. In Figure 3 (a) the infrared spectra observed for 4 types of sample (nylon 10/10 and its model compounds: N1010NN [$CH_3(CH_2)_8CONH(CH_2)_{10}NHCO(CH_2)_8CH_3$], N1010CC [$CH_3(CH_2)_9NHCO(CH_2)_8CONH(CH_2)_9CH_3$], and N10 [$CH_3(CH_2)_8CONH(CH_2)_9CH_3$]) are compared with each other. The positions predicted for several main bands are compared with the observed data as seen in Figure 3 (b). Prediction II is made by assuming all the methylene units as effective oscillators. Prediction I is based on the above-mentioned new concept. The latter gives a good agreement with the observed data. The methylene unit adjacent to the amide group is speculated to experience rather different thermal motion from that of the inner part of methylene sequence, resulting in the vibrational decoupling between them. Another possibility is a difference in electronic structure of methylene unit adjacent to the amide group from that of the inner part, giving different force constant and different vibrational frequency.

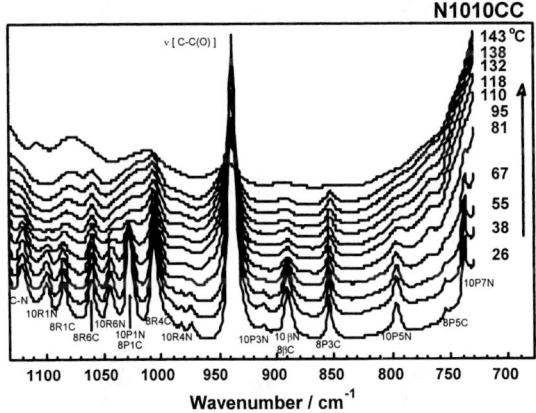

Fig. 4. Temperature dependence of infrared spectra of nylon 10/10 model compound (N1010CC, refer to the chemical formula given in the text).

The temperature dependence of these progression bands was measured. Above the Brill transition point, some of the progression bands disappeared and new progression bands appeared and increased in intensity as shown in Figure 4. At the same time the gauche bands were also detected. The generation of gauche bond shortens the long trans segment and the effective number of methylene units included in the trans-zigzag part becomes smaller. This change in methylene segmental length gives the change of phase difference and then the change of vibrational frequency. From the positions of the newly appeared bands, the average length of the thus shortened alkane segments could be estimated. In the case of Figure 4, for example, the trans-zigzag methylene segment is considered to change from $(CH_2)_{10}$ to $(CH_2)_7$ in average. In the Brill transition, the hydrogen bonds between the neighboring amide groups are not broken, as supported by the observation of hydrogen-bonded amide I band for example. Above the transition point the conformational disordering in the methylene segments is increased but under some constraints from the hydrogen bonds.

[1] M. Kobayashi, M. Sakashita M. *J. Chem. Phys.* **1992**, *96*, 748.
[2] S. Sasaki, K. Tashiro, M. Kobayashi, Y. Izumi, K. Kobayashi, *Polymer* **1999**, *40*, 7125.
[3] R. Hosemann, S. N. Bagchi, *"Direct Analysis of Diffraction by Matter"*, North-Holland, Amsterdam 1962.
[4] K. Tashiro, Y. Ueno, A. Yoshioka, M. Kobayashi, *Macromolecules* **2001**, *34*, 310.
[5] M. Kobayashi, K. Akita, H. Tadokoro, *Makromol. Chem.* **1968**, *118*, 324.
[6] K. Tashiro, A. Yoshioka, *Macromolecules* **2002**, *35*, 410.
[7] R. J. Brill, *Prakt. Chem.* **1942**, *161*, 49.
[8] T. Itoh, *Jpn. J. Appl. Phys.* **1976**, *15*, 2295.
[9] R. G. Snyder, J. H. Schachtschneider, *Spctrochimica Acta* **1963**, *19*, 85.

Macromol. Symp. **2003**, *201*, 65—75 65

Competing Short-Range and Long-Range Interactions in Block Copolymers: A Role of Connectivity in Polymer Science

T. Hashimoto,* K. Yamauchi, D. Yamaguchi, H. Hasegawa

Department of Polymer Chemistry, Graduate School of Engineering, Kyoto University, Kyoto 606-8501, Japan, E-mail: hashimoto@alloy.polym.kyoto-u.ac.jp

Summary: We explored a role of connectivity in polymer science by using two model systems: mixtures of two lamella-forming block copolymers with short and long chain lengths and triblock terpolymers of polyisoprene-*block*-polystyrene-*block*-poly(vinyl methyl ether). These two systems commonly have enriched competitions or interplays of short-range segmental interactions and long-range interactions arising from connectivity between block chains and their packing in the respective domains. The enriched interplays were found to give an intriguing cosurfactant effect for the block copolymer mixtures and interesting two-step microphase transitions for the triblcock terpolymers.

Kerwords: cosurfactant effects, microphase transition, mixtures of block copolymers, polyisoprene-*block*-polystyrene-*block*-poly(vinyl methyl ether), triblock terpolymers

Introduction

We would like to discuss "Short-range and Long-range Interactions" which can compete each other and which are important in self-assembly in block copolymer systems. For this purpose we select two model systems: a mixture of lamella-forming polystyrene-*block*-polyisoprene having a large difference in molecular weights. By presenting the experimental studies of these two systems, we hope we can highlight a role of connectivity in polymer sciene.

It is extremely interesting for us to realize that just a single covalent bond between one end of A and that of B polymers strongly alter the physics of these two-component polymer systems, i.e., mixtures of A and B denoted hereafter A/B and diblock copolymers designated hereafter A-B. Without this covalent bond, polymer mixtures A/B can achieve macrophase separation with a single interface at thermal equilibri-um. The important physics here is competing short-range

 DOI: 10.1002/masy.200351107

interactions between A-A, B-B and A-B segm-ents. On the other hand with this single covalent bond, A-B block copolymers can achieve microphase separation in which the bulk block copolymer is an assembly of monolayers of A and B blocks separated across the interface and is considered to be nothing other than interfaces themselves, if the monolayers are considered to be a part of the interface. It is very important to realize that the unique microphase-separated structure in neat block copolymers is created by an interplay of the short-range interactions inherent in the mixture and the long-range interactions arising from the block connectivity and the resulting packing effects of A and B in the respective microphases. The long-range interactions involve conformational entropy of block chains which is inherent and very important in polymer science. The interplay of these two interactions yields such long-range-ordered domain structures with various symmetries comprised of the nano-sized structural units,[1-3] such as spheres in body-centered and face-centered cubic lattice,[4] hexagonal cylinder, double gyroid network with $Ia\bar{3}d$ space group symmetry[5,6] and alternating lamellae for diblock copoymer melts.

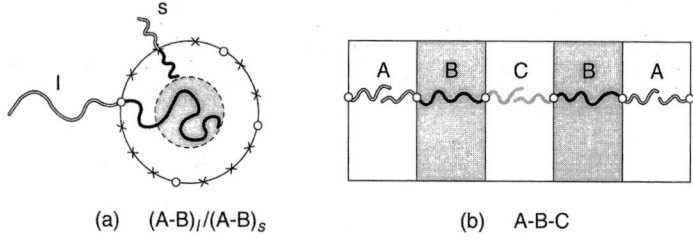

(a) $(A\text{-}B)_l/(A\text{-}B)_s$ (b)   A-B-C

Fig. 1. Schematic representation of the model systems used in this study. (a): mixtures of long and short block copolymers, (b): A-B-C triblock terpolymers.

We should further note the fact that there is enrichment of competitions and interplays in the short-range and long-range interactions for such systems as, for example, (a) mixtures of long and short block copolymers, and (b) A-B-C triblock terpolymers. In the mixtures of long and short block copolymers, designated respectively as l and s, as shown in Figure 1(a) with either identical or different block composition, the short-range interactions are the same as those in neat block copolymers but the long-range interactions between l and s will compete each other.

Furthermore, the short- and long-range interactions interplay each other. In the A-B-C triblock terpolymers, we have three kinds of competing short-range interactions between two different block sequences; A•B, B•C and C•A and long-range interactions due to A-B and B-C connectivity. The short- and long-range interactions interplay each other. We should note that these interplays and competitions are fundamental elements in polymer science and enrich varieties in nano-sized domain structures.

A Cosurfact Effect in a Mixture of Block Copolymers

Now let us discuss the self-assembly of mixtures of block copolymers. As we discussed already, we have the same short-range interactions between A and B as in neat A-B block copolymers. However, the long-range interactions of the block copolymers now have to compete each other, which should give new "delicate effects" on nano-sized patterns. The competing long-range interactions can give two cases: (a) one in which the junctions of two blocks share common interfaces and act as co-surfactants as shown in Figure 1(a), and (b) the other in which they share different interfaces, giving rise to macrophase separation into two different ordered phases. The macrophase separation between the two block copolymers having the same block compositions was experimentally discovered by Hashimoto and coworkers[7] and theoretically predicted by Matsen.[8]

Let us now demonstrate one intriguing example among various cosurfactant effects we encountered. We consider a mixture of lamellar forming polyisoprene-*block*-polystyrene (designated hereafter as PI-PS) block copolymers, one having total number-averaged molecular weight M_n of 100×10^3 and polystyrene (PS) volume fraction of 0.47 (Figure 2(a)) and the other having total M_n of 14.5×10^3 and PS volume fraction of 0.46 (Figure 2(c)). Surprisingly enough the 60/40 mixture in weight of the large/small molecular weight block copolymers shows up cylindrical domains as shown in Figure 2(b). The results are reproducible and confirmed also by SAXS. Please note that this cylinder is unique, because it has total weight fraction of cylinders as high as about 50%.

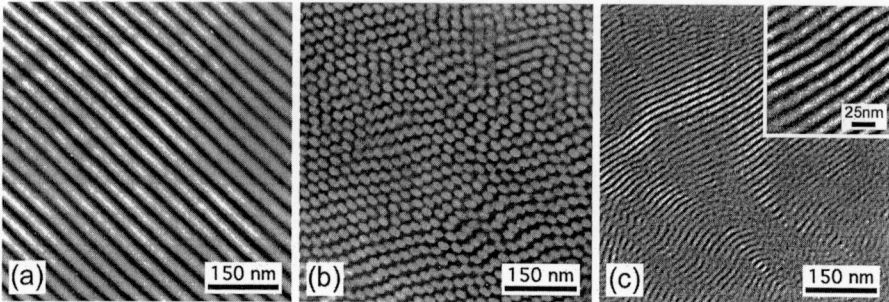

Fig. 2. TEM micrographs of neat PS-PI diblock copolymers: (a) PS(47K)-PI(53K), f$_{PS}$=0.47, (c) PS(7.7K)-PI(6.8K), f$_{PS}$=0.46, and its polymer blend: (b) (a)/(c) = 60/40 (wt%)blend.

Why can mixing the two blocks of nearly the same PS volume fraction give the cylinder? One important hint is that as far as the number of molecules is concerned, the small molecular weight block copolymer is a major component, occupying 72 mol% and the large molecular weight block copolymer is a minor component, occupying only 18 mol%. Thus we can interpret our result in such a way that the minority long block copolymer changed the curvature of the majority phase from lamellae to cylinder, due to the competing long-range interactions.

How can we intuitively understand this intriguing lamella-to-cylinder transition upon the mixing of the two blocks? The neat short block copolymer, which is the majority component and which is slightly asymmetric in volume of PS and PI blocks, may want to have the curved interface with a spontaneous curvature such as shown in Figure 3(a), if packing interactions from other block copolymer chains are negligible. However, in the real world, many-chain effects on packing do exist and do not allow the spontaneously curved interface as shown in part a, because it creates density dip in the PI phase and excess density in the PS phase. As a consequence we have a flat interface for the neat short block copolymers with such a composition as PS volume fraction of 0.46, and hence the lamellar morphology as shown in Figure 3(b). However, when the long nearly symmetric block copolymers are mixed by a small amount, this packing constraint encountered in the neat block copolymer may be relaxed and the interface may tend to have the spontaneous curvature driven by the slightly asymmetric short block copolymer as shown in Figure 3(c), even in the case where the asymmetry is very small. In this case the long block copolymers tend to fill the space unoccupied by the short slightly asymmetric block copolymers, therefore stabilizing the curved interface driven by the cosurfactant effect and the competing long range interactions of l and s. We found the lamella-to-cylinder transition induced by mixing the lamella-forming block copolymers can be explained theoretically based on the strong segregation theory proposed by Brishtein and her coworkers.[9-10]

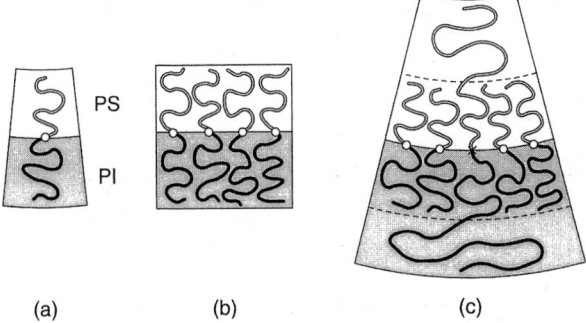

(a)　　　　　　　　　(b)　　　　　　　　　(c)

Fig. 3. Schematic illustrations of (a) a slightly asymmetric diblock copolymer molecule with the spontaneous curvature, (b) aggregation of the slightly asymmetric short diblock copolymers with a flat interface, (c) mixture of a large amount of slightly asymmetric short block copolymers and a small amount of long nearly symmetric block copolymers with the spontaneous curvature.

Phase Transitions in Triblock Terpolymers

Let us now move to the second topic of self-assembly of A-B-C triblock terpolymers. In A-B-C triblock terpolymers we have competing long-range interactions due to A-B and B-C connectivities as well as competing short-range interactions between A•B, B•C and C•A. Moreover, these long-range and short-range interactions interplay each other. The enriched competitions and interplays create rich variations of the ordered structures as illustrated in Figure 1 of our previous report.[11] Some of them have been actually found.[12-13] So far, however, most of the reported works have been directed toward finding various ordered structures developed by changing compositions or components of A-B-C. However, little works have been devoted to ordering process of these structures and phase transitions among various ordered structures.

If we can induce order-disorder transition and order-order transitions to the systems at will, we can attain a very rich morphology control. We like to present one of our recent studies along this line, concerning phase transitions with temperature for a given triblock terpolymer.

The triblock terpolymer we shall discuss here is polyisoprene-*block*-poly(deuterated styrene)-*block*-poly(vinyl methyl ether) (PI-DPS-PVME) having rather small block molecular weights of 14K, 3K and 3K for PI, poly(deuterated styrene) (DPS) and poly(vinyl methyl ether) (PVME) blocks, respectively. These blocks have interesting competing short-range interactions: (a) The interactions between PI and DPS and those between PI and PVME decrease with temperature so that they tend to go from the 2-phase to the 1-phase state with temperature; (b) On the contrary, the interactions between DPS and PVME increase with temperature so that they tend to go from the 1-phase to the 2-phase state with temperature. These interesting short-range interactions interplay with the long-range interactions coming from the PI-DPS connectivity and the DPS-PVME connectivity, which is anticipated to create interesting phase transitions with temperature.

Figure 4 shows morphology of the as-cast film at room temperature observed by transmission electron microscopy on the ultra thin sections stained with OsO_4. In part (a) we see a two-phase morphology composed of PI matrix stained dark by OsO_4 and unstained bright cylinders composed of DPS and PVME. At this stage we do not know whether or not DPS and PVME microphase-separate. Thus we further stained with aqueous phosphotungstic acid (PTA) which should selectively stain the PVME phase but not DPS phase. We observed no change with the PTA staining as shown in part (b). Therefore, we conclude that DPS and PVME are mixed together to form the cylinders hexagonally-packed in the PI matrix.

In the results shown in Figure 4(c), we exposed the ultrathin section stained with OsO_4 to water for 1 hour and then stained with PTA. We observe core-shell cylinders in which the unstained bright shell are PS phase and the stained dark cores are PVME. This result demonstrates the solvent (water)-induced microphase transition inside the nanocylinders in which PVME brushes are emanating from the vitrified DPS shells and swollen with water. Evaporation of water would create a channel of hole inside each PVME core. This nano-sized channel structure would be interesting for practical applications such as for membrane reactors, selective permeations, and so on.

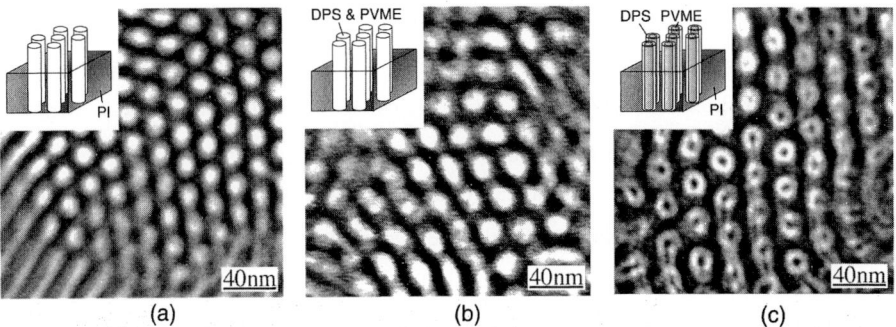

Fig. 4. TEM micrographs of as-cast PI-DPS-PVME film: (a) is the ultrathin section stained by OsO₄, (b) and (c) are the ultrathin sections first stained with OsO₄ and then PTA, the difference between (b) and (c) exists in the time exposed to purified water before staining with PTA: (b) 0 min. and (c) 60 min. The insertions are the model structures corresponding to each TEM micrograph.

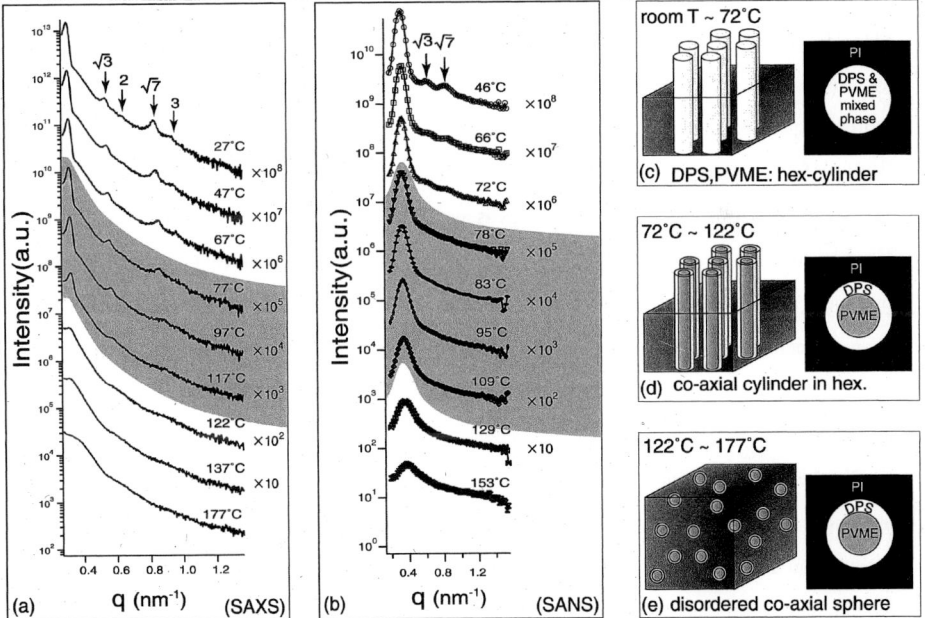

Fig. 5. In-situ SAXS (a) and SANS profiles (b) obtained at various temperatures and schematic models of microdomain structures in the temperature ranges between room temperature and 72°C (c), between 72 and 122°C (d), and between 122 and 177°C (e).

In order to explore such morphological change or phase transition *in-situ* at elevated temperatures, we should use scattering methods by which we can distinguish phases composed of these 3 components. For this purpose we used a combined small-angle neutron scattering (SANS) and small-angle X-ray scattering (SAXS). In SANS the scattering contrast of PI and PVME are almost same (2.69×10^9 and 3.5×10^9 cm/cm^3, respectively) and about 1/20 of DPS (6.49×10^{10} cm/cm^3) so that only DPS is highlighted against others. On the contrary in SAXS, the contrast between DPS and PVME are identical (3.41×10^{23} and 3.42×10^{23} electrons/cm^3, respectively), so that only PI is highlighted against others.

Figure 5 shows the SAXS (a) and SANS profiles (b) obtained in-situ at various temperatures and plotted with a logarithmic intensity scale and a linear scale for scattering vector q. At room temperature both the SAXS and SANS profiles show scattering peak at $1:\sqrt{3}:2:\sqrt{7}:3$ relative to the 1st-order peak position, suggesting hexagonally packed cylinders. This result is consistent with TEM observation of the two-phase cylindrical structure where DPS and PVME are mixed as shown in part (c): (a) With SAXS, DPS and PVME cannot be distinguished, so that the morphology looks the 2-phase structure with hexagonally packed cylinders is observed anyway; (b) With SANS, a high contrast of DPS with respect to PI is uniformly diluted with a low contrast of PVME, so that the observed morphology is again the 2-phase cylinder.

Upon raising temperature, the SANS profiles undergo a big change above 72°C in such a way that the scattering peaks become weak and broad, whereas the SAXS profiles show almost no change. Let us call this as *Transition 1*. Upon further raising temperature, the SAXS profiles now show a big change above 122°C in such a way that the peaks become weak and broad, whereas the SANS profiles show little change. Let us call this as *Transition2*. The SAXS and SANS profiles between *Transition 1* and *Transition 2* are distinguished from those below *Transition 1* and above *Transition 2* by shading in part (a) and (b) of Figure 5.

Based upon these results we can draw the following two conclusions. Conclusion 1: We obtain the two-step phase transition with temperature. In *Transition 1* at the low temperature, the phase transition occurs within the cylinders so that DPS and PVME respectively microphase-separate into shells and cores of the cylinders as shown in part (d). Note that this transition is obviously expected to cause a big change in SANS but no change in SAXS, as indeed observed in parts (b) and (a) above 72°C, respectively. In *Transition 2* at the high temperature, the core-shell cylinders transforme into core-shell spheres with liquid-like short-range order as shown in part (e), the detailed analysis of which will be reported elsewhere.[14] This transition causes a big change in SAXS. This transition may also cause a change in SANS. However the cylindrical shells may be thermally fluctuating and, therefore, give only a weak and broad SANS peak after the low-temperature *Transition 1*. Then no further drastic change upon this change of the structure is expected at higher temperature (*Transition 2*). We should note here that the low temperature transition occurs at the very low temperature of 72°C. If we have a DPS-PVME diblock copolymer with each block having M_n of 3K, respectively, it would microphase-separate only at much higher temperature, higher than at least 200°C.[15] Thus tethering one-end (DPS end in this case) of DPS-PVME block at the interface of PI microdomains enhances microphase separation between DPS and PVME. Note also that the microphase separation of PI and DPS is stable up to high temperature as high as 177°C. If we have PI-DPS diblock copolymer having M_n of PI and DPS of 14K and 3K, respectively, PI and DPS blocks will be demixed to form a disordered state at much lower temperature. Therefore, we can conclude that the tethering one end of DPS in PI-DPS at the interface of PVME microdomains again enhances the microphase separation between PI and DPS.

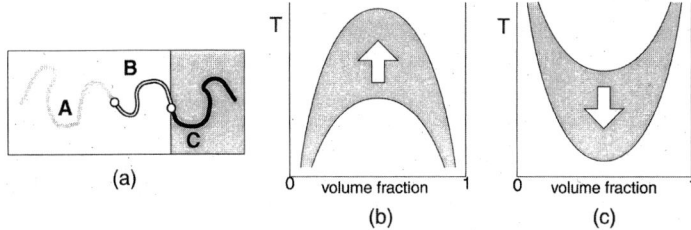

Fig. 6. Promoted microphase transition of A-B block chain in A-B-C triblock terpolymer with one end of B tethered on the interface of microdomains of C blocks (a). Part (b) and (c) show the cases where A-B has UCST-type interactions and LCST-type interactions, respectively.

Those tethering effects on the microphase separation as discussed above are the second conclusion of our studies and are summarized in Figure 6. The tethering one end of diblock copolymers as shown in part (a) where A-B diblock in A-B-C triblock terpolymer is tethered on the interface of C microdomains promotes microphase separation between A and B, raising UCST if A-B has UCST-type phase diagram or lowering LCST if B-C has LCST-type phase diagram as shown by the shaded regions in parts (b) and (c), respectively.

Concluding Remarks

We discussed the enriched short-range and long-range interactions in (1) mixtures of block copolymers having long and short chain lengths and (2) PI-DPS-PVME triblock terpolymers. In the mixture we elucidated an interesting cosurfactant effect originating from competing long-range interactions between the long and short block chains: The two lamella forming block copolymers mixed together to form special cylinders with a large PS volume fraction, as demonstrated by Figure 2 and interpreted by Figure 3. We note that the elastic free energy of block copolymer chains plays an important role here.

In the triblock terpolymer, we found interesting two-step phase transitions with temperature as summarized in Figure 5, and the enhanced microphase separation for the triblock terpolymer, as summarized in Figure 6. Thus even triblock terpolymers with a very small molecular weight undergo microphase separation.

We hope we could elucidate a role of connectivity in polymer science by using the two model systems as discussed above. Specifically the block connectivity plays an important role on molecular assembly of block copolymers through interplays and competitions of the short-range interactions and the long-range interactions discussed in this paper. The role of sequence distributions of different monomer units on molecular assembly and their static and dynamical properties and functions is one of the most important and fundamental problems in polymer science.

[1] M. Matsuo, T. Ueno, H. Horino, S. Chujyo, H. Asai, *Polymer*, **1968**, *9*, 425; T. Inoue, T. Soen, T. Hashimoto, and H. Kawai, *Macromolecules*, **1970**, *3*, 87

[2] H. Hasegawa, H. Tanaka, K. Yamasaki, T. Hashimoto, *Maromolecules*, **1987**, *20*, 1651

[3] M. W. Matsen, F. S. Bates, *Macromolecules*, **1996**, *29*, 1091

[4] Y.-Y. Huang, H.-L. Chen, T. Hashimoto, Macromolecules, in press and a paper in preparation.

[5] D. A. Hajduk, P. E. Harper, S.M. Grüner, C. C. Honeker, G. Kim, E.L. Thomas, *Macromolecules*, **1994**, *27*, 4063

[6] M. F. Schulty, F. S. Bates, K. Almdal, K. Martensen, *Phys. Rev. Lett.*, **1994**, *73*, 86

[7] T. Hashimoto, S. Koizumi, H. Hasegawa, *Macromolecules*, **1994**, *27*, 1562

[8] M. W. Matsen, *J. Chem. Phys.* **1995**, *103*, 3268

[9] D. Yamaguchi, T. Hashimoto, *Macromolecules*, **2001**, *34*, 6495

[10] Ju. V. Lyatskaya, E. B. Zhulina, T. M. Birshtein, *Polymer,* **1992**, *33*, 343

[11] K. Yamauchi, H. Hasegawa, T. Hashimoto, N. Köhler, K. Knoll, *Polymer,* **2002**, *43*, 3563

[12] Y. Mogi, K. Mori, Y. Matsushita, I. Noda, *Macromolecules*, **1992**, *25*, 5412

[13] R. Stadler, C. Auschra, J. Beckmann, U. Krappe, I. Voigt-Martin, L. Leibler, *Macromolecules,* **1995**, *28*, 3080

[14] K. Yamauchi, H. Hasegawa, T. Hashimoto, in preparation.

[15] T. Hashimoto, H. Hasegawa, T. Hashimoto, H. Katayama, M. Kamigaito, M. Sawamoto, M. Imai, *Macromolecules,* **1997**, *30*, 6819.

New Developments and Directions in the Area of Elastomers and Rubberlike Elasticity

J. E. Mark

Department of Chemistry and the Polymer Research Center, The University of
Cincinnati, Cincinnati, OH 45221-0172, USA
Email: markje@email.uc.edu

Summary: There are number of important developments in the area of elastomeric
polymers, including (i) network chains of controlled stiffness, (ii) model elastomers
(including dangling-chain networks), (iii) fluorosiloxane elastomers, (iv) new
thermoplastic elastomers, (v) other new elastomers, (v) bimodal network chain-length
distributions, (vi) cross linking in solution or in a state of deformation, and (vii) gel
collapse. Interesting elastomeric composites include those with (i) in-situ generated
ceramic-like particles, (ii) ellipsoidal fillers, (iii) clay-like layered fillers, (iv)
polyhedral oligomeric silsesquioxane (POSS) particles, (v) porous fillers, (vi)
elastomeric domains modifying ceramics, and (vii) controlled interfaces. New
characterization techniques are being developed for elastomers, and there have been
new developments in elasticity theory and in elastomer processing. Some examples
of societal aspects of relevance are (i) synthesis of elastomers in environmentally-
friendly solvents, (ii) biosynthesis, (iii) recyclability, (iv) improved adhesion to tire
cords, and (v) better barrier properties in anti-terrorism clothing. Educational topics
include curriculum development, and mobile laboratories for elastomer experiments
and demonstrations.

Keywords: bimodal distributions, biosynthesis, curriculum, elasticity theory,
elastomers, gel collapse, interfaces, mechanical properties, model networks,
recyclability, reinforcement

Elastomeric Polymers

Network chains of controlled stiffness. The primary interest here is to increase the melting

point of an elastomer such as poly(dimethylsiloxane) (PDMS) so that it undergoes strain-induced

crystallization. This crystallization is the origin of the superb mechanical properties of natural

rubber, and it results from the reinforcing effects of the crystallites. One way of stiffening

elastomeric chains such as PDMS is to put a meta or para phenylene group in the backbone.[1]

Model elastomers. Elastomers of this type are made by reacting functionally-terminated chains

with a multifunctional end-linking agent. Because of this preparative approach, much more

information on their structures is known than on those prepared using more uncontrolled

© 2003 WILEY-VCH Verlag GmbH & KGaA, Weinheim   DOI: 10.1002/masy.200351108

methods such as sulfur vulcanization or peroxide thermolysis. They also have unusually good mechanical properties, perhaps because of a reduced number of dangling chains.[2] The same synthesis approach has been used to put known numbers of dangling chains of known lengths into intentionally-imperfect networks.

Fluorosiloxane elastomers. Placing fluorine atoms into siloxane repeat units can be useful for increasing polysiloxane solvent resistance, thermal stability, and surface-active properties.[3-5]

New thermoplastic elastomers. There is a need to develop thermoplastic elastomers that are less expensive than the Kraton® styrene-butadiene-styrene triblock copolymers. The leading candidates are stereochemical copolymers of polypropylene, and chemical copolymers of ethylene and comonomers such as hexene-1.[6-8]

Some other new elastomers. One example of another interesting elastomeric material is a new hydrogenated nitrile rubber with good oil resistance and a wide service-temperature range.[9] Another is a type of "*baro*plastic" elastomer which parallels *thermo*plastic elastomers in that a pressure increase gives the desired softening required for processing instead of the usual temperature increase.[10]

Bimodal network chain-length distributions. End linking a mixture of very short chains with the much longer chains that are typical of elastomers gives networks with unusually good ultimate properties, including toughness. [11] This is illustrated in Figure 1.

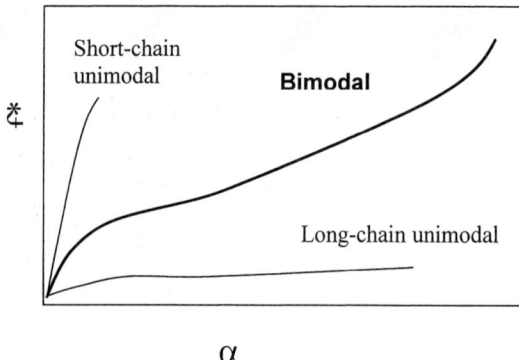

Fig. 1. Typical dependence of nominal stress against elongation for two unimodal networks having either all short chains or all long chains, and a bimodal network having some of both.

Cross linking in solution or in a state of deformation. Forming networks under these unusual conditions has a number of advantages, including the synthesis of elastomers exhibiting less stress relaxation, and stress-strain relationships that are closer to those expected from the simplest molecular theories.[2] Recent studies on networks cross linked in solution have focused on their unusually high extensibilities,[12,13] and changes in their extents of strain-induced crystallization.[14,15]

Gel collapse. Gels, which are networks swollen with a diluent, can be brought to the point where only small changes in a variable such as temperature, pH, ionic strength, etc. can bring about an abrupt shrinkage.[16-19] This is illustrated in Figure 2. The shrinkage occurs rapidly enough in fibers and films to be of interest with regard to producing switches, actuators, artificial muscle, and drug-delivery systems.

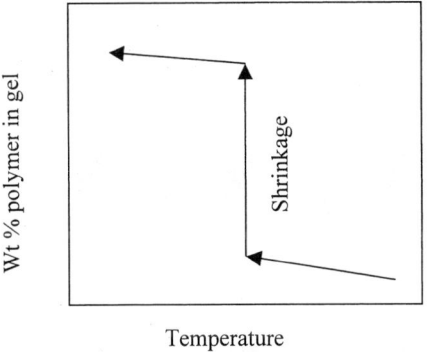

Fig. 2. A gel exuding solvent upon decrease in temperature, with the shrinkage representing "gel collapse".

Elastomeric Composites

In-situ ceramic-reinforced elastomers. One novel way of introducing ceramic-like fillers into a polymer is by the *in-situ* hydrolysis and condensation of an organosilicate, organotitanate, organozirconate, etc. to give silica, titania, zirconia, etc. These "sol-gel" techniques are not much used commercially thus far, because the processing is so different from the usual "*ex-situ*" incorporation of fillers after making them in a separate step, and then mechanically blending them into the polymer matrix.

Ellipsoidal fillers. Reinforcing fillers can be deformed from their usual approximately spherical shapes in a number of ways. For example, if the particles are a glassy polymer such as polystyrene, then deforming the matrix in which they reside above the glass transition temperature will convert them into ellipsoids.[20,21] Uniaxial deformations give prolate (needle-shaped) ellipsoids, and their axes will be in the direction of the deformation, as illustrated in Figure 3. Similarly, biaxial deformations give oriented oblate (disc-shaped) ellipsoids.

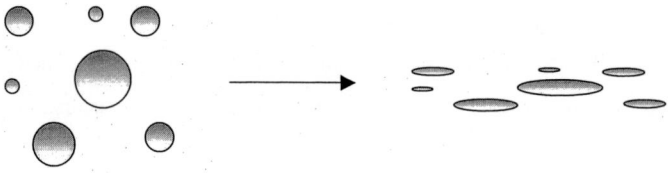

Fig. 3. Deformation of spherical particles into prolate ellipsoids.

One interest here is the anisotropic reinforcements such particles provide, and there have been simulations to better understand the mechanical properties of such composites. [22]

Clay-like filler. Exfoliating layered particles such as the clays, mica, and graphite is being used to provide very effective reinforcement of polymers at loading levels much smaller than in the case of solid particles such as carbon black and silica.[23-27] Other properties can also be substantially improved, including increased resistance to solvents, and reduced permeability and flammability.

Polyhedral oligomeric silsesquioxane (POSS) particles. These fillers are cage-like structures, and have been called the smallest possible silica particles. They typically contain between zero and eight organic functional groups per cage. The particles with no functional groups at all can be blended into polymers using the usual mixing or compounding processing, while those with one functional group can be attached to a polymer as side chains. Those with two functional groups can be incorporated into polymer backbones by copolymerization, and those with more than two can be used for forming cross linked networks.[28-32] Nanotubes are also of considerable interest in this regard.[33-35]

Porous fillers. Some fillers such as zeolites are sufficiently porous to accommodate monomers, which can then be polymerized. This threads the chains through the cavities, with unusually intimate interactions between the reinforcing phase and the host elastomeric matrix.[34] Because of the constraints imposed by the cavity walls, these confined materials show no glass transition temperatures. A typical structure of this type is illustrated in Figure 4.

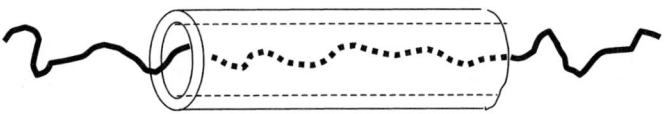

Fig. 4. Sketch of a polymer chain passing through the cavity of a zeolite.

In-situ elastomer-improved ceramics. The sol-gel technique used to precipitate ceramic-like fillers into an elastomer can also be used to precipitate elastomeric domains into a ceramic.[2] The main goal here is to improve the impact resistance of the resulting composite.

Composites with controlled interfaces. By choosing the appropriate chemical structures, chains that span filler particles in a polymer-based composite can be designed so that they are either durable, breakable irreversibly, or breakable reversibly.[36-38]

New Characterization Techniques

IR spectroscopy and birefringence[39] are being used to get new insights into the network chain orientation necessary for strain-induced crystallization. Also of importance are NMR, small-angle x-ray and neutron scattering, atomic force microscopy, Brillouin scattering,[40,41] and pulse propagation measurements.[40,42] In the last of these techniques, the delay in pulses passing through the network is used to obtain information on the network structure.

Theoretical Developments

Some of the most interesting advances in rubber elasticity theory are the various approaches being developed to take better account of chain entanglements.[2,43] In the "constraint" theories, the focus is on the way the constraints are placed within the network structure, some possibilities being illustrated in Figure 5.

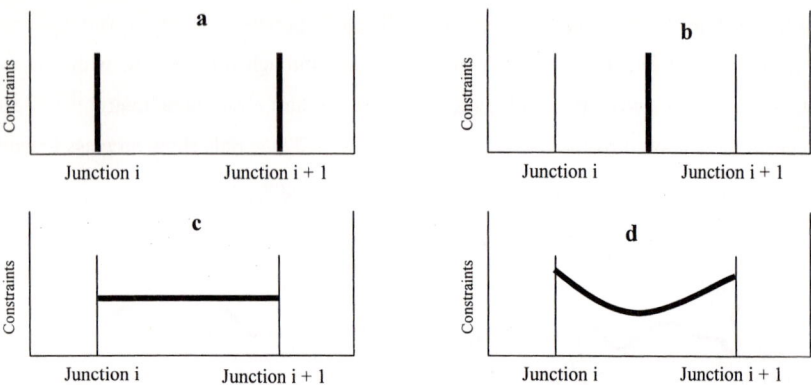

Fig. 5. Locations of constraints in the constraint theories: (a) at the junctions themselves, (b) at the mass centers of the network chains, (c) spread uniformly along the network chains, and (d) at possibly more refined locations based on future experimental information.

New Developments in Processing

Important topics in this area are the use of chaotic mixing to improve compounding,[44] and modeling that includes flow-induced crystallization during molding processes.

Societal Aspects

Of interest here are the possible synthesis of elastomers in environmentally-friendly solvents, and the understanding and exploitation of biosynthetic techniques.[45] Another environmental goal is recyclability.[46,47] Other topics much in the news currently are the improvement of safety aspects of tires (with an emphasis on more reliable bonding to tire cords), and better barrier properties in anti-terrorism protective clothing. Educational topics include curriculum development, and mobile laboratories for elastomer experiments and demonstrations.

Acknowledgments

It is a pleasure to acknowledge the financial support provided by the National Science Foundation through Grant DMR-0075198 (Polymers Program, Division of Materials Research), and by the Dow Corning Corporation.

[1]   J. E. Mark, B. Erman, "*Rubberlike Elasticity. A Molecular Primer*", Wiley-Interscience, New York 1988.
[2]   B. Erman, J. E. Mark, "*Structures and Properties of Rubberlike Networks*", Oxford University Press, New York 1997.
[3]   W. J. Bobear, in: "*Rubber Technology*", M. Morton, Ed., Van Nostrand Reinhold, New York 1973, p. 368.
[4]   H. Kobayashi, M. J. Owen, *Macromolecules* **1990** *23*, 4929.
[5]   D. V. Patwardhan, H. Zimmer, J. E. Mark, *J. Inorg. Organomet. Polym.* **1998** *7*, 93.
[6]   H. H. Brintzinger, D. Fischer, R. Mulhaupt, B. Reiger, R. M. Waymouth, *Angew. Chem. Ed. Engl.* **1995** *34*, 1143.
[7]   S. Mansel, E. Perez, R. Benavente, J. M. Perena, A. Bello, W. Roll, R. Kirsten, S. Beck, H.-H. Brintzinger, *Macromol. Chem. Phys.* **1999** *200*, 1292.
[8]   S. Lieber, H.-H. Brintzinger, *Macromolecules* **2000** *33*, 9192.
[9]   C. Wrana, K. Reinartz, H. R. Winkelbach, *Macromol. Mater. Eng.* **2001** *286*, 657.
[10]   M. H. Acar, J. A. Gonzales, A. M. Mayes, *Preprints, American Chemical Society Division of Polymer Chemistry, Inc.* **2002** *43(2)*, 55.
[11]   J. E. Mark, *Rubber Chem. Technol.* **1999** *72*, 465.
[12]   S. Kohjiya, K. Urayama, Y. Ikeda, *Kautschuk Gummi Kunstoffe* **1997** *50*, 868.
[13]   K. Urayama, S. Kohjiya, *Eur. Phys. J. B* **1997** *2*, 75.
[14]   J. Premachandra, J. E. Mark, *J. Macromol. Sci., Pure Appl. Chem.* **2002** *39*, 287.
[15]   J. Premachandra, C. Kumudinie, J. E. Mark, *J. Macromol. Sci., Pure Appl. Chem.* **2002** *39*, 301.
[16]   E. S. Matsuo, T. Tanaka, *Nature* **1992** *358*, 482.
[17]   K. Takemoto, R. M. Ottenbrite, M. Kamachi, Eds., "*Functional Monomers and Polymers*," Marcel Dekker, New York 1997.
[18]   Y. Tanabe, Ed., "*Macromolecular Science and Engineering. New Aspects*," Springer, New York 1999.
[19]   V. Y. Grinberg, A. S. Dubovik, D. V. Kuznetsov, N. V. Grinberg, A. Y. Grosberg, T. Tanaka, *Macromolecules* **2000** *33*, 8685.
[20]   S. Wang, J. E. Mark, *Macromolecules* **1990** *23*, 4288.
[21]   S. Wang, P. Xu, J. E. Mark, *Macromolecules* **1991** *24*, 6037.
[22]   M. A. Sharaf, A. Kloczkowski, J. E. Mark, *Polymer* **2002** *12*, 643.
[23]   E. P. Giannelis, R. Krishnamoorti, E. Manias, *Adv. Polym. Sci.* **1999** *138*, 107.
[24]   R. A. Vaia, E. P. Giannelis, *MRS Bull.* **2001** *26 (5)*, 394.
[25]   T. J. Pinnavaia, G. Beall, Eds., "*Polymer-Clay Nanocomposites*," Wiley, New York 2001.
[26]   Y. T. Vu, J. E. Mark, L. H. Pham, M. Engelhardt, *J. Appl. Polym. Sci.* **2001** *82*, 1391.
[27]   W. Zhou, J. E. Mark, M. R. Unroe, F. E. Arnold, *J. Macromol. Sci. - Pure Appl. Chem.* **2001** *A38*, 1.
[28]   J. D. Lichtenhan, J. Schwab, S. W. A. Reinerth, *Chem. Innov.* **2001** *31*, 3.
[29]   R. M. Laine, J. Choi, I. Lee, *Adv. Mats.* **2001** *13*, 800.
[30]   D. A. Loy, C. R. Baugher, D. A. Schnieder, A. Sanchez, F. Gonzalez, *Polym. Preprints* **2001** *42(1)*, 180.
[31]   K. J. Shea, D. A. Loy, *MRS Bull.* **2001** *26*, 368.
[32]   T. S. Haddad, A. Lee, S. H. Phillips, *Polym. Preprints* **2001** *42(1)*, 88.
[33]   H. Nakamura, Y. Matsui, *J. Am. Chem. Soc.* **1995** *117*, 2651.
[34]   H. L. Frisch, J. E. Mark, *Chem. Mater.* **1996** *8*, 1735.
[35]   S. J. Tans, M. H. Devoret, H. Dai, A. Thess, R. E. Smalley, L. J. Geerligs, C. Dekker, *Nature* **1997** *386*, 474.
[36]   B. T. N. Vu, J. E. Mark, D. W. Schaefer, *Preprints, American Chemical Society Division of Polymeric Materials: Science and Engineering* **2000** *83*, 411.
[37]   B. T. N. Vu, M. S. Thesis in Chemistry, University of Cincinnati, 2001.
[38]   D. W. Schaefer, B. T. N. Vu, J. E. Mark, *Rubber Chem. Technol.* **2002** *75*, 000.
[39]   L. Bokobza, N. Nugay, *J. Appl. Polym. Sci.* **2001** *81*, 215.
[40]   M. Sinha, Ph. D. Thesis in Physics, University of Cincinnati, 2000.
[41]   M. Sinha, J. E. Mark, H. E. Jackson, D. Walton, *J. Chem. Phys.* **2002** *117*, 2968.
[42]   M. Sinha, B. Erman, J. E. Mark, T. H. Ridgway, H. E. Jackson, *Macromolecules* submitted.
[43]   M. Rubinstein, S. Panyukov, *Macromolecules* **2002** *35*, 6670.
[44]   J. M. Ottino, F. J. Muzzio, M. Tjahjadi, J. Franjione, S. C. Jana, H. A. Kusch, *Science* **1992** *257*, 754.
[45]   T. Koyama, A. Steinbuchel, Eds., "*Biopolymers, Vol. 2: Polyisoprenoids*," Wiley-VCH, New York 2001
[46]   A. I. Isayev, S. H. Kim, V. Y. Levin, *Rubber Chem. Technol.* **1997** *70*, 194.
[47]   S. E. Shim, A. I. Isayev, *Rubber Chem. Technol.* **2001** *74*, 303.

Macromol. Symp. **2003**, *201*, 85—88

Supramolecular Aspects of Polymer Science: A Challenge for Solid State NMR

Hans Wolfgang Spiess

Max-Planck-Institute for Polymer Research, PO Box 3148, D-55021 Mainz, Germany

Summary: Recent developments in structural elucidation of supramolecular systems by advanced solid state NMR are described. Special emphasis is placed on hydrogen-bonded sytems and columnar stacks of aromatic moieties. In imidazole-based proton conductores spatially separated regions of high and low mobility are identified. In stacks of alkyl-substituted hexabenzocoronenes maximum charge carrier mobility is observed for crystal-like stacking of the discs.

Keywords: charge transport, NMR spectroscopy, proton transport, solid-state structure, supramolcular structure

Introduction

In advanced synthetic as well as in biological systems self-assembly of carefully chosen building blocks is of central importance.[1] Hydrogen bonding and aromatic π-π-interactions are key features of supramolecular chemistry. Despite being highly ordered on a local scale, such systems often do not crystallize. Therefore, their structures cannot be determined by conventional X-ray crystallography or neutron scattering. Alternatives are needed which should provide structural and dynamic information, preferably requiring only small amounts of as-synthesized samples. High resolution solid state ^{1}H-NMR can meet these requirements,[2] provided that sufficiently selective information can be extracted from the corresponding spectra.

Over the last years we have systematically pursued the development of solid state ^{1}H NMR for that purpose by combining fast Magic Angle Spinning (MAS) and Double-Quantum (DQ) NMR spectroscopy, which makes use of the homonuclear dipole-dipole coupling between protons.[3] The spectral resolution can be increased by use of similar methods exploiting

DOI: 10.1002/masy.200351109

heteronuclear ^{1}H-^{13}C dipole-dipole couplings.[4] These techniques have provided new insight in

- hydrogen bonded structures in the solid state[5]
- columnar stacking and molecular dynamics of discotics[6]
- chain order and translational motion in polymer melts[7]
- chain organization on surfaces.[8]

Solid state NMR probes the structure at the molecular level. Self-assembly, however, often leads to structures which are partially ordered on the nanometer scale. There, pulse EPR techniques[9] are better suited to determine their structure after introduction of spin labels. This technique has provided new insight in the nanostructure of polymers[10] including block copolymers with ionic groups.[11]

Hydrogen Bonded Systems

Protons in hydrogen bonded structures exhibit ^{1}H chemical shifts well away from the aliphatic peaks. Therefore, high resolution ^{1}H solid state NMR is ideally suited for investigating such systems as noted early on in our study of the unusual hydrogen bonded structures in benzoxazine dimers.[5] These dimers are of interest as model compounds for a new class of phenolic materials, the polybenzoxazines prepared by Ishida and coworkers.[12] These systems combine several favorable properties, such as near-zero shrinkage on polymerization as well as low water absorption, which are attributed to their favorable hydrogen-bonding. Application of improved methods have meanwhile allowed us to locate the protons in these compounds with high precision[13] and evidence for helical structures has been found in benzoxazine oligomers.[14] Further examples are collected in a recent review.[15]

Imidazole Based Proton Conductors

Recently, the new techniques were applied to study proton conductivity.[16] The materials of interest, ethylene oxide tethered imidazole heterocycles[17] (Imi-nEO), are characterized by variable temperature experiments, as well as 2D homonuclear DQ NMR and 2D exchange spectroscopy.[2] Quantum chemical calculations provide a full assignment and understanding of the ^{1}H chemical shifts, based on a single-crystal structure obtained for Imi-2EO. Three types of hydrogen bonded N-^{1}H resonances are observed. Double quantum NMR experiments identify those hydrogen-bonded protons that are mobile on the time scale of the experiment, and thereby, are able to participate in charge transport. In addition, evidence for locally ordered domains within all the Imi-nEO materials is provided. Disordered (mobile) and ordered components in Imi-2EO dramatically differ in their ^{1}H spin-lattice relaxation times. At lower temperatures and in absence of excess protons 2D NOESY spectra[2] show no evidence of chemical exchange processes between the ordered and disordered domains. These results indicate that the highly ordered regions of the materials do not (or only poorly) contribute to proton conductivity, which is rather taking place in the disordered regions.

Photonic Materials

Advanced solid state NMR is also highly suited for elucidating columnar stacking and its relation to function.[6] Solid columnar discotic and liquid crystalline (LC) phases formed by alkyl-substituted hexabenzocoronene (HBC) mesogens have been investigated by a combination of X-ray scattering and a variety of advanced solid state NMR methods. Correlations between chemical structure, molecular packing and dynamics are established. Maximum charge carrier mobility is observed for a crystal-like stacking of the discs in the column with optimized π-π-interactions.[18]

Self-assembly of fluorinated tapered dendrons can drive the formation of supramolecular liquid crystals with promising optoelectronic properties from a wide range of organic materials. Attaching conducting organic donor or acceptor groups to the apex of the dendrons leads to supramolecular nanometer-scale columns that contain in their cores π-stacks of donors, acceptors or donor–acceptor complexes exhibiting high charge carrier mobilities. Moreover, functionalized dendrons and amorphous polymers carrying compatible side groups co-assemble so that the polymer is incorporated in the centre of the columns through donor–acceptor interactions and exhibits enhanced charge carrier mobilities. The mutual arrangement of the functional groups was determined by solid state NMR.[19]

Conclusions

The development of the technique is far from being complete. Major advances are expected in the years to come. This includes in particular increased sensitivity through inverse detection in heteronuclear solid state NMR. Thus solid state NMR is envisaged to become an indispensable tool in the field of supramolecular polymer chemistry.

Supported lipid bilayers were prepared of the phospholipid DMPC, 1,2-dimyristoyl-*sn*-glycero-3-phosphocholine using the method of vesicle fusion.[3] These surfaces were "starved" of polymer by depositing a small amount, too small to saturate them. Fig. 1 shows an experiment in which, after polymethacrylic acid (PMA) was allowed to adsorb to ≈ 5% of the saturated amount adsorbed, the solution of polymer was replaced by pure buffer solution. Thereafter the total mass adsorbed was constant (top panel) but the intensity ratio of the charged and uncharged carboxylic groups adsorbed continued to decrease (bottom panel). In this circumstance where surface coverage was limited to much less than a monolayer, one expects the polymer to adopt a flat "pancake" conformation at equilibrium. We suppose that the spreading of adsorbed chains, i.e. the increase of segment-surface contacts with the dipolar headgroups, produced the observed changing ratio of charged to uncharged ionized groups. The puzzle was to understand why the rate of spreading is so slow, as local diffusion times of lipids within DMPC bilayers are far more rapid than this.

The molecular weight of the spreading chains was varied over a wide range (a factor of 30) and this kinetic process was found to show no dependence on molecular weight. This is reminiscent of the N-independent spreading rate recently discovered when liquid droplets spread on a viscoelastic surface. The usual strong viscosity dependence of the spreading rate, for cases where the surface is a hard solid, is lost. The rate-limiting step is observed instead to be viscoelastic response of the underlying material. In this present system of supported phospholipid bilayer, measurements of the infrared dichroism of the headgroup gave evidence of direct coupling between the two processes, adsorption and surface reconstruction. Extrapolating from the theoretical prediction by others that polymer adsorption stiffens a membrane locally, we conjecture that the surface reorganization process involved changes of local curvature and local stiffness while the polymer chains unwound from the three-dimensional solution conformation towards the two-dimensional pancake conformation.[3]

Diffusion of Small Molecules Embedded Within Adsorbed Polyelectrolytes

The directed flow of fluids contained within small channels constitutes an emerging theme of modern chemistry and materials science. New applications are proliferating, for example in the fields of microfluidics, chemical analysis, and protein crystallization. Most prior work has concerned channels whose cross-section dimensions are on the order of micrometers to hundreds of micrometers. It is also attractive to consider the potential for directed flow of molecules contained within molecularly-thin layers.

The scheme we have selected to prepare molecularly-thin layers consists in using adsorbed polymers, which has the advantage of simplicity. Whereas for initial studies we have used polymers that coated a solid surface uniformly, it is easy to envision methods to patterned arrays of adsorbed polymers, for example by lithography or stamping a solid surface to render it selectively adsorbing. In this way we envision forming polymer wires or polymer ribbons, with size limited by the resolution available using lithography or stamping. Once such channels have been formed, we envision directing flow of molecules embedded within them under the stimulus of electric field or temperature gradients.

An alternative approach to produce molecularly-thin channels might consist of placing two extended atomically-smooth flat surfaces in close proximity, in a generalization of the surface forces technique. This is simple when dealing with curved surfaces but becomes problematical and laborious when it is intended that the surfaces be parallel over linear dimensions of millimeters. As another alternative one can envision more costly methods, such as fabricating nanometer-sized channels using advanced methods of silicon machining such as electron beams, which would have the advantage of higher spatial resolution.

Initial studies focused on the dynamics of the complex system of a polycation, quaternized poly-4-vinylpyridine (QPVP), which was allowed to adsorb to freshly cleaved mica, whose surface in aqueous environment is negatively charged. By changing the ionic strength as a variable in the adsorption condition, the structure of QPVP layer was tuned and the mobility of the probe molecules was measured accordingly using fluorescence correlation spectroscopy (FCS).[4]

Such features of small (probe) molecules diffusing within surface channels one macromolecule thick raise a number of interesting questions that require further study. For example, how will the diffusion of the probe depend on the ionic strength of the solution? By studying this, the relationship between the mobility of the probe and the strength of the electrostatic interaction with the QPVP segment can be discovered. Another important factor is that the probe molecule can not only probe the structure and dynamics of the polyelectrolyte layer, but also can represent the mobility of the solvent molecules in the monolayer. By choosing dye molecules with different size as well as different chemical structures such as polarity and hydrophobicity, the dynamics of the solvent inside such a surface modification structure will be revealed. In addition to these questions that are science-motivated, the direction of flow by application of external fields may find technological application.

Acknowledgements

This work was supported by the U.S. Department of Energy, Division of Materials Science, under Award No. DEFG02-91ER45439 and DEFG02-02ER46019 through the Frederick Seitz Materials Research Laboratory at the University of Illinois at Urbana-Champaign.

[1] S. A. Sukhishvili, Y. Chen, J. Müller, K. Schweizer, E. Gratton, and S. Granick, *Nature* 2000, 406, 146.
[2] S. A. Sukhishvili, Y. Chen, J. Müller, E. Gratton, K. Schweizer, and S. Granick, *Macromolecules* 2002, 35, 1776.
[3] A. F. Xie and S. Granick, *Nature Materials* 2002, 1, 129.
[4] J. Zhao and S. Granick, unpublished experiments.

Macromol. Symp. **2003**, *201*, 95—102

Surface Mobility in Monodisperse Polystyrene Films

Tisato Kajiyama,* Daisuke Kawaguchi, Keiji Tanaka

Department of Applied Chemistry, Faculty of Engineering,
Kyushu University, Fukuoka 812-8581, Japan

Summary: Surface dynamics in monodisperse polystyrene films was examined by lateral force microscopy in conjunction with dynamic secondary ion mass spectroscopy. Glass transition temperature, T_g, at the surface was markedly lower than the corresponding bulk T_g. Also, it was shown that polymer chains present at the surface could diffuse even at a temperature below the bulk T_g. The surface depth, in which molecular motion was activated, was of the order of 5 nm.

Keywords: dynamic secondary ion mass spectroscopy, diffusion, glass transition temperature, lateral force microscopy, surface, polystyrenes

Introduction

Surfaces and interfaces of polymeric materials play an important role in many technological applications such as lubrication, adhesion, biomaterials, etc. To design highly functionalized polymeric materials, the systematical understanding of aggregation states and physical properties in the vicinity of surface and interfacial layers, which differs from those in the interior bulk region, is quite important. However, surface physical properties, especially molecular motion, has been far from clear for the moment.

So far we have systematically studied surface molecular motion in monodisperse polystyrene (PS) films by mainly using scanning force microscopy.[1,2] As a main result, it was elucidated that molecular motion at the film surface was activated in comparison with that in the interior bulk region. This activation of surface mobility has been explained in terms of chain end effect[3] and reduced cooperativity.[4] Based on the aforementioned situation, it seems reasonable to infer that surface chains are movable in a relatively large distance even at a temperature below the bulk glass transition temperature, T_g^b, as long as the temperature is higher than the surface glass transition temperature, T_g^s. However, a little information about such has been known. The objective of this study is to reveal directly chain diffusion at the surface, based on the time evolution of the interfacial width of PS bilayers.

DOI: 10.1002/masy.200351111

Experimental

Monodisperse PSs were synthesized by an anionic polymerization using *sec*-butyllithium and methanol as an initiator and a terminator, respectively. PS films were coated from toluene solutions onto silicon wafers by a spin-coating method. The film was dried at 296 K for more than 24 h and then annealed at 393 K for 24 h under vacuum. The film thickness evaluated by ellipsometric measurement was approximately 200 nm. The surface relaxation behavior of the PS film was examined by using lateral force microscopy (LFM; SPA 300 HV, Seiko Instruments Industry Co., Ltd.) with an SPI 3800 controller. The applied force to the cantilever was set to be 10 nN in a repulsive force region. T_g^b was determined by differential scanning calorimetry (DSC8230, Rigaku Co., Ltd.). For interdiffusion experiment, bilayer films composed of monodisperse PS and deuterated PS (dPS) were prepared by a floating method.[5] The interfacial broadening of the bilayers by annealing was examined by dynamic secondary ion mass spectroscopy (DSIMS; SIMS 4000, Atomika Analysetechnik GmbH). The incident beam of oxygen ions with 4keV and ca. 30 nA was focused onto a 200 μm x 200 μm area of the specimen surface. The incident angle was 45 deg.

Results and Discussion

First of all, to discuss about surface molecular motion with the segmental scale, T_g^s in the PS films was examined. Since the manifestation of frictional force of polymeric solids is closely related to their viscoelastic properties, it is possible to examine surface molecular motion of the polymeric solids by using LFM.[2,6] That is, when the energy dissipation increases at the surface due to molecular motion, lateral force increases. Hence, it can be postulated that lateral force alteration with measuring temperature is essentially similar to the temperature dependence of dynamic loss modulus or mechanical loss tangent.

Figure 1 shows the lateral force versus temperature curves for the PS films with number-average molecular weight, M_n, of 4.9k and 140k at the scanning rate, v, of 1 μm•s⁻¹. The ordinate is normalized by the peak value of lateral force to show how the lateral force varies with temperature in the vicinity of a transition region. The lateral force-temperature curves shown in Figure 1 make it clear that each surface transition temperature for the PS films with M_n of 4.9k and 140k is much lower than the corresponding T_g^b of 348 and 376 K, respectively, based on the onset of DSC curves. Thus, it seems reasonable to conclude that the segmental motions at PS film surfaces are more activated than that in the internal bulk phase.

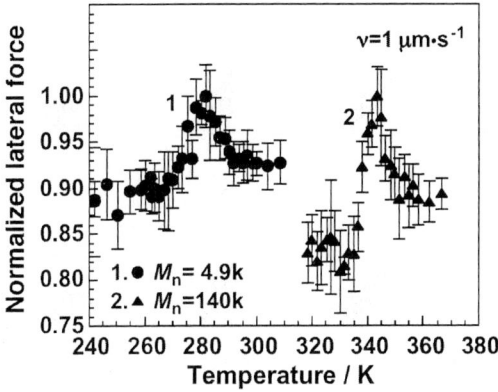

Fig. 1. Temperature dependence of lateral force for the PS films with M_n of 4.9k and 140k at the scanning rate of 1 μm·s^{-1}.

An onset temperature on the lateral force-temperature curve, that is, the temperature at which the magnitude of lateral force starts to increase, can be empirically defined as T_g^s.[4] Figure 2 shows T_g^s so obtained as a function of M_n, and T_g^b determined by DSC are also plotted for a comparison. The arrow beside the ordinate denotes our room temperature, which is abbreviated as "R.T.". It is clear that T_g^s exhibits the stronger M_n dependence than T_g^b. It is of interest to note that the surface is already in a glass-rubber transition or rubbery state at room temperature in case of a M_n smaller than approximately 40k.

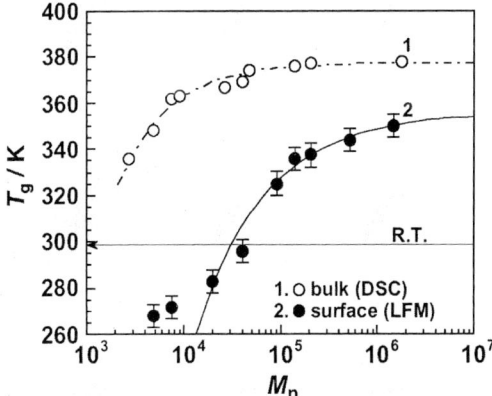

Fig. 2. Molecular weight dependences of T_g^b and T_g^s for the PS films. The broken and solid curves are drawn in the context of the power law analysis.

We now turn to surface perpendicular diffusion in PS films. This can be examined by using a bilayer film composed of two different components, in which both surfaces stand face-to-face. Figure 3 shows the typical DSIMS profile of proton and deuterium and carbon ions for the (PS/dPS) bilayer. Here, one component of the bilayer was labeled by deuterium to detect experimentally where the interface was. Since the intensity of the carbon ion C^+ was almost constant through the bilayer, it is clear that the steady-state etching proceeded during the measurement. The abscissa of the etching time was simply converted to the depth from the surface on the assumption of a constant sputtering rate through the bilayer, which was pre-examined using the dPS film with a known thickness.

A measured concentration profile by DSIMS is generally broadened from an ideal one owing to an atomic mixing effect. The broadening of the measured SIMS profile was quantified by the instrument function Δz_g corresponding to the depth resolution. Figure 4 shows our definition of the interfacial thickness; (a) the deuterium ion intensity profile $I_{D+}(z)$ through the interface and (b) the derivative of $I_{D+}(z)$ by the distance from the center of the interface. Assuming that the $dI_{D+}(z)/dz$ can be expressed by Gaussian function, the Δz_i (i = g or m), where the Δz_m denotes the measured

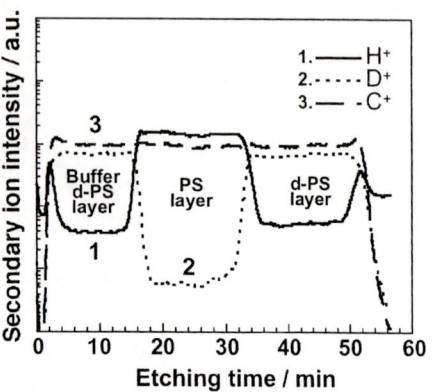

Fig. 3. Typical DSIMS profile of proton and deuterium and carbon ions for the (PS/dPS) bilayer.

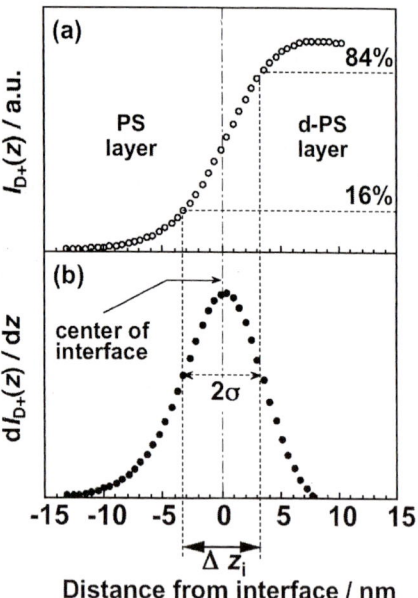

Fig. 4. Our definition of the interfacial thickness, Δz_i (i=g or m) from DSIMS profile; (a) the deuterium ion intensity profile, $I_{D+}(z)$, through the interface, and (b) the derivative of $I_{D+}(z)$ by the distance from the center of the interface. The Δz_g and the Δz_m were defined as twice standard deviation of Gaussian function.

apparent width of the bilayer interface, is defined as twice the standard deviation of Gaussian function, corresponding to the depth range where I_{D+} arises from 16 to 84 % of the maximum value. Given that the measured and ideal profiles are expressed by Gaussian functions as shown in Figure 4, the real interfacial thickness Δz is given in terms of the Δz_m and the Δz_g.[7]

$$\Delta z = (\Delta z_m^2 - \Delta z_g^2)^{1/2} \qquad (1)$$

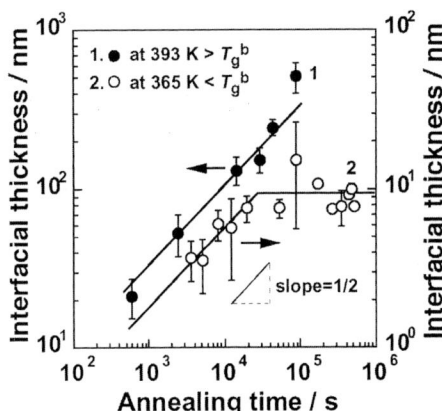

Fig. 5. Double-logarithmic plots of interfacial thickness versus annealing time for (PS/dPS) bilayer annealed at 393 K and 365 K. Slope of 1/2 is drawn in the context of the Fickian diffusion.

Figure 5 shows the time evolution of the interfacial thickness for the (PS/dPS) bilayer by annealing as a function of temperature. Each elemental film possesses M_n of 29k. The choice of this M_n was based on Figure 2, showing that the T_g^s of the PS film with M_n of 29k was approximately 290 K. Since the discrepancy between the T_g^b and the T_g^s is relatively large, it is easy to regulate the polymer diffusion only into the surface region. At 393 K above the T_g^b, the interfacial thickness proportionally increased to the 1/2 power of the annealing time, t. This result is in good accordance with the Fickian diffusion, because the calculated reptation time is approximately 120 s. On the other hand, at 365 K, which was above the T_g^s and below the T_g^b, the interfacial thickness reached a constant value of (9.6 ± 2.5) nm after $t = 2 \times 10^4$ s, although the initial interfacial evolution could be described in terms of Fickian diffusion. This result implies that there is a gradient of the glass transition temperature in the surface region, and makes it clear that the polymer interdiffusion across the bilayer interface definitely occurs even at a temperature below its T_g^b. Since the bilayer interface was originally composed of two film surfaces, a half of the evolved interfacial width corresponds to the

100

surface region of one film, in which polymer diffusion was attained. Thus, it seems reasonable to claim that the surface layer of 4.8 nm thick is in a glass-rubber transition state or even in a rubbery one at 365 K. It is of interest to note that the thickness of this surface layer is almost the same as the radius of gyration of an unperturbed PS chain with M_n of 29k, 4.5 nm. The thickness of the surface layer being in the glass-rubber transition state should strongly depend on the temperature and the molecular weight. A more conclusive study based on this point of view will be reported shortly.

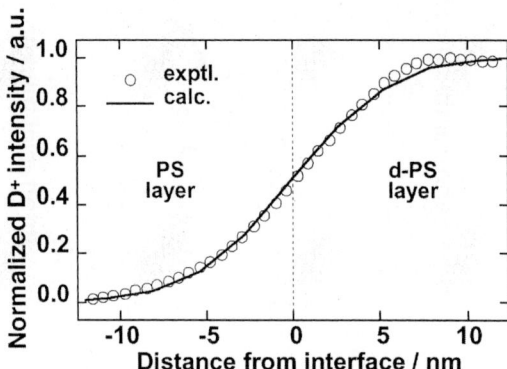

Fig. 6. Depth profile of deuterium ion, D$^+$, at the bilayer interface annealed at 365 K and for 72 h. The ordinate is normalized by the maximum intensity of D$^+$ ion. The solid curve indicates the best fitted one by using eq. (3) with the D_{eff} of 1.8 x 10^{-7} μm^2•s^{-1} as a fitting parameter.

Next, the effective diffusion coefficients D_{eff}s at 393 and 365 K are extracted from the SIMS depth profiles. The concentration profile $C(z')$ of deuterium ion along the direction normal to the surface is given by[8]

$$C(z') = 0.5[1 - erf\left\{\frac{z'}{\sqrt{4D_{eff}t}}\right\}]$$ (2)

where z' is the distance from the center of the interface. However, the DSIMS profile obtained by the experiment is convoluted with the instrument function, as mentioned before. In that case, the apparent concentration profile can be expressed as follows;[8]

$$C_{app}(z') = 0.5[1 - erf\left\{\frac{z'}{\sqrt{a^2 + 4D_{eff}t}}\right\}]$$ (3)

where a is the apparent broadening factor and thus the instrument function of 7.4 nm was used for the a. Figure 6 shows the typical depth profile of normalized deuterium ion at the bilayer

© 2003 WILEY-VCH Verlag GmbH & KGaA, Weinheim

interface annealed at 365 K for 72 h. The solid curve is the best-fit one with a D_{eff} as a fitting parameter by using Eq. (3). The D_{eff}s at 393 and 365 K were estimated to be $(1.8 \pm 0.7) \times 10^{-7}$ and $(7.5 \pm 5.8) \times 10^{-10}$ $\mu m^2 \cdot s^{-1}$, respectively.

It has been widely accepted that the temperature dependence of D_{eff} follows the Vogel relationship.[9,10]

$$\log\left(\frac{D_{eff}}{T}\right) = A - \frac{B}{T - T_{\infty}} \tag{4}$$

where A and B are constants, and besides T_{∞} is the temperature at which long-range motion of the polymer completely ceases, that is, $(T_g\text{-}50)$ K. When eq. (4) is applied to our results, the determination of T_{∞} is somewhat complicated. What was examined in this study was the surface diffusion, and then it is not clear which T_g, $T_g{}^s$ or $T_g{}^b$, should be taken. Since in the case of the annealing at 393 K, the chain diffusion could take place into the internal bulk phase, the $T_g{}^b$ was used for the determination of T_{∞}, resulting in the T_{∞} of 326 K. Using Green and Kramer's equation and value for A and B,[10] respectively, the D_{eff} at 393 K was calculated to be 9.8×10^{-8} $\mu m^2 \cdot s^{-1}$. This value was in good agreement with the measured D_{eff} of $(1.8 \pm 0.7) \times 10^{-7}$ $\mu m^2 \cdot s^{-1}$ in this study. On the contrary, the annealing temperature of 365 K was below the $T_g{}^b$. Hence, the $T_g{}^s$ of 264 K was used for the determination of the T_{∞}, namely, the T_{∞} of 214 K. In that case, the calculated D_{eff} at 365 K was 1.8×10^{-2} $\mu m^2 \cdot s^{-1}$, which was totally different from the D_{eff} by the experiment. For the moment, we have surmised that the inconsistency of the calculated value by eq. (4) with the experimental D_{eff} is based on the ill-chosen value of the T_{∞}, which might be underestimated. This explanation can be understood by taking into account that the T_{∞} is gradually increased along the direction normal to surface and then eventually reached the bulk value of 326 K due to the gradient of the glass transition temperature at the surface region.

Conclusions

Surface dynamics in the PS film was studied, from the segmental scale to center of mass diffusion, by LFM in conjunction with DSIMS. $T_g{}^s$ in the PS films was much lower than the corresponding $T_g{}^b$, and ist M_n dependence was more remarkable at the surface. The time evolution of the PS bilayer interface was examined at temperatures below and above the $T_g{}^b$. At 393 K above the $T_g{}^b$, the interfacial thickness monotonically increased with the annealing time, obeying the Fickian diffusion. In contrast, in the case of the annealing at 365 K being between the T_g^s and the T_g^b, the interfacial thickness was remained to be a constant after the initial interfacial evolution. This result apparently indicates that the polymer interdiffusion can be attained due to the enhanced molecular mobility in proximity to the interface, which used to be in the film surface region, even at a temperature lower than the T_g^b as long as the annealing temperature was higher than the T_g^s.

Acknowledgements

We greatly thank Prof. Atsushi Takahara, Kyushu University, for his warm encouragement and helpful discussion. This work was in part supported by a Grant-in-Aid for Scientific Research (A) (#13355034) from the Ministry of Education, Culture, Sports, Science, and Technology, Japan.

[1] T. Kajiyama, K. Tanaka, I. Ohki, S.-R. Ge, J.-S. Yoon, A. Takahara, *Macromolecules* **1994**, *27*, 7932.
[2] (a) K. Tanaka, A. Taura, S.-R. Ge, A. Takahara, T. Kajiyama, *Macromolecules* **1996**, *29*, 3040; (b) T. Kajiyama, K. Tanaka, A. Takahara, *Macromolecules* **1997**, *30*, 280.
[3] (a) K. Tanaka, X. Jiang, K. Nakamura, A. Takahara, T. Kajiyama, T. Ishizone, S. Nakahama, *Macromolecules* **1998**, *31*, 5148; (b) N. Satomi, K. Tanaka, A. Takahara, T. Kajiyama, T. Ishizone, S. Nakahama, *Macromolecules* **2001**, *34*, 8761.
[4] (a) T. Kajiyama, K. Tanaka, N. Satomi, A. Takahara, *Macromolecules* **1998**, *31*, 5150; (b) K. Tanaka, A. Takahara, T. Kajiyama, *Macromolecules* **2000**, *33*, 7588.
[5] D. Kawaguchi, K. Tanaka, A. Takahara, T. Kajiyama, *Macromolecules* **2001**, *34*, 6164.
[6] J. A. Hammerschmidt, W. L. Gladfelter, G. Haugstad, *Macromolecules* **1999**, *32*, 3360.
[7] S. J. Whitlow, R. P. Wool, *Macromolecules* **1991**, *24*, 5926.
[8] P. M. Hall, J. M. Morabito, N. T. Panousis, *Thin Solid Films*, **1977**, *41*, 341.
[9] P. F. Green, E. J. Kramer, *Macromolecules*, **1986**, *19*, 1108.
[10] K. A. Welp, R. P. Wool, S. K. Satija, S. Pispas, J. Mays, *Macromolecules*, **1998**, *31*, 4915.

Nanostructures Formed by Combination of Nanotube and Polymer Chain

Kohzo Ito,* Takeshi Shimomura, Yasushi Okumura

Graduate School of Frontier Sciences, University of Tokyo, Hongo, Bunkyo-ku, 113-8656 Tokyo, Japan
E-mail: kohzo@k.u-tokyo.ac.jp

Summary: We have investigated polymeric supramolecular systems of inclusion complexes between molecular nanotube and polymer chains theoretically and experimentally. This system indicates inclusion-dissociation behavior much sharper than the inclusion complex between cyclic molecule and small compounds. And we propose some functional supramolecules utilizing the combination of the nanotube or nanoring and polymer chains such as insulated molecular wire and topological gel.

Keywords: conducting polymer, cyclodextrin, gel, inclusion complex, nanotube, polyaniline, supramolecule

Introduction

Various organic supramolecules with unique structures have attracted great interests of many scientists.[1-5] As an example, Harada et al. prepared a polyrotaxane supramolecule in which cyclodextrin (CD) molecules of cyclic form were threaded on a polymer chain with bulky ends[6,7] and then synthesized molecular nanotubes with the diameter smaller than carbon nanotubes by crosslinking the adjacent CD units in the polyrotaxane.[8] This molecular nanotube, highly soluble in several kinds of solvents such as water, has a constant inside diameter (0.45nm) and a longitudinal length of submicron order controllable by varying the length of the polyrotaxane.

Owing to the infinitesimal inside diameter of the molecular nanotube, a polymer chain included in the nanotube has an extended conformation, such as a planar zigzag one, with no degrees of freedom other than a translational motion along its longitudinal axis. Therefore, the inclusion of a polymer chain into a molecular nanotube is entropically unfavorable and promoted by attractive interaction such as hydrophobic one between the chain and the nanotube. In other

© 2003 WILEY-VCH Verlag GmbH & KGaA, Weinheim   DOI: 10.1002/masy.200351112

words, heating results in the dissociation of the polymer chain from the nanotube with recovery of the intrinsic entropy of random coiled conformation. Moreover the conformational entropy and inclusion interaction are roughly evaluated to be proportional to the length of the nanotube and polymer chain. Consequently, the free energy changes drastically with inclusion or dissociation between a long molecular nanotube and polymer chain, which leads to phase transitional behavior.

We have recently investigated the inclusion complex formation between molecular nanotubes and polymer chains theoretically[9,10,14] and experimentally.[11-13,16-19] As mentioned above, this polymeric supramolecular system indicates the inclusion-dissociation behavior more cooperative and sharper than the inclusion complex formation between cyclic molecule and small molecular compounds.[10] And we have proposed some functional supramolecules utilizing the combination of nanotube or nanoring and polymer chains such as insulated molecular wire[16,19] and topological gel.[18] In this report, we will introduce these polymeric supramolecular systems.

Insulated Molecular Wire

Much attention has recently been paid to molecular devices.[20,21] Conjugated conducting polymers were regarded as a promising candidate for molecular wire connecting among electrodes and functional organic molecules such as molecular diode. However, it is quite difficult to actually use conducting polymers for molecular wire and measure the conductivity of a long single chain of conducting polymer. This is mainly because conducting polymer chains generally form fibrils entangled complicatedly. Certainly, some conducting polymer chains with alkyl side chains are soluble in organic solvents and can be isolated in very dilute solution. However, the polymer chain then forms coiled conformation because of the large conformational entropy. The coiled conformation consists of randomly distributed trans and gauche configurations. The trans configuration, namely, the planar zigzag or coplanar one, delocalised electron due to π conjugation while the gauche configuration breaks the conjugation to form defect. This indicates that high conductivity is not expected in the coiled conformation of the conducting polymer but in the all trans configuration, rodlike conformation. Consequently, to use a conducting polymer chain for molecular wire, we have to isolate a single chain and extend it to rodlike conformation. Incidentally, we can somewhat stretch a conducting polymer chain by adding poor solvent or doping, but then the extended polymer chain rapidly aggregates to become insoluble.

To resolve the difficulty in using the conducting polymer for molecular wire, we applied the inclusion complex formation with cyclodextrin (CD) to a conducting polymer chain.[16] What is the advantage of covering the conducting polymer chain with CDs? The coverage should reduce the attractive interaction between polymer chains considerably because CD has hydrophilic outside and highly soluble in various organic solvents. This means that one can easily isolate a single conducting polymer chain by covering with CDs. Next it is also expected that the conducting polymer chain forming the inclusion complex with CDs would be confined to rodlike conformation because the inside diameter of CD is extremely small. This means that the polymer chain has all trans, planar zigzag or coplanar configuration. Therefore, the π conjugation system should spread over a whole chain of the conducting polymer, which results in high conductivity. This inclusion complex has a molecular wire as an axis, which is covered with insulating cyclic molecules. Therefore it can be regarded as insulated molecular wire. Incidentally, there were some other approaches to form insulated molecular wires such as the chemical coupling of several conjugated monomer units threaded through β-CD,[22] the complex formation of camphorsulfonic acid and polianiline[23] and so on.

A conducting polymer used in the inclusion complex formation is emeraldine base polyaniline (PANI), which is highly soluble in n-methyl-2-pyrrolidone (MP) and has the average contour length of 300nm. Other soluble conducting polymers have too bulky side chains such as hexyl one to be included into the fine nanotube. We mixed MP solution of PANI with aqueous solution of α-, β- and γ-CDs at various temperatures. Then blue precipitation appeared in only the solution with β-CD at low temperature below ca. 275K. It has been reported that very high concentration of rodlike inclusion complexes yields precipitation, which is therefore used as an evidence of the inclusion complex formation. Accordingly, the experimental results suggested that only β-CD formed the inclusion complex with PANI because of a close fit in size. This is consistent with a report that aniline molecule, monomer of PANI, forms the inclusion complex with β-CD only. Furthermore, it is another important point of the experimental results that the precipitation was observed at low temperature. This is in qualitative agreement with the theoretical prediction that the inclusion behavior should be promoted at low temperature as mentioned before.

Next we investigated the structure of the inclusion complex by the electric birefringence spectroscopy and STM. The electric birefringence appears in solution containing rodlike molecules.[16] The experimental results of the electric birefringence showed that large electric birefringence was observed in the solution of β-CD and PANI at low temperature below ca. 275K and drastically decreased down to zero with increasing temperature. This means that β-CD and PANI forms rodlike inclusion complex at lower temperature while β-CD is dissociated from PANI at higher temperature. On the other hand, the STM image of HOPG substrate, on which low-temperature solution of β-CD and PANI was dropped and spincoated, showed a rodlike structure caught by an exfoliation of step on HOPG. The length is almost equal to the average contour length 300nm of PANI and the height is close to the outside diameter of β-CD. Consequently, we concluded that the rodlike structure was identified as the inclusion complex of β-CD and PANI, namely, the insulated molecular wire.

Moreover, we investigated the insulation effect of β-CD on oxidization of PANI by iodine.[17] It was reported that when iodine was added to MP solution of emeraldine base PANI, the solution color changed from blue to violet owing to the oxidization of PANI by iodine. Accordingly, the optical absorption spectroscopy determines whether PANI is oxidized or not. When we added iodine to solution of β-CD and PANI at 275K, the solution color did not change although the solution color of PANI alone at 275K changed to violet by addition of iodine. This indicates that β-CD fully covers PANI and prevents oxidization of PANI by iodine at low temperature. Next we heated up the solution of the insulated molecular wire and iodine to 288K. Then the solution color shifted to violet ca. 4 hours after heating. This suggests that β-CD was slowly dissociated from PANI at 288K and then PANI was oxidized by iodine. Namely, PANI is not oxidized by iodine as long as β-CD covers PANI perfectly.

Very recently, we formed another insulated molecular wire consisting of the α-cyclodextrin nanotube and PANI. In this case, we observed some precipitation of the nanotube and PANI even at room temperature.[19] This indicates that the inclusion interaction between the nanotube and PANI is stronger than that of β-CD and PANI. Figure 1 shows the AFM image of the insulated molecular wire on a mica substrate where the mixture of the nanotube and PANI was spin-coated. The height of the rodlike structure is nearly equal to the outside diameter of the nanotube. As mentioned above, PANI could not form inclusion complex with α-CD. This may seem to be inconsistent with the experimental results that the nanotube consiting of α-CD includes PANI.

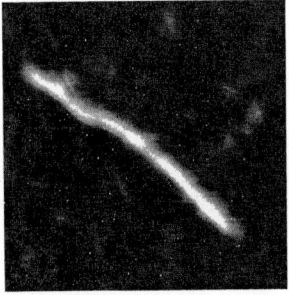

Fig. 1. AFM topographic image (200nmX200nm) of an insulated molecular wire formed by the α-CD nanotube and PANI. The length is almost equal to the average contour length 300nm of PANI and the height is close to the outside diameter of α-CD.

However, it was reported that the α-CD nanotube included some low molecular compounds such as diphenylhexatriene, which could not form the inclusion complex with α-CD. [24] This is ascribed to some structural change by the cross-linkage of adjacent α-CDs. The same change is considered to occur in the present case of the inclusion complex between the α-CD nanotube nad PANI.

Topological Gel

Polymeric system shows self-assembled higher-order structures formed by a variety of intramolecular and intermolecular interactions. The structures and physical properties in nanoscale exert a great influence on the macroscopic properties. Very recently, we have reported a novel kind of gel other than the conventional physical and chemical gels.[18] The gel has high modulus, transparency and swellability arising from the nanoscale peculiar structure.

First of all let me introduce the hystorical background of the topological gel. Such a supramolecular gel with a polyrotaxane architecture was first formed by Harada's group in 1993.[25] They formed a physical gel having crosslinking points due to the hydrogen bond between α-CDs threaded onto polyethylene glycol (PEG). In 2000, Yui et al. synthesized biodegradable hydrogel having crosslinking points of the polyrotaxane of α-CD and PEG.[26] The biodegradation of the bulky end groups of the polyrotaxane resulted in the liquefaction of the gel, so that the hydrogel is applicable to the regenerative medicine. At almost the same time, we formed the topological gel having figure-of-eight crosslinks moveing freely in a polymer network.

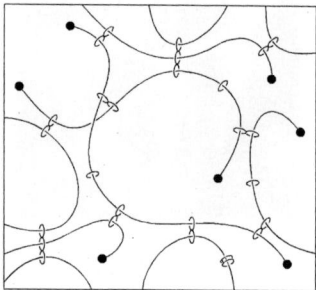

Fig. 2. Schematic diagram of the topological gel. The gel has figure-of-eight crosslinks freely moving in a polymer network.

We first synthesized a polyrotaxane in which a PEG chain with large molecular weight (Mw=20,000 and more) is sparsely included by α-CD. By chemically cross-linking α-cyclodextrins contained in the polyrotaxanes in solutions, we got transparent gels with good tensibility, low viscosity and large swellability in water. In this gel, the polymer chains with bulky end groups are neither covalently cross-linked like chemical gels nor attractively interacted like physical gels, but are topologically interlocked by figure-of-eight cross-links. It is expected that the figure-of-eight cross-links can pass the polymer chains freely to equalize the 'tension' of the threading polymer chains just like pulleys. Therefore, the nanoscopic heterogeneity in structure and stress may be automatically relaxed in the gel. Then we call this topological gel by figure-of-eight cross-links a 'topological gel'.

On tensile deformation, the polymer chains in the chemical gel are broken gradually due to the heterogeneous polymer length between fixed cross-links. On the other hand, the polymer chain in the polyrotaxane gel can pass through the figure-of-eight cross-links acting like pulleys to equalize the tension of the polymer chains cooperatively. Note that the equalization of tensions can occur not only in a single polymer chain, but also among adjacent polymers interlocked by the figure-of-eight cross-links. We call this 'pulley effect'. The physical properties of the topological gel are supposed to mainly result from the pulley effect.

The topological gel is a real example of the sliding gel theoretically considered so far and can be regarded as the third gel other than the chemical and physical gels, where the polymer network is interlocked by topological restrictions. The concept of the topological gel is important not only in the creation of high-performance gels or rubbers, but also as a new framework of artificial molecular motors based on the sliding motion just like the actin and the myosin.

Conclusion

We have so far introduced the polymeric supramolecules of the inclusion complex between the molecular nanotubes and polymer chains. The fundamental feature of the polymeric inclusion complex is transitional behavior different from the inclusion complex between small molecules. We have proposed some applications of the polymeric inclusion complex such as the insulated molecular wire and the topological gel. They are expected to develop into nanoscale electronic devices and nanomachines.

It may be stressed that there are many tools consisting of a string and ring or tube. This means that they are excellent combination and produce novel functions which are not given by each of them. On the other hand, we have not sufficiently applied it to functional materials in nanoscopic scale yet. This report indicates that we can control the nanoscopic structure and macroscopic properties by utilizing the combination between a polymer chain and nanoring or naotube.

Acknowledgements

We are indebted to Prof. Akira Harada for introducing us to this new field and helpful advice about synthesis of the molecular nanotube. This study was partly performed through Special Coordination Funds of the Ministry of Education, Culture, Sports, Science and Technology of the Japanese Government and partly supported by CREST of JST (Japan Science and Technology).

110

[1] G. Wenz, *Angew Chem., Int. Ed. Engl.* **1994**, *33*, 803.

[2] D. Philp and J. F. Stoddart, *Angew Chem., Int. Ed. Engl.* **1996**, *35*, 1154.

[3] A. Harada, *Coordination Chemistry Reviews* **1996**, *148*, 115.

[4] J. –P. Sauvage, D. Dietrich-Buchcker, "*Molecular Catenanes, Rotaxanes and Knots*", Wiley-VCH, 1999.

[5] D. A. Tomalia and I. M. J. Fréchet, "*Dendrimers and Other Dendritic Polymers*", John Wiley and Sons, Chichester UK, 2001, p.3.

[6] A. Harada and M. Kamachi, *Macromolecules* **1990**, *23*, 2821.

[7] A. Harada, J. Li and M. Kamachi, *Nature* **1992**, *356*, 325.

[8] A. Harada, J. Li and M. Kamachi, *Nature* **1993**, *364*, 516.

[9] Y. Okumura, K. Ito and R. Hayakawa, *Phys. Rev. Lett.* **1998**, *80*, 5003.

[10] Y. Okumura, K. Ito and R. Hayakawa, *Polym. Adv. Technol.* **2000**, *11*, 815.

[11] E. Ikeda, Y. Okumura, T. Shimomura, K. Ito and R. Hayakawa, *J. Chem. Phys.* **2000**, *112*, 4321.

[12] Y. Okumura, E. Ikeda, T. Shimomura, K. Ito and R. Hayakawa, *Rep. Prog. Polym. Phys. Jpn.* **1997**, *40*, 95.

[13] M. Saito, T. Shimomura, Y. Okumura, K. Ito and R. Hayakawa, *J. Chem. Phys.* **2001**, **114**, 1.

[14] Y. Okumura and R. Hayakawa, *Phys. Rev. E* **1999**, *59*, 3823.

[15] Y. Okumura, K. Ito, R. Hayakawa and T. Nishi, *Langmuir* **2000**, *26*, 10278.

[16] K. Yoshida, T. Shimomura, K. Ito and R. Hayakawa, *Langmuir* **1999**, *15*, 910.

[17] T. Shimomura, K. Yoshida, K. Ito and R. Hayakawa, *Polym. Adv. Technol.* **2000**, *11*, 837.

[18] Y. Okumura and K. Ito, *Adv. Mater.* **2001**, *13*, 485.

[19] T. Shimomura, K. Yoshida, K. Ito and R. Hayakawa, *Polym. Adv. Technol.* **2000**, *11*, 837.

[20] D. T. McQuade, A. E. Pullen and T. M. Swager, *Chem. Rev.* **2000**, *100*, 2537.

[21] E. W. Meijer and A. P. H. J. Schenning, *Nature* **2002**, *419*, 353.

[22] P. N. Taylor, M. J. O'Connell, L. A. McNeill, M. J. Hall, R. T. Aplin and H. L. Anderson, *Angew Chem., Int. Ed. Engl.* **2000**, *39*, 3456.

[23] H. Kosonen, J. Ruokolainen, M. Knaapila, M. Torkkeli, R. Serimaa, W. Bras, A. P. Monkman, G. ten Brinke and O. Ikkala., *Synth. Mett.* **2001**, *121*, 1277.

[24] G. Li and L. B. Mcgown, *Science* **1994**, *264*, 249; Y. Kawaguchi, H. Arai, M. Okada, M. Kamachi and A. Harada, *Polym. Prep. Jpn.* **1999**, *48*, 436.

[25] J. LI, A. Harada and M. Kamachi, *Polym. J.* **1993**, *26*, 1019.

[26] J. Watanabe, T. Ooya, N. Yui, *J. Artif. Organs.* **2000**, *3*, 1136.

Development of Multivalent Macromolecular Ligands for Enhanced Detection of Biological Targets

Kalle Levon,* Bin Yu

Polymer Research Institute, Polytechnic University, Six Metrotech Center, NY 11201, USA
E-mail: klevon@poly.edu

Summary: Macromolecular conjugates enable simultaneous binding of multiple ligands on one biological entity and these polyvalent interactions can be collectively stronger than the corresponding monovalent ligands. We have synthesized macromolecules and conjugated them with a lectin (Helix Pomatia lectin, HPA), and an antibody, both with shown affinities to certain bacteria. The binding ability was studied by flow cytometry and the results showed that the affinity of the biomolecules was greatly enhanced due to the polyvalent effect.

Keywords: bacteria detection, biomolecules, macromolecules, multivalent, polyvalent

Introduction

Polyvalent interactions are a characteristic phenomenon in biology. Because these interactions can be collectively much stronger than the corresponding monovalent interactions, they would lower the needed concentration of biologial ligand molecules in sensor applications and thus increase the stability parameters of the sensors. Recently, one of the most extensively studied polyvalent interaction system has been inhibition of the adhesion of influenza virus A to erythrocytes.[1-7] The inhibition concentration of polymeric ligand could be 10^7-10^8 times lower than that of monomeric ligands. Whitesides *et al* investigated the inhibition of agglunation of erythrocytes by influenza virus using polyacrylamides bearing pendant alfa-sialoside groups and showed that the cooperative binding was shown to be effective in the inhibition (1). Influenza virus binds tightly to cell with the recognition of sialic acid (SA) groups on the cell surface by a viral surface protein, hemagglutinin (HA). The binding is considered to be due the polyvalent interactions as solubilized HA binds only weakly to methyl alfa-sialoside. The extent of inhibition was shown to be a function of the conformation of the polymer and thus the function to the monomer size and concentration. For instance with high degree of substitution, the

DOI: 10.1002/masy.200351113

polymer exhibits extended conformation and is not effective in the inhibition. When the functional monomer concentration was below 0.05, no inhibition was observed. The concentration range 0.2-0.6 showed strong inhibition and with the concentration above 0.6, the polymer with extended conformation was not shown to be effective anymore. It was confirmed that the initial affinity of the ligand molecule is an essential starting requirement as polymer with non-specific interactions was not able to create inhibition. The quantitative affinity experiments did though show that in the polyvalent polymer, the affinity is weaker than in the individual molecules as the dissociation constant of the complex did not increase but actually showed less tight binding. The polyvalence was explained to be due to entropically enhanced binding expresses as enhanced probability, similar as in a chelate effect. Additional reasons for the inhibition were mentioned to be steric (colloidal) stabilization to prevent hemagglutination and possibly viscosity or the aggregation of the particles.

In the above studies, the biomolecules for conjugation were small molecules; sialic acids, mono- or disaccharides. Macromolecular conjugation with large molecules, such as antibodies or lectins would be much more difficult. Not only because the conjugating chemistry will become complicated due to the multifunctional groups in these large multifunctional molecules, but a blocking of the binding site can also be possible. Recently, Whitesides conjugated a dodecameric peptide (HTSTYWWLDGAP) into a polyacrylamide side chain and found that the inhibitiom ability was much higher than the monomeric peptide.[8]

We have investigated the surface composition of *B anthracis* and screened potential ligand molecules for the detection of these harmful bacteria. In this work we report that the conjugation of antibodies, lectins and heptapeptides with polymethacrylamide chains.

Experimental

Materials: Lectin, Helix Pometia (HPA) was purchased from Sigma. The intact BD8 antibody was supplied by John Kearney in University of Alabama at Birmingham. Oregon Green 488 succinimide and 4'-(aminomethyl)fluoresein were purchased from Molecular Probe Co. N-(2-hydroxylpropyl) methacrylamide (HPMA) and N- (3-aminopropyl) methacrylamide (APMA) were supplied by Polyscience, Inc.. Acryoyl-poly(ethylene glycol)-NHS (monomer 2) was purchased from Shearwater Polymers, Inc.

Monomer synthesis: NaOH (8g, 0.2mol) was slowly added with stirring to an aqueous 6-aminocaproic acid (13.1g, 0.1 mol, in 40 ml H_2O) in an ice bath. Methacroyl chloride (10.5g, 0.1mol) was dropped into the above solution. After additional stirring at ice bath, the reaction mixture was neutralized by diluted HCl to pH 2.0. $CHCl_3$ was used to extract the final product. Removal of the solvent gave a solid. The crude product (4g) was directly mixed with N-hydroxysuccinimide (NHS, 2.3g) in dry dioxane. DCC (4.1g) was added to catalyze the reaction. The reaction was monitored by TLC. The product was purified by column separation using hexane-ethyl acetate (1:1, v/v) as the eluent. Total yield: 65% mp.76-77°C. H-NMR ($CHCl_3$) δ (ppm) 5.68($\underline{H}$C=C), 5.29($\underline{H}$C=C), 3.42(N-$\underline{H}_2$C-C), 2.64 (OC-$\underline{H}_2$C-C), 1.96 ($\underline{CH}_3$-C=), 1.4-1.8(-$\underline{CH2}$-$\underline{CH2}$-$\underline{CH2}$-).

The copolymerization of HPMA and APMA was initiated by 2-hydroxy-4'- (2-hydroxyethoxy)-2-methylpropiophenone (HHMP) with the irradiation of UV light at room temperature. The product was further reacted with predetermined amount of Oregon Green 488 succinimide, subsequently (about 2 hours) with excess of SMCC to give a copolymer with maleiimide functional groups and fluorescent chromospheres.

The radical copolymerization was carried out followingly: The two monomers, HPMA and the monomer (described above), and the initiator (AIBN) were mixed in dry acetone. After degassed, the polymerization solution was maintained at 60°C under N_2 for 24 hours. The copolymer was washed with cold acetone and further purified by dissolving in methanol and precipitated in acetone. This copolymer was stored in an inert environment at -20°C. Predetermined amount of 4'-aminomerthylfluorescein reacted with the polymer in dry DMF.

Oregon Green 488 succinimide (5mg) in DMSO (0.5ml) was added to a HPA lectin (10mg) or BD8 antibody (3mg) in sodium bicarbonate buffer solution(pH 8.3). After 2 hours, the reaction mixture was dialysised against sodium bicarbonate buffer solution(pH 8.3).

HPA conjugation: Oregon Green 488 labeled HPA was conjugated into polymer via the reaction of NH_2 groups in HPA and the NHS activated carboxyl groups. This amide formation was done by adding DMF solution of polymer into HPA in PBS buffer (pH 7.4) at 4°C. After overnight reaction, the conjugate was dialyzed against PBS buffer for 48 hrs. Antibody conjugation: BD8 antibody conjugate was obtained similarly with HPA lectin conjugate, by the reaction of polymer with Oregon Green 488 labeled BD8 antibody.

Charaterization ^{1}H NMR was recorded on a 400MZ Varian Unity spectrometer. GPC was performed on Waters using PBS buffer as eluent. Fluorescence spectra were obtained with a Perkin-Elmer spectrometer. HPA lectin and antibodies concentration in conjugates were determined by the protein assay. The peptide concentration in conjugates was measured by the method described in literature.[10]

Results

The polymer synthetized had three different functionalities. First, certain monomer (HPMA) composition was to provide water solubility. The second monomer composition for to provide attachment of the fluorescent lable and the third was designed for the attachment of the ligand itself. An example of the polymer synthesized for the conjugation (experimental details described in the experimental section) is shown in Scheme 1.

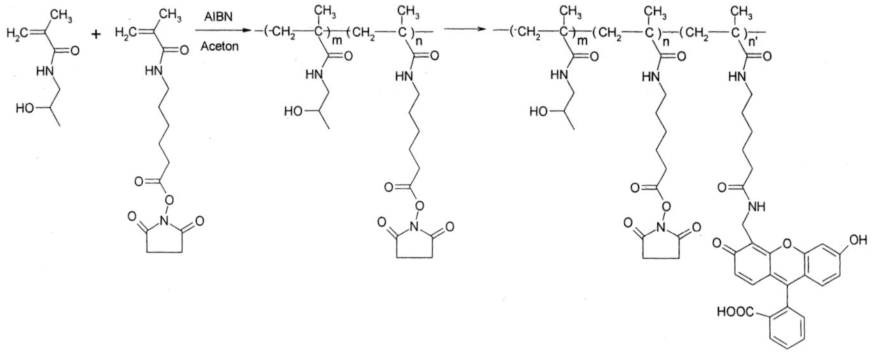

Scheme 1.

Before the evaluation of the possible binding with the developed conjugates, a control experiment with unconjugated polymers was designed to confirm that the polymers do not create non-specific binding. In flow cytometry experiments, one follows the population of labeled species (y-axis) versus the changes in the fluoresence intensity. As shown in Figure 1, shifts in the x-axis can not be observed and thus the polymer does not bind the either bacteria.

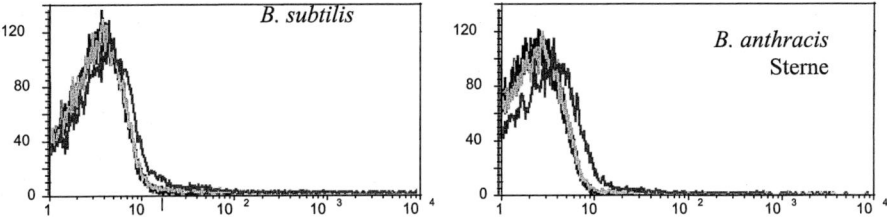

Fig. 1. Flow cytometry results on the non-specific binding of the polymers without the specific biological ligands.

HPA lectin is known to have affinity to selected sugar molecules. The binding experiments show that the labeled lectin interacts with *Bacillus thuriengiensis* at the highest concentration (Figure 2 A). The binding experiments with the lectin-polymer conjugate show a great enhancement as seen in Figure 2. The polymer conjugate (B) shows much higher binding compared to the labeled lectin alone (A). This not only confirms that sugars exist on the surface of the spores but also that polvalence increases the binding.

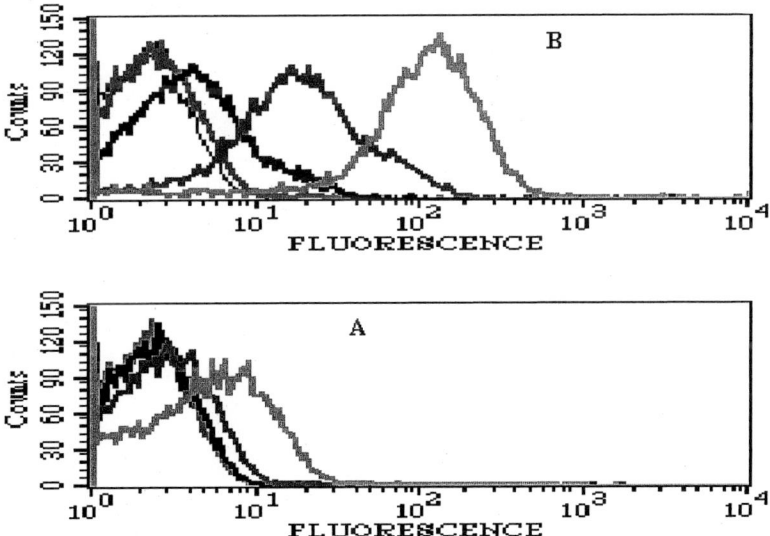

Fig. 2. Flow cytometry results on A. Lectin (HPA) and B. HPA-polymer conjugate with *B thuriengiensis.*

Lectin-sugar interactions are common in nature and lectins are large proteins so the use of these ligands are not optimal for the development of sensors. We were able to apply antibodies which had been developed in the presence of the exosporium of *B. anthracis*. We conjugated the antibody molecules as explained in the experimental section. Neither the antibody nor the antibody-polymer conjugate showed any binding to *B. subtilis* spores (the lowest picture in Figure 3). The antibody alone showed interaction with *B. anthracis* Sterne spores but only in the high concentrations. The antibody conjugates show the enhancement in a significant manner (the middle picture in Figure 3), not only the binding intensity has increased but the interactions can be observed at much lower antibody concentrations. The use of macromolecular conjugates is a potential method to build both multi- and polyvalent components for the detection of biological warfare agents.

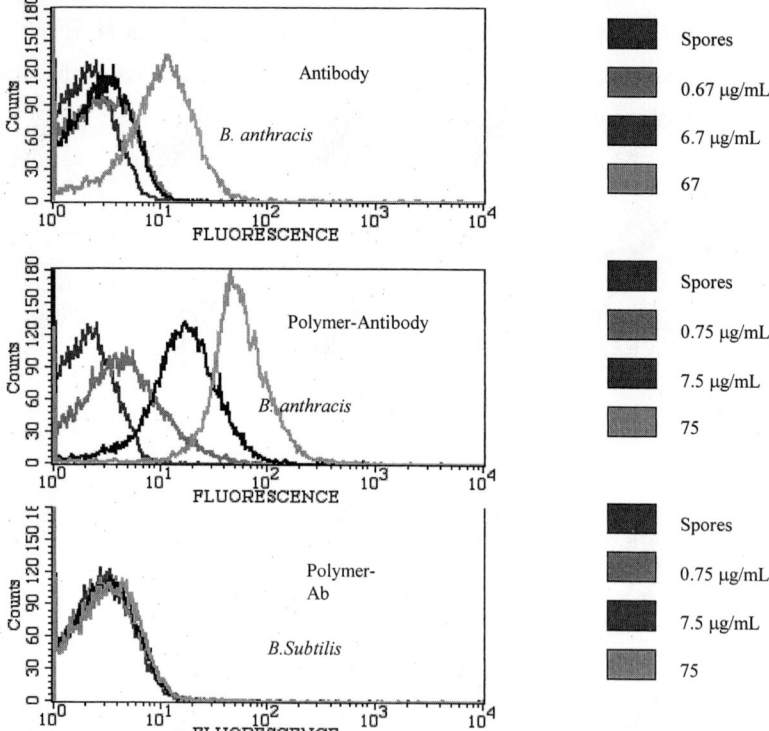

Fig. 3. Flow cytometry results for the anitbody and antibody-polymer conjugates.

[1] (a) M. Mammen, S. Choi, G. Whitesides, *Angew. Chem. Int. Ed.*, 37, 2754 (1998), (b) S. Borman, *C&EN*, 78 (41), October 9, 2000, and c) G. Whitesides *et. al.* J. Med. Chem., 1994, 37, 3419-3433

[2] (a) A. Spaltenstein, G. M. Whitesides, *J. Am. Chem. Soc.*, 113, 686-687 (1991) (b) M. Mammem, K. Helmerson, R. Kishore, S. Choi, W. D. Phillips, G. W. Whitesides, *Chem. Biol.*, 3, 757-763 (1996)

[3] M. Baek, R. Roy, *Biomacromolecules*, 1, 768-770 (2000)

[4] (a) K. H. Mortell, R. V. Weatherman, L. L. Kiessling, *J. Am. Chem. Soc.*, 118, 2297-2298 (1996) (b) K. H. Mortell, M. Gingras, L. L. Kiessling, *J. Am. Chem. Soc.*, 116, 12053-12054(1994) (c) E. J. Gordon, W. J. Sanders, L. L. Kiessling, *Nature*, 392, 30-31 (1998) (d) C. W. Cairo, J. E. Gestwicki, M. M. Kanai, L. L. Kiessling, *J. Am. Chem. Soc.*, 124, 1615-1619 (2002)

[5] J. Wang, X. Chen, W. Zhang, S. Zacharek, Y. Chen, P. Wang, *J. Am. Chem. Soc.*, 121, 8174-8181(1999)

[6] (a) H. Kamitakahara, T. Suzuki, N. Nishigori, Y. Suzuki, O. Kanie, C. Wang, *Angew. Chem. Int. Ed.*, 37, 1524-1528 (1998) (b) R. H. Kramer, J. W. Karpen, *Nature*, 395, 710-713 (1998) (c) P. I. Kitov, *Nature*, 403, 669-672 (2000)

[7] M. Mourez, R. S. Kane, J. Mogridge, S. Metallo, P. Deschatelets, B. R. Sellman, G. W. Whitesides, R. J. Collier, *Nature Biotechnology*, 19,958-961 (2001)

[8] Y. Wang, F. Chang, Y Zhang, N. Liu, S. Gupta, M. Rusckowski, D. J. Hnatowich, *Bioconjugate Chem.*, 12, 807-816 (2001)

[9] R.B. Sashidhar, A. K. Capoor, D. Ramana, *J. Immunol. Methods*, 167, 121-127 (1994)

[10] (a) Z. R. Lu, S.Q. Gao, P. Kopeckova, J. Kopecek, *Bioconjugate Chem.*, **11**, 3-7 (2000). (b) A. David, P. Kopeckova, A. Rubinstein, J. Kopecek, *Bioconjugate Chem.*, 12, 890-899 (2001)

© 2003 WILEY-VCH Verlag GmbH & KGaA, Weinheim

Optical Absorption and Photoluminescence Properties of the PPV Nanotubes and Nanowires

Kyungkon Kim,[1] *Sae Chae Jeoung,*[2] *Jinwoo Lee,*[3] *Taeghwan Hyeon,*[3]
Jung-Il Jin*[1]

[1] Division of Chemistry and Molecular Engineering and Center for Electro- and Photo-Responsive Molecules, Korea University, Seoul 136-701, Korea
[2] Spectroscopy Laboratory, Korea Research Institute of Standards and Science, Taejon 305-600, Korea
[3] School of Chemical Engineering and Institute of Chemical Processes, Seoul National University, Seoul 151-742, Korea

Summary: We measured optical absorption and time resolved photoluminescence decay properties of the PPV nanotubes and nanowires which were prepared by CVD polymerization using templates. When compared with bulk PPV films, their nano objects showed different optical properties, long photoluminescence decay time and higher photoluminescence efficiencies.

Keywords: nanotubes; nanowires

Introduction

The present world-wide, explosive interests in nanotechnology is arousing renewed attention to organic conductors because of their potentials in a wide variety of applications, especially in nanodevices. Among many organic conductors, poly(*p*-phenylenevinylene) (PPV) [1, 2]is unique in that the polymer can be prepared by many synthetic routes and possesses many interesting electrical and optical properties including photo- and electroluminescence.[3] In general, nano sized materials show much different properties from those of macroscopic bulk state.[4] In order to find proper applications of nano structured materials in the nano devices, it is essential to understand structure-properties relationship of nano dimension materials. Recently[5, 6], we reported that PPV nanotubes and nanorods can be easily prepared by performing the chemical vapor deposition (CVD) polymerization of α, α'-dichloro-*p*-xylene inside of pores of organic or inorganic templates. We have extended our studies of PPV nano objects to examine their optical properties and time resolved photoluminescence decay behavior.

© 2003 WILEY-VCH Verlag GmbH & KGaA, Weinheim   DOI: 10.1002/masy.200351114

Experimental

In the synthesis of PPV nanotubes and nanowires we used commercially available nano porous alumina membrane (nominal pore diameter: 200 nm; Whatman, England) and mesoporous silica[7] (3.2, and 10 nm pore diameter), respectively. The detailed CVD polymerization and PPV nano object preparation procedure can be found in our previous report.[5] For the preparation of PPV nanowires, we increased deposition zone temperature up to 100°C to facilitate penetration of activated monomer molecules into the small pores of the template. After CVD polymerization of PPV inside pores of mesoporous silica, the product was washed with THF several times to remove PPV formed outside of the pores. PPV films were prepared on the quartz substrate by CVD polymerization.

Measurement of UV-vis absorption and photoluminescence (PL) spectra of the PPV nanotubes was performed after the alumina membrane was dissolved by 3 M NaOH solution. Isolated PPV nanotubes were dispersed in $CHCl_3$ using ultrasonification. In the case of PPV nanowires, UV-vis absorption and PL spectra were obtained for the mesoporous silica template containing the nanowires. In order to reduce the scattering from the template, glycerol / propanol index matching solution was used.[8]

Absolute quantum yield of the dispersed PPV nano objects was obtained by using coumarin 307 as a reference material. Experimental setup for the time resolved PL decay can be found elsewhere.[9]

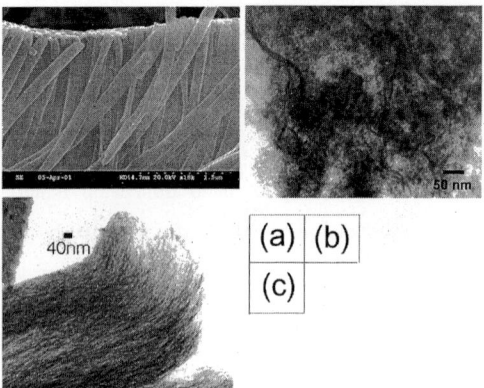

Fig. 1. SEM and TEM images of PPV nanotubes and nanowires: (a) PNT-200 (SEM) (b) PNW-10 (TEM) (c) PNW-4 (TEM).

Result and Discussion

Figure 1 shows SEM and TEM images of PPV nanotubes and nanowires prepared on the inner surface of the pores of alumina membrane and inside the pores of the mesoporous silica, respectively. The nominal pore diameter of the alumina membrane was 200 nm and the pore sizes of the mesoporous silicas were 10 and 3.2 nm. The shape of the prepared nano object was dependent on the size of the template used. When we used alumina membrane as a template, we obtained PPV nanotubes. In contrast, mesoporous silica produced PPV nanowires. The diameter and wall thickness of PPV nanotubes (PNT-200) obtained from the alumina membrane was 285 ± 25 nm and 28 ± 3 nm, respectively. Although it is difficult to determine the exact diameter of the PPV nanowires due to low resolution of TEM images, diameter of PPV nanowires obtained using 3.2 (PNW-4) and 10 nm (PNW-10) pore diameter mesoporous silica were about 4 and 10 nm, respectively.

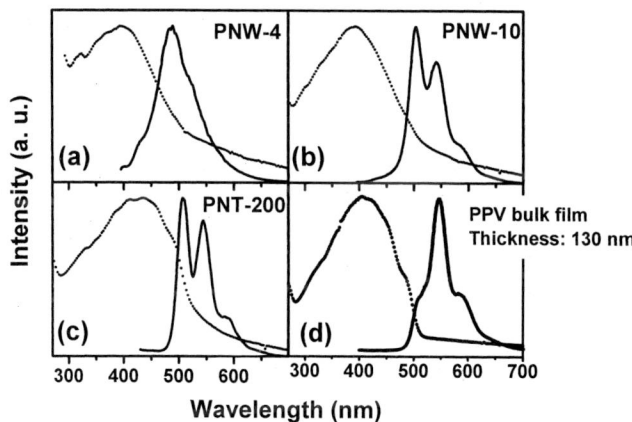

Fig. 2. Comparison of UV-vis absorption and photoluminescence spectra of PPV nanotubes and nanowires.

Figure 2 compares optical properties of the prepared nano objects. When compared with absorption spectrum of the bulk PPV film, overall shapes of other spectra are very similar to that of the bulk film with no significant shift in their absorption maxima positions. Since PPV molecules are in an inhomogeneous environment in the pores of alumina membrane and mesoporous silica, spectral broadening was observed. This phenomenon was also earlier observed in the UV-vis absorption spectrum of MEH-PPV prepared in an aligned mesoporous silica.[8, 10] On the other hand, there are observed significant differences among the PL spectra of the samples. The PL peak from the $S_1 \rightarrow S_0$ 0-0 transition, i. e., highest electronic energy transition, increases in energy as the pore size of template is reduced. Especially, in the case of PNW-4, the peak from the 0-0 transition was shifted to the shortest wavelength (higher energy) side. This is due to reduced reabsorption by neighboring chains of the emitted light originated from the 0-0 transition because of complete isolation of the nano particles suspended in the solvent medium that consist of much fewer polymer chains compared to bulk films and also due to much reduced interchain contacts between the polymer chains. We could observe an increased intensity of PL originated from the 0-0 transition when we decreased the thickness of the bulk PPV film, which can be explained from viewpoint of reabsorption.

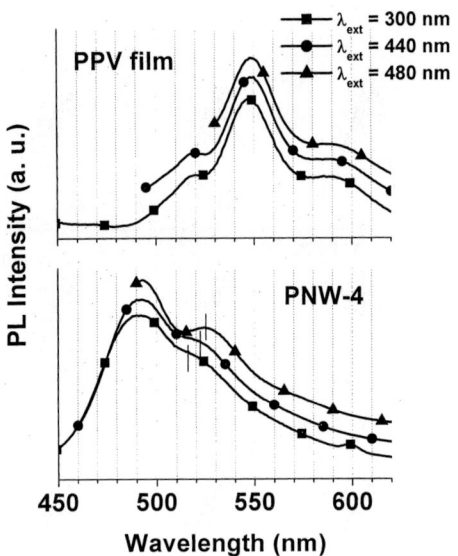

Fig. 3. Comparison of photoluminescence spectra of PPV films and nanowires at different excitation wavelengths.

Another observation made is that the PL from the PNW-4 red-shifted substantially as the excitation wavelength is increased from 300 nm to 480 nm (Figure 3). In contrast, as we can see from Figure 3, the position of the PL spectra of the PPV films, however, did not change when we changed the excitation wavelength, because the rate of intermolecular energy transfer (few ps) is more rapid than radiative recombination in the PPV chain. When PPV chains are in the pores of the mesoporous silica, they are separated by the walls of the mesoporous silica, which remarkably reduces the chances for energy transfer among the polymer chains. In other words, the rate of the intermolecular energy transfer is reduced. As a result, the higher energy excitation at 300 nm could not be transferred to the lower energy sites and PL from the $S_1 \rightarrow S_0$ 0-1 transition is observed at 517 nm which is shorter than 525 nm that is observed when the excitation wavelength was 480 nm.

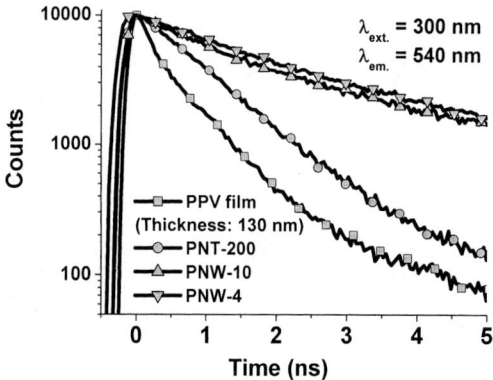

Fig. 4. Photoluminescence decay of PPV film, nanotubes and nanowires.

This also could be confirmed by the time resolved PL decay of the PPV nano objects and films (Figure 4). And the results of the spectral analysis are summarized in Table 1. As one can see from Figure 4, the PL decay for the PPV film was fastest and PNT-200 revealed the second fastest decay in PL. They both exhibit a fast initial decay followed by a slower decay. But decay rate of the PNT-200 (τ_1: 310 ps, τ_2: 880 ps) is much slower than that of the bulk film (τ_1: 87 ps, τ_2: 480 ps). Usually, impurity and structural defects of polymer chains are the non-radiative decay sites and most polymers have these kinds of sites. When polymer chains are contacted to each other, the probability of interchain energy transfer to the radiative sites or non-radiative decay increases, which causes a fast radiative decay for the bulk film. PNT-200, PNW-4 and

PNW-10 are different from true solutions of polymers. They, however, form nano domains in the dispersed solution and, thus, they are much closer to dilute solutions than bulk films. It means that they are far apart enough to prevent interchain energy transfer to non-radiative decay sites, which, in turn, would result in a long radiative decay when compared to that of bulk films. Especially, PNW-4 and PNW-10 show exceptionally long radiative decay rates (τ_3 in Table 1) and the decay time parameter values are 2.2 and 2.4 ns, respectively. It is quite likely that as soon as PNW-4 and PNW-10 are exposed to excitation beam, polymer chains outside of the pores are first excited and then generated excitons migrate to the lower energy sites along the chains which are isolated in solution-like environment by the walls of the mesoporous silica.[10] This appears to be the origin of the longest radiative decay time (τ_3 in Table 1). The τ_3 value is comparable to that of a diluted solution of MEH-PPV[11] and a diluted blend film of PPV oligomer and PMMA.[11] Therefore, this kind of isolation effect would result in a high PL quantum efficiency for the PPV nano objects.[12, 13] In fact, the value of PL quantum efficiency increases in reverse order of pore size of the template used as shown in Table 2. The maximum PL efficiency obtained was 22.7% (PNW-4) which is twice the value for the bulk PPV film.

Table 1. Photoluminescence decay times of PPV film, nanotube and nanowire.

	τ_1	τ_2	τ_3
	Decay time (ratio) (ns)	Decay time (ratio) (ns)	Decay time (ratio) (ns)
Bulk film (130 nm thick)	0.087 (0.83)	0.48 (0.17)	-
PNT-200	0.31 (0.86)	0.88 (0.14)	-
PNW-10	0.17 (0.46)	0.75 (0.33)	2.4 (0.21)
PNW-4	0.27 (0.36)	0.92 (0.46)	2.2 (0.18)

Table 2. Absolute PL quantum efficiencies of PPV film and PPV nano objects.

	Quantum efficiency (%)
PPV film	12.0[14]
PNT-200	14.8
PNW-10	18.2
PNW-4	24.7

Conclusion

In conclusion, we measured optical absorptions and time resolved PL properties of PPV nanofilms, nanotubes, and nanowires which were prepared by CVD polymerization method using various nano sized templates. In comparison with bulk PPV film, the PL peak from the S_1 $\rightarrow$ S_0 0-0 transition increased for the PPV nano objects. And in the case of PNW-4, a spectral shift to short wavelength was observed. Also their PL decay times increased as the pore size of the template was reduced. For the case of the PL decay of PNW-4 and PNW-10, exceptionally long radiative decay time was observed. The much slower radiative decay time is in accord with absolute PL quantum yield. All the results could be interpreted as a result of the prevention of interchain interactions throuth the isolation of the PPV chains in form of nanosized domanins.

Acknowledgment

This work was supported by the Korea Science and Engineering Foundation through the Center for Electro- and Photo- Responsive Molecules, Korea University. Kyungkon Kim is the recipients of Brain Korea 21 assistantship supported by the Ministry of Educations and Human Resources Development.

[1]   C. Kvarnstrom, A. Ivaska, in *Habdbook of Conductive Molecules and Polymers, Vol.*, ed. (Ed.:H. S. Nalwa), John Wiley & Sons, New York, **1997**, Chapter 9.

[2]   C. S. Moratti, in *Handbook of Conducting Polymers,* 2nd ed. (Eds.T. A. Skothtim, R. E. Elsenbaumer, J. R. Reynolds), Marcel Dekker, New York, **1998**, Chapter 13.

[3]   J. H. Burroughes, D. D. C. Bradley, A. R. Brown, R. N. Marks, K. Mackay, R. H. Friend, P. L. Burn, A. B. Holmes, *Nature* **1990**, *347*, 539.

[4]   S. A. Empedocles, R. Neuhauser, K. Shimizu, M. G. Bawendi, *Adv. Mater.* **1999**, *11*, 1243.

[5]   K. Kim, J.-I. Jin, *Nano Letters* **2001**, *1*, 631.

[6]   K. Kim, G. Zhong, J.-I. Jin, J. H. Park, S. H. Lee, D. W. Kim, Y. W. Park, W. Yi, *ACS Symp. Series in press.*

[7]   A. Stein, B. J. Melde, R. C. Schroden, *Adv. Mater.* **2000**, *12*, 1403.

[8]   J. Wu, A. F. Gross, S. H. Tolbert, *J. Phys. Chem. B* **1999**, *103*, 2374.

[9]   Y. H. Kim, S. C. Jeoung, D. Kim, S. J. Chung, J.-I. Jin, *Chem. Mater.* **2000**, *12*, 1067.

[10]   T.-Q. Nguyen, J. Wu, V. Doan, B. J. Schwartz, S. H. Tolbert, *Science* **2000**, *288*, 652.

[11]   C. M. Heller, I. H. Campbell, B. K. Laurich, D. L. Smith, D. D. C. Bradley, P. L. Burn, J. P. Ferraris, K. Mullen, *Phys. Rev. B: Condens. Matter* **1996**, *54*, 5516.

[12]   R. C. Smith, W. M. Fisher, D. L. Gin, *J. Am. Chem. Soc.* **1997**, *119*, 4092.

[13]   T. W. Lee, O. O. Park, J. H. Yoon, J. J. Kim, *Adv. Mater.* **2001**, *13*, 211.

[14]   U. Lemmer, R. F. Mahrt, Y. Wada, A. Greiner, H. Bassler, E. O. Gobel, *Appl. Phys. Lett.* **1993**, *62*, 2827.

Macromol. Symp. **2003**, *201*, 127—134

Self-Amplifying Sensory Materials: Energy Migration in Polymer Semiconductors

Kenichi Kuroda, Timothy M. Swager*

Department of Chemistry, Massachusetts Institute of Technology, Cambridge, MA, USA 02139
E-mail: tswager@mit.edu

Summary: Signal amplification for ultra-sensitive detection has been achieved by energy migration in conjugated semiconducting polymeric assemblies. Critical to optimizing this effect is the synthesis of non-aggregate polymers, the multi-dimensional directional transport of excited states (excitons), and extending the intrinsic excited state lifetime of conjugated polymers. We developed new water-soluble non-ionic conjugated polymers for use in biosensory applications, which can be used to provide highly sensitive/specific ultra-trace detection that is immune to specificity problems that plauge ionic conjugated polymers.

Keywords: conjugated polymers, energy migration, non-aggregate polymers, semiconducting polymeric assemblies, signal amplification

Introduction

Fluorescent semiconductive polymers are powerful tools to create ultra-sensitive sensory materials and their unique electronic properties provide a new transduction capablity in sensory detection schemes.[1,2] We have focused on conjugated polymers with rigid rod-like structures consisting of aromatic groups connected directly or with double or triple bonds, where π-electrons are delocalized along the polymer chain. The origin of the amplification that enables these materials to produce ultra-sensitive sensors is the extended electronic structures that create energy bands (a conduction band and a valence band). Excited states (excitons) can move through these energy bands and in this way the conjugated polymer behaves as a molecular wire for the transport of excitons. The excitons travel along these molecular wires by a combination of Förster (long range dipolar interaction through space) and Dexter (strong electronic coupling by short range interaction) transport mechanisms, and the sensory detection event is realized when the excitons encounter a energy trap at a receptor site that is activated by a bound analyte molecules.[3] In most schemes, the excitons are quenched at an occupied receptor sites, and thus the fluorescence is diminished. In our schemes, the key factor leading to the high sensitivity of conjugated polymeric sensors is the migration of excitons through the material. Therefore, our efforts have focused on increasing the efficiency of excition transport, which in turn increases the overall exciton diffusion length and thus the probability of the exciton encountering a receptor site occupied by an analyte molecule. According to this principle, we have designed new sensory systems based upon conjugated polymers, particularly poly(arylene ethynylene) derivatives (PAEs). In our quest for greater sensitivity, we have gained new insight into different electronic properties of semiconducting polymers important to exciton transport.

Non-aggregated Polymers

In our fundamental studies of the optical properties of PAE-aggregates using Langmuir Blodgett film techniques, we have shown that aromatic planer structures of PAE's backbone stack and aggregate strongly and, unless prevented by steric factors, display strong intermolecular electronic coupling to give non-emissive self-quenched materials.[4, 5, 6] In order to prevent this unfavorable event for exciton transport and thereby keep high quantum yields, we have designed pentiptycene[7] and rotaxane[3, 8] containing PAEs (Polymer 1 and 2) that prevent strong interactions between polymer chains that leads to a non-emissive solid state. Pentiptycene-based PAE (Polymer 1) showed a high fluorescence quantum yield in the film and a strong response to trinitro-toluene (TNT) molecules by fluorescence quenching[7]. In addition to non-aggregation of the polymers, this structure provides cavities that accommodate small aromatic molecules and enhance TNT binding in the films. Recently, this pentiptycene monomer was incorporated in cationic PAE particles for the purpose of DNA detection in water by fluorescence quenching.[9]

Polymer 1   Polymer 2   Polymer 3

In contrast to the non-aggregation approach, chemosensors that make use of fluorescence self-quenching of PAEs have also been developed.[10] In this case PAEs were designed with pendant crown ether groups (Polymer 3). The 15-crown-5 ethers do not display selectivity for specific alkali metal ions. However polymer 3 is endowed with specificilty due to the fact that 15-crown-5 ethers engages in multivalent (2:1) binding of potassium ions, however with sodium and lithium they display only 1:1 complexes. The resultant intermolecular multivalent process produces a highly discriminating sensor for potassium ions, wherein the K^+ ions pull the polymer chains together and induce the formation of non-emissive aggregates.

Multi-Dimensional Directional Energy Migration

When the conjugated polymer sensor is used in dilute solution, movement of excitons is limited within a single polymer chain and the migration can be regarded as a one-dimensional random walk along the polymer chain. This results in the excitons visiting the same spot on the chain multiple times during the lifetime and spending majority of its lifetime close to where it was created. This inefficiency of energy migration in one-dimension reduces the chance for excitons to sample more receptors, which is needed to increase the probability of encountering analyte molecules to give a sensory response. Energy migration is enhanced when in a thin film where the exciton can move from one polymer to another by Förster and/or Dexter transport mechanisms and may diffuse beyond the length of individual polymer chains. As the exciton's ability to sample different receptor sites is very fast relative to the off-rate of the bound analytes, increasing dimensionality and the number of receptor sites an exciton can visit during its lifetime increases the amplification. For example a two-dimensional film behaves as a sheet transporting excitons rather than just a wire.

The enhancement in thin films is best highlighted by the extraordinary sensory properties of pentiptycene-based PAE (Polymer 1). This material is the basis of an ultrasensitive detector for TNT that displays orders of magnitude (10^4-10^5) more sensitivity than any other explosive-detection systems.[7] However the excitons still have finite diffusion lengths, and to increase diffusion lengths 3-dimentional multi-layer polymer films have been designed, wherein the excitons are directed though polymer layers to transduction sites. As mentioned earlier, fluorescence of PAE solid films (and indeed most conjugated materials) tend to decrease due to formation of non-emissive aggregates. Therefore the rigid three-dimensional structures that prevent strong interaction with neighboring polymer chains (Polymer 1 and 2) are necessary. By inducing a directional (vectorial) migration of excitons in polymer films with an energy gradient, energy migration is made more efficient.[8]

130

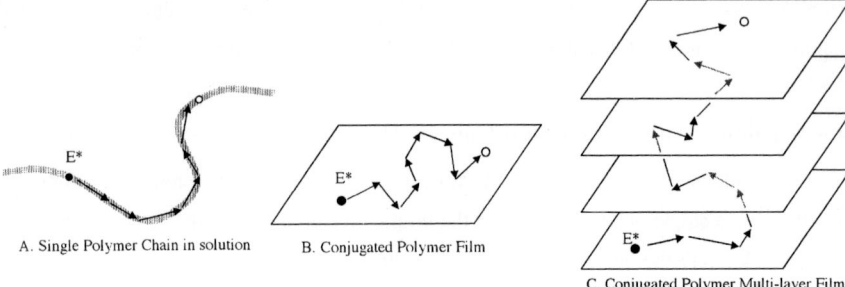

A. Single Polymer Chain in solution

B. Conjugated Polymer Film

C. Conjugated Polymer Multi-layer Film

Fig. 1. Schematic presentation of multi-dimensional random walk of exciton (E*): (A) along a isolated polymer chain in solution (one-dimension), (B) on a conjugated polymer thin film (two-dimention), (C) within a conjugated polymer multi-layer film (three-dimension). In all cases the forward and backward random walks are present. In the single polymer chain the exciton spends much more of the time retracing its path. This is less likely in thin films and can be further minimized by creating an energy gradient in multi-layer films.

Lifetime Modulation

Another approach for increasing sensitivity is to extend the intrinsic lifetime of excited state of PAEs. Based upon what we have determined regarding the mechanisms of energy transfer in PAEs,[11] larger lifetimes of excited states leads to longer exciton diffusion lengths, which increase the chance that excitons will encounter receptor sites bound with analyte molecules. The lifetimes of PAEs have been extended by integrating larger two-dimensional polycyclic aromatic structures into the polymer backbone. Larger lifetimes have been realized in triphenylene[11] or dibenzo[g,p]chrysene[12] based-polymers (Polymer **4** and **5**) as compared with simple (phenylene-ethynylene) polymers, with similar fluorescence quantum yields.[11]

Polymer **4**

Polymer **5**

Light Harvesting

Fluorescence resonance energy transfer (FRET) is a widely used method to probe biological macromolecular conformations in proteins and DNA and also detect trace amounts of analyte molecules and biomolecular recognition events. In FRET schemes, specific emission can be observed depending on the pair of chromophores. A limitation of fluorescence quenching detection schemes is that a multitude of potential interfering species can cause quenching and the detection signal may not always be specific to the quencher. As we mentioned above, multilayers of conjugated polymers have been designed for the directional transportation of excited states. The ability to control the direction of energy migration in PAEs provides a number of opportunities to construct excellent chemosensors based on FRET schemes. The key feature is that excitons are directionally transported toward an acceptor to which the excited state energy is transferred. The acceptor can receive energy effectively from multiple polymer segments, which have been transported over large distances; so light-harvesting FRET produces an amplified emissive signal from the acceptor. We have demonstrated this principle to produce a turn-on chemosensor involving FRET between a chromophore and PAE.[13] Polymer thin films were constructed by layer-by-layer deposition of cationic PAE (Polymer 6) and anionic polyacrylamide (Polymer 7) modified with aminofluorescein, a chromophore widely used in biological and biosensory schemes, and the emission signal from the fluorescein is amplified. This study demonstrated that the PAE film harvests light and transfers it efficiently to a small dye molecule (fluorescein). Multi-layer polymer assemblies that consist of three different PAEs (Polymer 2, 8, 9) having different energy gaps were also designed.[14] The bandgaps decrease from the bottom layer (Polymer 8) to middle layer (Polymer 2) and to surface layer (Polymer 9) so that the film has a vectrical energy gradient. In this case, the polymers have fulfilled the FRET requirement of spectral overlap of a donor emission and an acceptor absorption. The excited state of the bottom layer, Polymer 8, migrates directionally to the polymer layers with lower bandgaps toward surface via Förster mechanisms, and dominant emission is from the surface polymer (Polymer 9) with the smallest band gap. This directional migration of excitons enforced by an energy gradient makes exciton transport length longer and thus energy migration more efficient because multiple visits of the same site are reduced. This result also revealed that we can extend the exciton diffusion in multilayer conjugated polymer assemblies, and this is beneficial to development of not only chemosensors, but also organic electronics and electrooptical devices.

Polymer 6   Polymer 7   Polymer 8   Polymer 9

Conjugated Polymers for Biosensors

The sensory schemes discussed above also have great potential for producing highly sensitive biosensory materials. In particular, water-soluble conjugated polymers can find many applications for producing ultra-sensitive biosensors that function in aqueous environments for diagnostic use in medicine, biotechnology, and security applications, with the latter being greatly needed to counter the threat of bio-weapons. Indeed conjugated polymer-based biosensors in thin film form and/or in organic media have received considerable attention.[2] One of the major problems for conjugated polymeric materials in these applications is the intrinsically hydrophobic nature of the conjugated system that gives rise to poor water-solubility. This strong hydrophobic nature of polymer backbone generally results in aromatic π-π stacking which limits their solubility in water. This strong interaction between polymer chains enhances the formation of non-emissive aggregates that reduce sensitivity. To overcome this difficulty, ionic conjugated polymers have been utilized in biosensory schemes.[2,17,18,19] Ionic groups such as sulfonate,[15] carboxylate,[16] or ammonium groups[17] in the sidechains provide strong hydration and electrostatic repulsions between polymer chains that promote water solubility of the conjugated polymers. By making use of the fact that these ionic conjugated polymers (conjugated polyelectrolytes) can electrostatic complex with ionic molecules, new detection schemes have been proposed and examined. Chen et al.[18] have recently made use of our amplification principles and have demonstrated that fluorescence of water-soluble anionic PPV derivative with sulfonate group on the side chain (MPS-PPV) was quenched efficiently by methylviologen (MV^{2+}) as shown in our earlier studies. The strong fluorescence quenching of MPS-PPV results from formation of electrostatic complex of the anionic polymer and the cationic quencher. In this scheme the fluorescence of MPS-PPV is quenched by B-MV that binds electrostatically to the polymer and the fluorescence of the polymer was recovered removing the quencher through avidin binding. Also building upon our principles MPS-PPV has been shown to be quenched electron transfer protein, cytochrome c (cyt c), reported by Fan et al.[19] and DNA detection using a cationic conjugated polymer (CCP) was reported by Gaylord et al.[17] The latter scheme makes use of our earlier FRET method and a fluorescein-labled neutral peptide nucleic acid (PNA-C*) that has complementary sequence with target DNA was paired with DNA/CCP hydrides by nucleic acid matching. In this process, target DNA served as a specific interaction to connect a fluorescent probe (PNA-C*) and CCP. This complexation brings the fluorecein in the vicinity of CCP and thus results in signal amplification from the fluoescein by FRET.

Methylviologen (MV^{2+})   MPS-PPV   CCP

R: -(CH$_2$)$_6$NMe$_3$ I

Polymer 10   Polymer 11

In spite of these laboratory demonstrations ionic conjugated polymers have severe limitations for use in biosensor schemes because (1) the solution conditions (pH, ionic strength, temperature) have to be adjusted to prevent polymer aggregation and (2) electrostatic interaction between ionic polymers and biomolecules such as proteins and DNA are non-specific and therefore will reduce specificity for target molecules. To overcome these problems we have synthesized non-ionic water-soluble conjugated polymers, which promise to constitute a platform for producing biosensors with higher specificity. Polymer 10 was designed to be surrounded by dendritic side chains with a number of hydroxyl group and amide bonds.[20] This structure was designed to shield the hydrophobic polymer backbone from water and promote the solubility in water. Polymer **10** is completely water-soluble, wehereas less hydrophilic Polymer **11** has a limited solubity in water.

In summary, we have developed fluorescent conjugated polymers and their assemblies as a versitle platform from which to produce ultra-sensitive sensory materials. The high sensitivity in detection was obtained by intricate designs of chemical structures directed at producing high quantum yield fluorescent polymers, extended lifetimes, and structured films for efficient exciton transport in the materials. These design principles are presently being extended by ourselves and others to biosensor and bioconjugates with semiconducting polymers.

[1] T. M. Swager, *Acc. Chem. Res.*, **1998**, 31, 201; T. M. Swager, J. H. Wosnick, *MRS Bulletin*, **2002**, June, 446; J. H. Wosnick, T. M. Swager, *Curr. Opin. Chem. Bio.*, **2000**, 4, 715.
[2] D. T. McQuade, A. E. Pullen, T. M. Swager, *Chem. Rev.*, **2000**, 100, 2537
[3] Q. Zhou, T. M. Swager, *J. Am. Chem. Soc.* **1995**, 117, 7017; Q. Zhou, T. M. Swager, *J. Am. Chem. Soc.* **1995**, 117, 12593
[4] J. Kim, T. M. Swager, *Nature*, **2001**, 411, 1030,;
[5] J. Kim, I. A. Levitsky, D. T. McQuade, T. M. Swager, *J. Am. Chem. Soc.* **2002**, 124, 7710.
[6] D. T. McQuade, J. Kim, T. M. Swager, *J. Am. Chem. Soc.* **2000**, 122, 5885
[7] J.-S. Yang, T. M. Swager, *J. Am. Chem. Soc.* **1998**, 120, 5321; J.-S. Yang, T. M. Swager, *J. Am. Chem. Soc.* **1998**, 120, 11864; J. C. Cumming, C. Aker, M. Fisher, M. Fox, M. J. la Grone, D. Reust, M. G. Rockley, T. M. Swager, E. Towers, V. Williams, *IEEE Transactions on Geoscience and Remote Sensing* **2001**, 39 (6), 1119.
[8] I. A. Levitsky, J. Kim, T. M. Swager, *J. Am. Chem. Soc.* **1997**, 121, 1466.
[9] J. H. Moon, R. Deans, E. Krueger, L. F. Hancock, *Chem. Comm.* **2003**, 104.
[10] J. Kim, D. T. McQuade, S. K. McHugh, T. M. Swager, *Angew. Chem. Int. Ed.*, **2000**, 39, 3868.
[11] A. Rose, C. G. Lugmire, T. M. Swager, *J. Am. Chem. Soc.* **2001**, 123, 11298.
[12] S. Yamaguchi, T. M. Swager, *J. Am. Chem. Soc.* **2001**, 123, 12087.
[13] D. T. McQuade, S. K. McHugh, T. M. Swager, *J. Am. Chem. Soc.* **2000**, 122, 12389
[14] J. Kim, D. T. McQuade, A. Rose, Z. Zhu, T. M. Swager, *J. Am. Chem. Soc.* **2001**, 123, 11488.
[15] S. Shi, E. Wudl, *Macromolecules*, **1990**, 23, 2119
[16] H. Häger, W Heitz, Macromol. Chem. Phys., **1998**, 199, 1821
[17] B. S. Gaylord, A. J. Heeger, G. C. Bazan *Proc. Natl. Acad. Sci. U.S.A.*, **2002**, 99, 10954.
[18] L. Chen, D. W. McBranch, H. –L. Wang, R. Helgeson, F. Wudl, D. G. Whitten, *Proc. Natl. Acad. Sci. U.S.A.*, **1999**, 96, 12287.
[19] C. Fan, K. W. Plaxco, A. J. Heeger, *J. Am. Chem. Soc.* **2002**, 124, 5642.
[20] K. Kuroda, T. M. Swager, *Chem. Comm.* **2003**, 26.

Macromol. Symp. **2003**, *201*, 135—142

Macromolecular Helicity Inversion of Poly(phenylacetylene) Derivatives Induced by Various External Stimuli

Katsuhiro Maeda, Kazuhide Morino, Eiji Yashima*

Department of Molecular Design and Engineering, Graduate School of Engineering, Nagoya University, Chikusa-ku, Nagoya 464-8603, Japan

Summary: Unique macromolecular helicity inversion of stereoregular, optically active poly(phenylacetylene) derivatives induced by external achiral and chiral stimuli is briefly reviewed. Stereoregular, *cis-transoidal* poly(phenylacetylene)s bearing an optically active substituent, such as (1*R*,2*S*)-norephedrine (poly-**1**) and β-cyclodextrin residues (poly-**2**), show an induced circular dichroism (ICD) in the UV-visible region of the polymer backbone in solution due to a predominantly one-handed helical conformation of the polymers. However, poly-**1** undergoes a helix—helix transition upon complexation with chiral acids having an *R* configuration, and the complexes exhibit a dramatic change in the ICD of poly-**1**. Poly-**2** also shows the inversion of macromolecular helicity responding to molecular and chiral recognition events that occurred at the remote cyclodextrin residues from the polymer backbone; the helicity inversion is accompanied by a visible color change. A similar helix—helix transition of poly((*R*)- or (*S*)-(4-((1-(1-naphthyl)ethyl)carbamoyl)phenyl)acetylene) is also briefly described.

Keywords: chiral, helix, helix—helix transition, induced circular dichroism, polyacetylenes

Introduction

In polymer and supramolecular chemistry, significant attention has been paid to developing synthetic polymers and polymeric assemblies that adopt a helical conformation even in solution not only to mimic nature but also to develop novel chiral materials for the use as liquid crystals, enantioselective catalysts, and chiral selectors.[1-4] To date, several synthetic helical polymers have been prepared, which can be classified into two types with respect to the nature of the helical conformation; one is a stable (or static) helical polymer even in solution and the other is a dynamic helical polymer. These synthetic helical polymers exhibiting an optical activity due

DOI: 10.1002/masy.200351116

to the one-handed helicity can be prepared either by polymerization of optically active monomers[2-4] or by asymmetric polymerization of achiral or prochiral monomers with chiral catalysts or initiators.[1] Therefore, the helix-sense can be controlled thermodynamically by chiral substituents covalently bonded to the polymer main chain or kinetically during the polymerization. Besides these helical polymers, we recently succeeded in developing a conceptually new helical polymer, which is an induced helical polymer. A predominantly one-handed helicity can be induced on optically inactive polymers including polyacetylenes bearing various functional groups upon complexation with optically active small molecules capable of interacting with the functional groups.[5-7] Details of these helical polymers have been thoroughly reviewed elsewhere.[1-4,8]

Chart 1. Structures of synthetic helical polymers that exhibit a helix—helix transition.

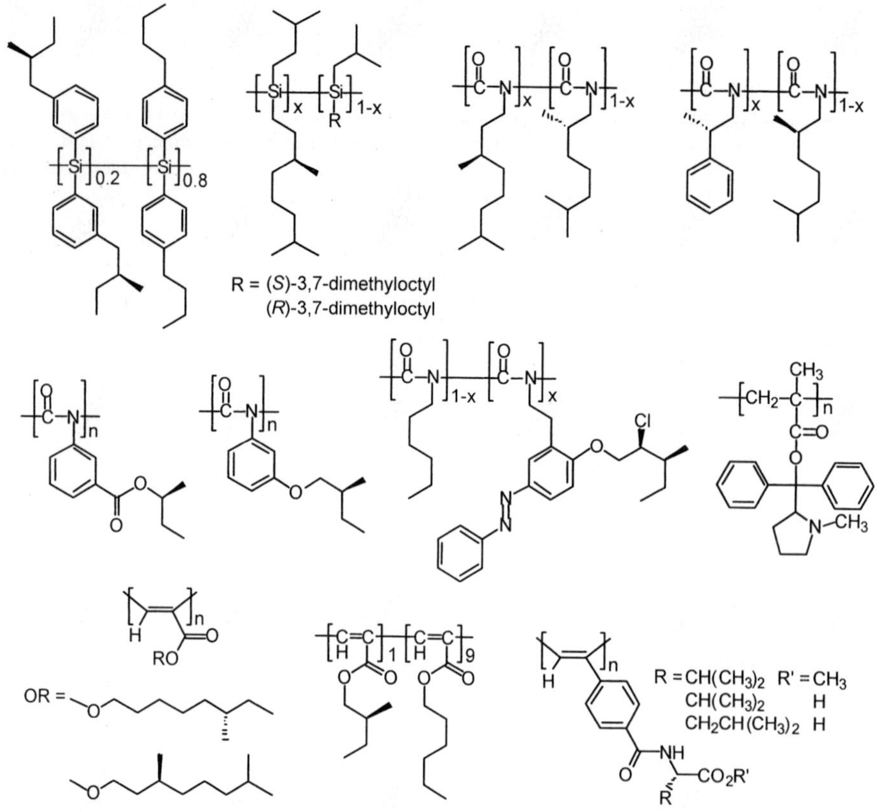
R = (S)-3,7-dimethyloctyl
(R)-3,7-dimethyloctyl
OR =
R = CH(CH3)2 R' = CH3
CH(CH3)2 H
CH2CH(CH3)2 H

Another interesting feature of helical polymers is their helix-helix transition. As shown in Chart 1, several synthetic polymers[9] as well as biopolymers[10] that undergo a helix-helix transition by various external achiral stimuli such as temperature, solvent, light, and the addition of acids, have been reported. In this paper, we describe unique helix—helix transitions of *cis-transoidal* poly(phenylacetylene)s bearing an optically active substituent (poly-**1**—**3**) (Chart 2) responding to external achiral and chiral stimuli. To the best of our knowledge, synthetic helical polymers exhibiting a helix-helix transition by external chiral stimuli have not yet been reported except for our examples.[11-13]

Chart 2. Structures of poly-**1**—poly-**3**.

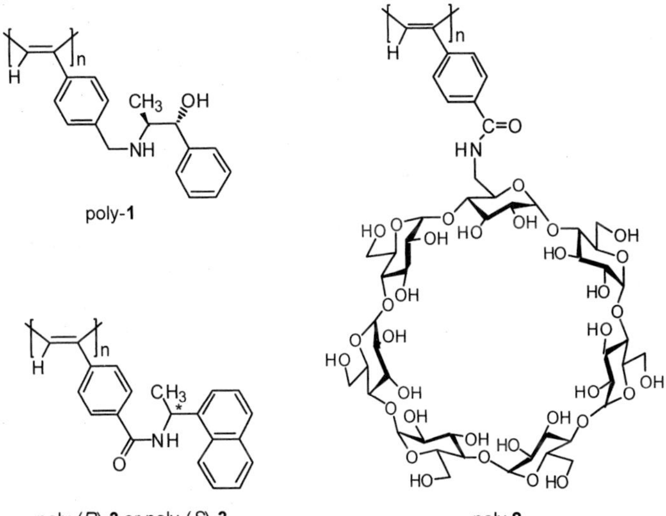

poly-**1**

poly-(*R*)-**3** or poly-(*S*)-**3**

poly-**2**

Helix—Helix Transition of Poly(phenylacetylene)s by External Stimuli

A *cis-transoidal* poly(phenylacetylene) having an optically active amino alcohol residue derived from norephedrine as the side group (poly-**1**) is optically active and exhibits a characteristic ICD in the UV-visible region in dimethyl sulfoxide (DMSO) due to a predominantly one-handed helical conformation of the polymer (Figure 1). However, the poly-**1**

138

undergoes a transition from one-helix to another induced by diastereomeric complexation with optically active acids such as mandelic acid (4).[11] As shown in Figure 1, the addition of (R)-4 to the poly-1 solution in DMSO induces drastic changes in the ICD of poly-1 to give an almost mirror image at [(R)-4]/[poly-1] = 50. On the other hand, in the presence of excess (S)-4, the ICD of poly-1 hardly changed. These results indicate that the poly-1 undergoes a helix—helix transition upon complexation with (R)-4 and that the optically active 4 can be used to regulate the helix-sense of poly-1.

In chloroform, an equimolar amount of chiral carboxylic acids having an R configuration was sufficient for an almost complete inversion of the ICD of poly-1 because ion association between poly-1 and carboxylic acids in less polar chloroform might be stronger than that in polar DMSO. This helix inversion behavior can be applicable to predict the absolute configuration of chiral acids because only acids with an R configuration bring about a helix—helix transition of poly-1.

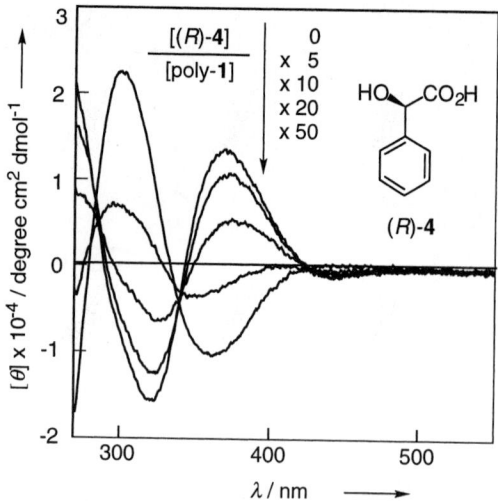

Fig. 1. CD spectral change of poly-1 (1 mg/mL) with (R)-4 in DMSO at room temperature.[11]

An optically active *cis-transoidal* poly(phenylacetylene) bearing a β-cyclodextrin (β-CyD) residue as the side group (poly-**2**) also exhibits an interesting macromolecular helicity inversion accompanied by a visible color change responding to various external stimuli.[12] Poly-**2** shows an ICD in the UV-visible region in DMSO and alkaline water, due to the prevailing one-handed helical conformation because the polymer has an optically active β-CyD unit. However, at high temperatures, the ICD pattern dramatically changed in DMSO, and the sign inverted accompanied by a color change from red to yellow. This indicates that the polymer underwent a helix—helix transition due to changing the temperature. Poly-**2** also exhibited a similar color change in the presence of an increasing amount of water and alcohols accompanied by inversion of the Cotton effect at 25°C. Moreover, the transition temperature can be controlled by tuning the solvent compositions.

It is well-known that β-CyD possesses a chiral hydrophobic cavity to form inclusion complexes with a variety of organic molecules that fit the cavity size. Poly-**2** was found to show a similar helicity inversion of the main chain with a color change through inclusion complexation with guest molecules such as 1-adamantanol and (-)-borneol into the cyclodextrin cavity.

The macromolecular helicity of poly-**2** can also be switched by complexation with chiral molecules. For instance, poly-**2** exhibited a color change (from yellow-orange to red) with a negative first Cotton effect sign in the presence of excess (*S*)-1-phenylethylamine ((*S*)-**5**) in DMSO-alkaline water (3/7, v/v), while the solution color remained yellow with a positive first Cotton effect sign in the presence of excess (*R*)-**5** (Figure 2). The chirality of **5** might induce the macromolecular helicity inversion of poly-**2**, which can be observed by the naked eye. The present system may be based on a change in the tunable helical pitch arising from the helicity inversion and can be used for the construction of conceptually new chiral materials as chiral sensors and chiral selectors.

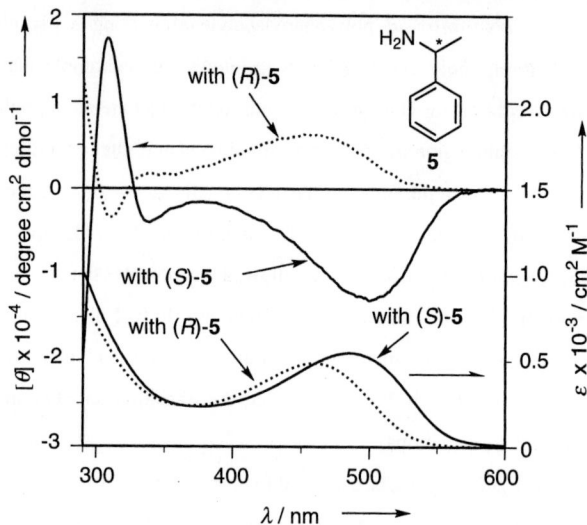

Fig. 2. CD and absorption spectra of poly-**2** (1 mg/mL) in an alkaline water (pH 11.7)-DMSO (7/3, v/v) in the presence of (*R*)-**5** (dotted line) and (*S*)-**5** (solid line) at 25 °C.[12]

A similar stimuli-responsive, macromolecular helicity inversion was also observed for poly((*R*)- or (*S*)-(4-((1-(1-naphthyl)ethyl)carbamoyl)phenyl)acetylene) (poly-(*R*)-**3** and poly-(*S*)-**3**) by changing the external conditions, such as temperature, solvent, or by interacting with optically active small molecules such as (*R*)- and (*S*)-1-(1-naphthyl)ethylamine (**6**).[13] A typical helix inversion of poly-(*R*)-**3** responding to chiral interaction with amines is described below.

Figure 3 shows the CD spectra of poly-(*R*)-**3** in the absence and presence of (*R*)- and (*S*)-**6** ([**6**]/[poly-**3**] = 400) in DMF. The ICD of poly-(*R*)-**3** changed with an excess of (*R*)-**6** and gave an almost mirror image. The ICD of poly-(*R*)-**3** also changed in the presence of the same amount of (*S*)-**6**, but the Cotton effect sign was the same as that of poly-(*R*)-**3**. A similar inversion of the Cotton effect sign was also observed for poly-(*S*)-**3** in the presence of (*R*)- and (*S*)-**6**. Thus, optically active **6** can be used to regulate the helix-sense of poly-**3**.

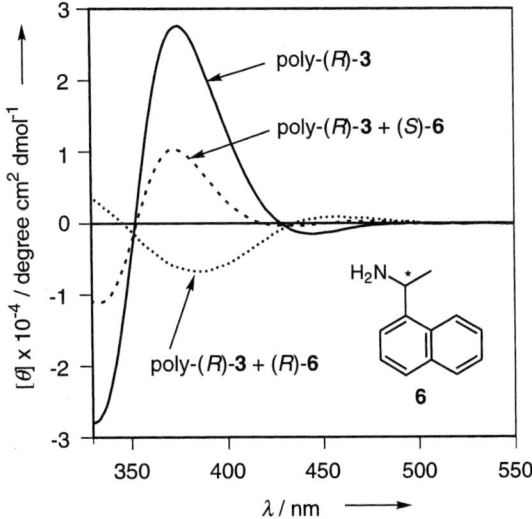

Fig. 3. CD spectra of poly-(R)-**3** (1 mg/mL) in DMF in the absence (solid line) and presece of (R)-**6** (dotted line) and (S)-**6** (broken line) ([**6**]/[poly-(R)-**3**] = 400) at room temperature.[13]

Conclusion

We have demonstrated in this article that the macromolecular helicity of some helical polyacetylenes can be switched by external chiral stimuli as well as achiral ones. Although helical polymers exhibiting a helix—helix transition by chiral stimuli are still rare, we expect that related helical polyacetylenes bearing other chiral functional substituents would also respond to various chiral small molecules, thus exhibiting a unique helix—helix transition.

Acknowledgment

This work was partially supported by Grant-in-Aid for Scientific Research from Japan Society for the Promotion of Science and the Ministry of Education, Culture, Sports, Science, and Technology, Japan.

142

[1] (a) Y. Okamoto, T. Nakano, *Chem. Rev.* **1994**, *94*, 349. (b) T. Nakano, Y. Okamoto, *Chem. Rev.* **2001**, *101*, 4013.

[2] M. M. Green, N. C. Peterson, T. Sato, A. Teramoto, R. Cook, S. Lifson, *Science* **1995**, *268*, 1860.

[3] (a) A. E. Rowan, R. J. M. Nolte, *Angew. Chem. Int. Ed.* **1998**, *37*, 63. (b) J. J. L. M. Cornelissen, A. E. Rowan, R. J. M. Nolte, N. J. A. M. Sommerdijk, *Chem. Rev.* **2001**, *101*, 4039.

[4] M. Fujiki, *Macromol. Rapid. Commun.* **2001**, *22*, 539.

[5] (a) E. Yashima, T. Matsushima, Y. Okamoto, *J. Am. Chem. Soc.* **1995**, *117*, 11596. (b) E. Yashima, T. Nimura, T. Matsushima, Y. Okamoto, *J. Am. Chem. Soc.* **1996**, *118*, 9800. (c) E. Yashima, Y. Maeda, Y. Okamoto, *Chem. Lett.* **1996**, 955. (d) E. Yashima, T. Matsushima, Y. Okamoto, *J. Am. Chem. Soc.* **1997**, *119*, 6345. (e) E. Yashima, K. Maeda, Y. Okamoto, *Nature* **1999**, *399*, 449. (f) M. A. Saito, K. Maeda, H. Onouchi, E. Yashima, *Macromolecules* **2000**, *33*, 4616. (g) K. Maeda, H. Goto, E. Yashima, *Macromolecules* **2001**, *34*, 1160. (h) K. Maeda, S. Okada, E. Yashima, Y. Okamoto, *J. Polym. Sci., Part A: Polym. Chem.* **2001**, *39*, 3180. (i) H. Onouchi, K. Maeda, E. Yashima, *J. Am. Chem. Soc.* **2001**, *123*, 7441.

[6] E. Yashima, K. Maeda, T. Yamanaka, *J. Am. Chem. Soc.* **2000**, *122*, 7813.

[7] M. Ishikawa, K. Maeda, E. Yashima, *J. Am. Chem. Soc.* **2002**, *124*, 7448.

[8] E. Yashima, *Anal. Sci.* **2002**, *18*, 3.

[9] (a) Y. Okamoto, T. Nakano, E. Ono, K. Hatada, *Chem. Lett.* **1991**, 525. (b) G. Maxein, R. Zentel, *Macromolecules* **1995**, *28*, 8438. (c) K. Maeda, Y. Okamoto, *Macromolecules* **1998**, *31*, 5164. (d) K. K. L. Cheuk, F. Salhi, J. W. Y. Lam, B. Z. Tang, *Polym. Prep.* **2000**, *41(2)*, 1567. (e) K. Hino, K. Maeda, Y. Okamoto, *J. Phys. Org. Chem.* **2000**, *13*, 361. (f) K. S. Cheon, J. V. Selinger, M. M. Green, *Angew. Chem. Int. Ed.* **2000**, *39*, 1482. (g) M. Fujiki, *J. Am. Chem. Soc.* **2000**, *122*, 3336. (h) J. R. Koe, M. Fujiki, M. Motonaga, H. Nakashima, *Chem. Commun.* **2000**, 389. (i) M. Fujiki, J. R. Koe, M. Motonaga, H. Nakashima, K. Terao, A. Teramoto, *J. Am. Chem. Soc.* **2001**, *123*, 6253. (j) H. Nakako, R. Nomura, T. Masuda, *Macromolecules* **2001**, *34*, 1496. (k) A. Teramoto, K. Terao, Y. Terao, N. Nakamura, T. Sato, M. Fujiki, *J. Am. Chem. Soc.* **2001**, *123*, 12303.

[10] (a) H. Toriumi, N. Saso, Y. Yasumoto, S. Sasaki, I. Uematsu, *Polym. J.* **1979**, *11*, 977. (b) J. Watanabe, K. Okamoto, K. Satoh, K. Sakajiri, H. Furuya, *Macromolecules* **1996**, *29*, 7084.

[11] E. Yashima, Y. Maeda, Y. Okamoto, *J. Am. Chem. Soc.* **1998**, *120*, 8895.

[12] E. Yashima, K. Maeda, O. Sato, *J. Am. Chem. Soc.* **2001**, *123*, 8159.

[13] K. Morino, K. Maeda, E. Yashima, *Macromolecules* in press.

Macromol. Symp. **2003**, *201*, 143—154

Supramolecular Polymers: From Scientific Curiosity to Technological Reality

A.W. Bosman,[1] L. Brunsveld,[2] B.J.B. Folmer,[2] R.P. Sijbesma,[2] E.W. Meijer*[2]

[1]SupraPolix Research Center, EUTECHpark, Horsten 2, 5612 AX Eindhoven, The Netherlands
[2]Laboratory of Macromolecular and Organic Chemistry, Eindhoven University of Technology, P.O. Box 513, 5600 MB Eindhoven, The Netherlands

Summary: Supramolecular polymers[1] are introduced as a new approach to come to materials in which the repeating units are not connected by covalent bonds but by specific secondary interactions. Self-complementary quadruple hydrogen bonded structures with high association constants are presented as easy to synthesize fragments in supramolecular polymers. Some of the many possibilities of equilibrium polymers are discussed, while it is shown that these supramolecular polymers can obtain materials properties normally only obtained with macromolecules.

Keyword: equilibrium polymers, hydrogen bonding, macromolecules, supramolecular polymers, ureidopyrimidinone units

Introduction[1]

Synthetic polymeric materials are amongst the most important classes of new materials introduced in the previous century. They are primarily used for construction purposes, but also electronic and biomedical applications are at the forefront of science and technology. It was only after the pioneering work of Staudinger, that it became evident that polymeric properties in both solution and solid state are the result of the macromolecular nature of the molecules. A large number of repeating units are covalently linked into a long chain and the entanglements of the macromolecular chains are responsible for many of the typical polymer properties.[2] Before macromolecules were generally accepted, the majority of scientists was convinced that polymer properties were the result of the colloidal aggregation of small molecules or particles.

The impressive recent progress in supramolecular chemistry,[3] paved the way to design polymers and polymeric materials that lack the macromolecular structure. Instead, highly directional secondary interactions are used to assemble the many repeating units into a polymer array.[4]

 DOI: 10.1002/masy.200351117

144

Polymers based on this concept hold promise as a unique class of novel materials, because they combine many of the attractive features of conventional polymers with properties that result from the reversibility of the bonds between monomeric units. Architectural and dynamic parameters that determine polymer properties, such as degree of polymerization, lifetime of the chain and its conformation are a function of the strength of the non-covalent interaction, which can reversibly be adjusted. This results in materials that are able to respond to external stimuli in a way that is not possible for traditional macromolecules.[1] These aspects of supramolecular polymers have led to a recent surge in attention for this promising class of compounds[5-7] and have stimulated our group to bring together materials science and supramolecular chemistry. The cartoon in Figure 1 shows the required directionality in the supramolecular interactions as compared with the historical and the current macromolecular view on polymers.

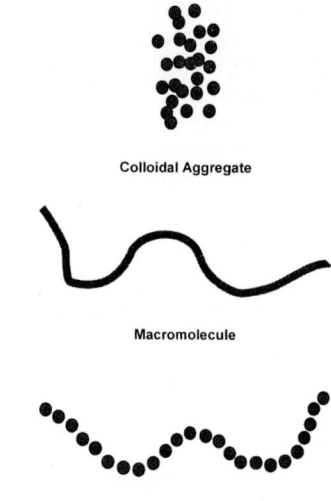

Colloidal Aggregate

Macromolecule

Supramolecular Polymer

Fig. 1. Supramolecular polymers as a redesign of the historical view of polymeric materials and compared to Staudinger's macromolecular concept.

The impressive recent progress in supramolecular chemistry,[8] paved the way to design polymers and polymeric materials that lack the macromolecular structure. Instead, highly directional secondary interactions are used to assemble the many repeating units into a polymer array.[9] Polymers based on this concept hold promise as a unique class of novel materials, because they combine many of the attractive features of conventional polymers with properties that result from the reversibility of the bonds between monomeric units. Architectural and dynamic parameters that determine polymer properties, such as degree of polymerization, lifetime of the chain and its conformation are a function of the strength of the non-covalent interaction, which can reversibly be adjusted. This results in materials that are able to respond to external stimuli in a way that is not possible for traditional macromolecules.[1] These aspects of supramolecular polymers have led to a recent surge in attention for this promising class of compounds[10-12] and have stimulated our group to bring together materials science and supramolecular chemistry. The cartoon in Figure 1 shows the required directionality in the supramolecular interactions as compared with the historical and the current macromolecular view on polymers.

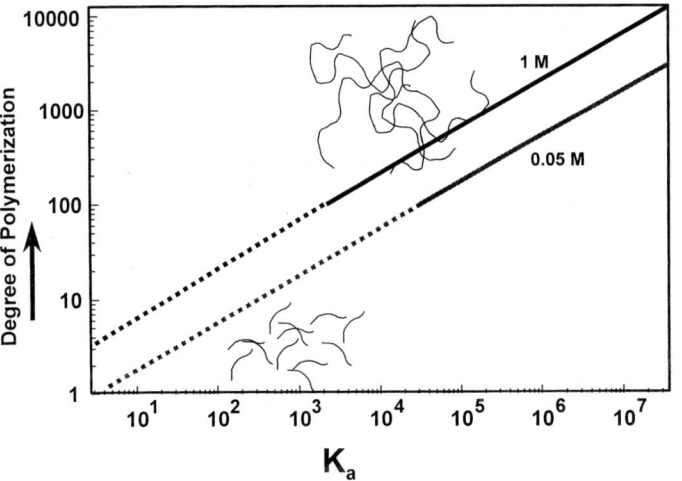

Fig. 2. A theoretical plot of the relation between association constant and degree of polymerization as a function of concentration of the bifunctional monomer using a simple isodesmic association function.

In supramolecular polymers, which are formed by the reversible association of bifunctional monomers, the average degree of polymerization (DP) is determined by the strength of the end group interaction.[13] The degree of polymerization is obviously dependent on the concentration of the solution and the association constant and a theoretical relationship is given in Figure 2. To obtain polymers with a high molecular weight, a high association constant between the repeating units is a prerequisite. In analogy with covalent condensation polymers, the chain length of supramolecular polymers can be tuned by the addition of monofunctional "chain stoppers".[14] This also implies that impurities containing only one function will have a strong influence on the maximal DP, since they will act as chain stopper. Hence, as in traditional polymer synthesis, the purification of the monomers is extremely important to obtain high molecular weights.

What Defines a Supramolecular Polymer?

The term *supramolecular polymer* is rather popular and used for a variety of different structures, utilizing secondary (or supramolecular) interactions between chains or for the construction of polymer chains. In the following definition, the polymers that have secondary interactions between macromolecular chains only, are ignored, since all polymers possess either hydrogen bonding (nylons), dipole-dipole (polyesters) or London-dispersion interactions (polyethylene), that determine their materials properties. Hence, supramolecular polymers are defined as those polymers that are made out of repeating units held together by other bonds than just covalent bonds. In these supramolecular polymers, however, there is a strong interplay between intra- and interchain interactions and the cooperativity between the two has a large impact in overall strength of the bonding as well as in the ultimate properties (comparable to the cooperativity between the hydrogen bonding and base-pair stacking in DNA). In order to make a rather stable individual supramolecular polymer chain, the bonding within the chain should be significantly stronger than the interactions between the chains. If this requirement is not included in the definition, many structures that lack a covalent architecture can be called polymers. This is nicely demonstrated by looking at terephthalic acid; in its crystalline form the molecules are connected in a long chain by the dimerization of the carboxylic acids by hydrogen bonding. However, upon melting or dissolution these chains are disrupted, due to the lack of strength of the acid dimer under these conditions. Furthermore, it is evident that such crystals do not exhibit any useful polymeric materials properties and the supramolecular structure formed is not only the result of the hydrogen bond dimerization but also due to crystal packing. Many other examples of this kind can be found in the recent literature.

Consequently, the balance between strong unidirectional association and uncontrolled multidirectional association (or gelation) is one of the major aspects in the design of supramolecular polymers. There are three main categories of supramolecular polymers: coordination polymers, polymers using π-π interactions, and hydrogen-bonded polymers (see Figure 3). However, only the latter are able to yield polymeric materials with a large variety of interesting bulk properties, because of their general flexibility to tune the strength of the bonding by external stimuli. As a result of this progress in supramolecular polymers, it has been stated recently that "polymer chemistry comes full circle".[3]

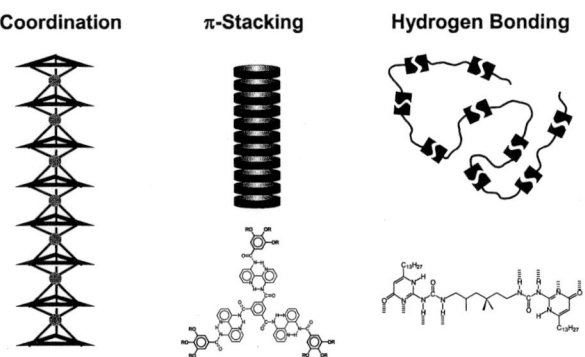

Fig. 3. The three main categories of making polymers without covalent bonding between the repeating units, as subdivided on the type of secondary interaction.

Supramolecular Polymers Based on Hydrogen Bonding[15]

Although hydrogen bonds between neutral organic molecules are not among the strongest non-covalent interactions, they hold a prominent place in supramolecular chemistry due to their directionality and versatility.[16-18] The relationship between the degree of polymerization and the strength of the non-covalent interaction between monomers in a supramolecular polymer (see Figure 2) implies that multiple hydrogen bonds with a high association constant are required to obtain significant degrees of polymerization.

Combining several hydrogen bonds in a functional unit is a valuable tool to increase the strength of this interaction. Moreover, employing a particular arrangement of the hydrogen bonding sites enhances its specificity. The strength of single hydrogen bonds basically depends on the nature of donor and acceptor, although it is influenced to a large extent by the solvent. Association strength between multiple hydrogen bonding units obviously depends on the same factors, as well as on

148

the number of hydrogen bonds. It has also been shown that the particular arrangement of neighboring donor and acceptor sites is an additional factor, which significantly affects the strength of the complexation. This phenomenon was first recognized for the association of linear arrays of 3 hydrogen bonding sites: whereas complexes between the common ADA-DAD motif exhibits an association constant of around 10^2 M^{-1} in chloroform, this value is around 10^4 M^{-1} in complexes with a DAA-DDA motif, while AAA and DDD arrays exhibit association constants exceeding 10^5 M^{-1}. Detailed calculations by Jorgenson[19,20] showed that this effect is due to differences in secondary interactions between these motifs. In the complexes, diagonally opposed sites repel each other electrostatically when they are of the same kind (both donor or both acceptor), while disparate sites attract each other. In the DDD-AAA motif the number of attractive secondary interactions is maximized, while in the ADA-DAD motif the number of repulsive interactions is at its largest.

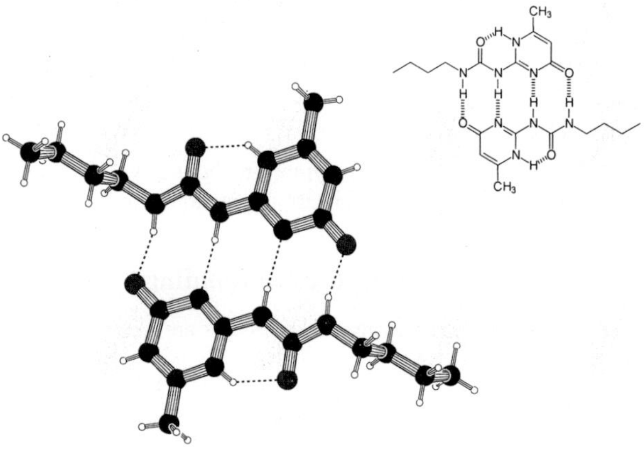

Fig. 4. X-ray structure showing the H-bonding pattern between two ureidopyrimidinone units.

Very stable complexes can be obtained when quadruple hydrogen bonding units are employed.[21-25] Aspects of multiple hydrogen bonding units that are of special importance with respect to application in supramolecular polymers are the self-complementarity of DADA and DDAA arrays, and the possibility of tautomerism. The latter may lead to loss of complexation when complementarity is lost, or when a DDAA array tautomerizes to a DADA array with a higher number of repulsive secondary interactions. We have reported on self-complementary quadruple

H-bonding units based on mono-ureido derivatives of diamino-triazines[24] (DADA-array) with a dimerization constant of $K_{dim}= 2\times10^4$ M^{-1} and hydrogen bonding units based on 2-ureido-4[1H]-pyrimidinones (DDAA, see Figure 4), that dimerize in chloroform with an association constant of $K_{dim}=6\times10^7$ M^{-1}.[25,26]

Application of hydrogen bonding units as associating end-groups in difunctional or multifunctional molecules results in the formation of supramolecular polymers with varying degrees of polymerization (DP). The early examples of hydrogen bonded supramolecular polymers rely on units, that associate using single, double or triple hydrogen bonds all having association constants below 10^3 M^{-1}. In isotropic solutions, the DP of these polymers is expected to be low. The development of the ureidopyrimidinone functionality (UPy), a synthetically very accessible quadruple hydrogen bonding unit with a very high association constant, has helped enormously to open the way for the exploration of all aspects of supramolecular polymers.

The ureidopyrimidinone unit can be made in a one-step procedure from commercially available compounds,[24,26] and dimerizes with an association constant of 6×10^7 M^{-1} in CDCl$_3$. Difunctional compound (7), possessing two of these ureidopyrimidinone units forms a stable and long polymer chain in solution as well as in the bulk (Figure 5).[27,28]

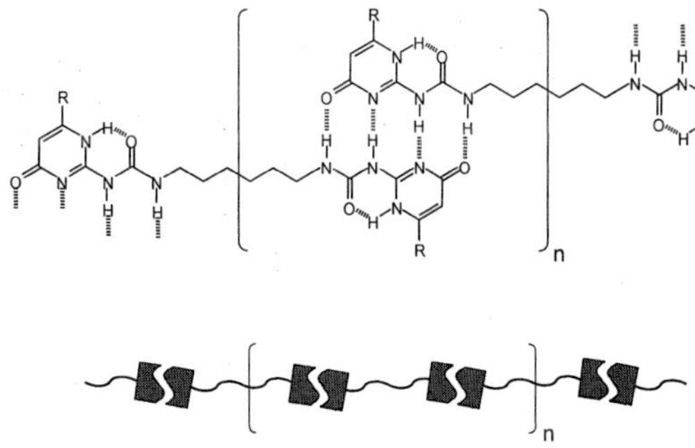

Fig. 5. Polymeric assembly of a bifunctional ureidopyrimidinone derivative.

Dissolving a small amount of this low molecular weight compound in chloroform, results in solutions with a high viscosity. It can be calculated that polymers with chain lengths of the order of 10^6 Dalton can be formed when highly purified monomers are used. The presence of monofunctional impurities is expected to lead to a dramatic reduction in DP, because they will act as "chain stoppers". In fact, deliberate addition of small amounts of monofunctional compounds results in a sharp drop in viscosity, proving the reversibility and unidirectionality of association. The reversibility of the linkages between the building blocks is instrumental in the development of materials that change their properties in response to environmental changes, so called 'smart materials'. Application of a light sensitive monofunctional compound yielded a material from which the degree of polymerization in solution could be tuned by UV-irradiation.[29]

Although the supramolecular polymers based on bifunctional ureidopyrimidinone derivatives in many ways behave like conventional polymers, the strong temperature dependence of their mechanical properties really sets them apart from macromolecular polymers. At room temperature, the supramolecular polymers show polymer-like viscoelastic behavior in bulk and solution, whereas at elevated temperatures liquid-like properties are observed. These changes are due to a three-fold effect of temperature on the reversible polymer chain. Due to the temperature dependence of the K_a value of UPy association, the average DP of the chains is drastically reduced at elevated temperatures. Simultaneously, faster dynamics of the scission-recombination process leads to faster stress relaxation in an entangled system. These two effects occur in addition to the temperature dependent stress relaxation processes that are also operative in melts or solutions of conventional polymers.

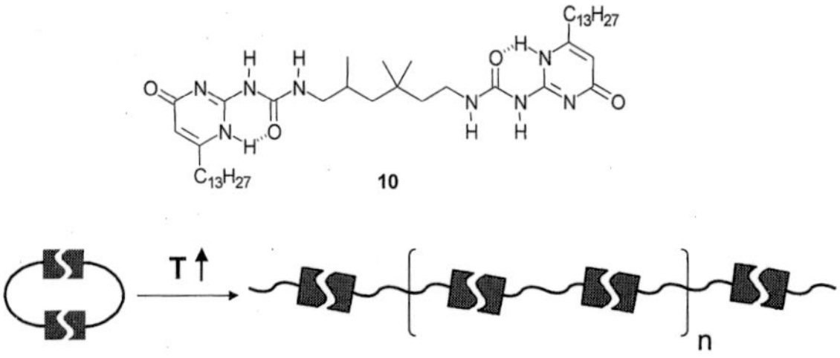

Fig. 6. Entropy driven ring-opening polymerization of the cyclic dimer of **10**.

Similar to the behavior in the melt, solution viscosities of UPy-based supramolecular polymers are also strongly temperature dependent. Recently a very surprising inversion of the normal temperature dependence of the solution viscosity was observed in solutions of preorganized difunctional compounds (**10**), which form a mixture of linear polymer chains and cyclic dimers (Figure 6).[30] The thermodynamic parameters of this equilibrium are such that polymerization is favored at higher temperatures. As a result, the viscosity of a 145 mM chloroform solution of the compound was observed to increase by a factor of 3.9 when the temperature was increased from 255 to 323 K. Entropy-driven polymerizations are rare, and the unexpected effect in this system is the first time it was observed in a reversible synthetic system.

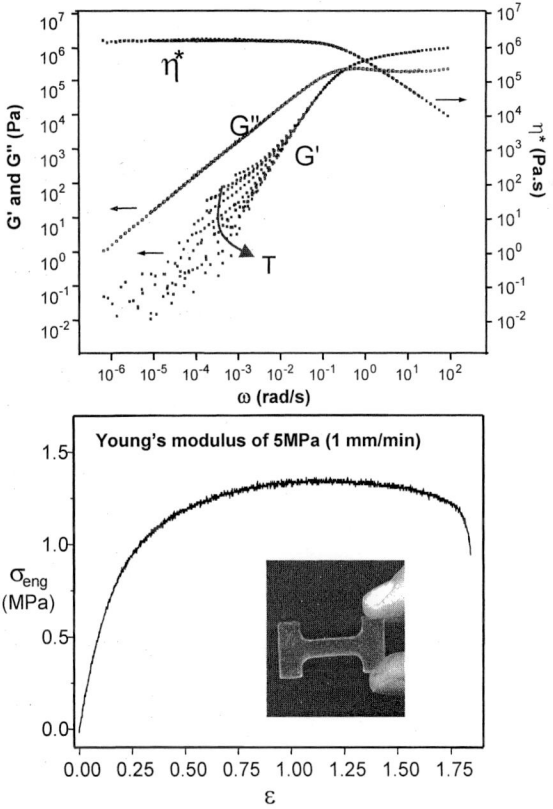

Fig. 7. Mechanical spectra of the supramolecular polymer based on a poly(ethylene/butylene). Top: master curves of the storage (G') and loss modulus (G'') are obtained by the superposition of curves obtained at different temperatures, with 40 °C as a reference temperature; the complex viscosity (η^*) versus frequency is given as well. Bottom: stress-strain measurement revealing the soft-rubberlike behavior.

The quadruple hydrogen bonded unit has been further employed in the chain extension of telechelic polysiloxanes,[31] poly(ethylene/butylenes), polyethers, polyesters and polycarbonates.[32] In these compounds, the material properties were shown to improve dramatically upon functionalization, and materials were obtained that combine many of the mechanical properties of conventional macromolecules with the low melt viscosity of organic compounds (Figure 7). Consequently, in this strategy the gap is closed between polymers and oligomers using the best of both worlds. Especially in the field of conjugated polymers for plastic electronic devices, expectations for future applications of this strategy are high.[33] Recently, Coates et al. used a vinyl-substituted UPy-unit to be part of an olefin polymerization using the Brookhart catalyst. With small amounts of UPy-units incorporated, the polyolefins showed thermoplastic elastomeric properties.[34]

The reversibility of supramolecular polymers adds new aspects to many of the principles that are known from condensation polymerizations. For example, a mixture of different supramolecular monomers will yield copolymers, but it is extremely simple to adjust the copolymer composition instantaneously by adding an additional monomer. Moreover, the use of monomers with a functionality of three or more, will give rise to network formation. However, in contrast to condensation networks, the 'self-healing' supramolecular network can reassemble to form the thermodynamically most favorable state, thus forming denser networks.[35]

Although the 'virtual' molecular weight and lifetime of supramolecular polymers and networks based on strong hydrogen bonding functionalities is extremely high, low creep resistance is an intrinsic property of these materials. Strong inter-chain interactions, especially in crystalline domains can be employed to reduce creep, leading to thermoplastic elastomers with enhanced processability. With the facile synthetic accessibility of these self-complementary UPy-units at hand, it is expected that many novel material properties can be obtained.

Conclusions and Outlook

Ten years ago, the first supramolecular polymers were seen a scientific curiosities. Nowadays, this field of research is generating several technologically important applications. Progress in supramolecular chemistry has made it possible to assemble small molecules into polymer arrays, and the created structures possess many of the well-known properties of "traditional" macromolecules. Due to the reversibility in the bonding, these supramolecular polymers are under thermodynamic equilibrium and their properties can be adjusted by external stimuli. Hydrogen bonded systems have shown to become of technological relevance and surpass the state of being scientific curiosities only. A large variety of applications are feasible, especially since the chosen approach can also be used for the modification of telechelic oligomers or to modify existing polymers. Moreover, the possibility to tune the properties by changing the relative ratio of UPy-monomer in the copolymer feed seems very attractive, while hybrids between blocks of macromolecules and supramolecular polymers are easy to prepare. Therefore, novel thermoplastic elastomers, superglues, hotmelts and tunable polymeric materials are within reach (Figure 8).

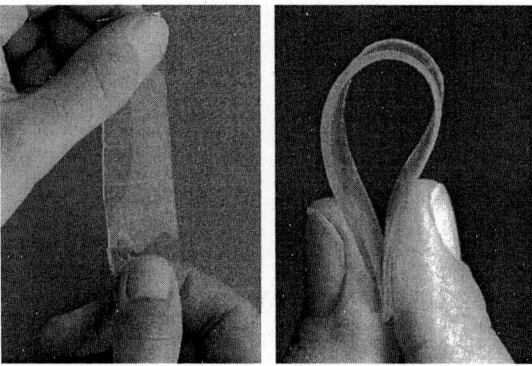

Fig. 8. Supramolecular polymer materials created with the ureidopyrimidone unit.

Acknowledgments

The authors like to acknowledge the many contributions of and discussions with the members of the Laboratories of Macromolecular and Organic Chemistry at the Eindhoven University of Technology. The University group acknowledges the financial support of the University, the Dutch National Science Foundation NWO and DSM Research.

154

[1]   a) Brunsveld, L.; Folmer, B. J. B.; Meijer, E. W. *MRS Bulletin* **2000**, *25*, 49. b) Brunsveld, L.; Folmer, B. J.
        B.; Meijer, E. W.; Sijbesma, R.P. *Chem. Rev.* **2001**, *101*, 4071.
[2]   Staudinger, H. Die Hochmolekulare Organische Verbindungen, (Springer, Berlin, 1932).
[3]   Lehn, J.-M. Supramolecular Chemistry (VCH, Weinheim, 1995).
[4]   Zimmerman, N.; Moore, J.S. and Zimmerman, S.C. *Chem. Ind.* **1998**, 604.
[5]   Supramolecular Polymers; Ciferri, A., Ed.; Marcel Dekker, Inc: New York, 2000.
[6]   Moore, J. S. *Curr. Opin. Colloid Interface Sci.* **1999**, *4*, 108.
[7]   Zimmerman, N.; Moore, J. S.; Zimmerman, S. C. *Chem. Ind.* **1998**, 604.
[8]   Lehn, J.-M. Supramolecular Chemistry (VCH, Weinheim, 1995).
[9]   Zimmerman, N.; Moore, J.S. and Zimmerman, S.C. *Chem. Ind.* **1998**, 604.
[10]   Supramolecular Polymers; Ciferri, A., Ed.; Marcel Dekker, Inc: New York, 2000.
[11]   Moore, J. S. *Curr. Opin. Colloid Interface Sci.* **1999**, *4*, 108.
[12]   Zimmerman, N.; Moore, J. S.; Zimmerman, S. C. *Chem. Ind.* **1998**, 604.
[13]   Bruce Martin, R. *Chem. Rev.* **1996**, *96*, 3043-3064.
[14]   Flory, P. J. Principles of Polymer Chemistry, Cornell University Press: Ithaca, New York, 1953.
[15]   a) Sherrington, D. C.; Taskinen, K. A. *Chem. Soc. Rev*, **2001**, *30*, 83. b) Schmuck, C.; Wienand, W. *Angew.
        Chem. Int. Ed.* **2001**, *40*, 4363.
[16]   Krische, M. J.; Lehn, J.-M. Struct. Bond. **2000**, *96*, 3.
[17]   Sijbesma, R. P.; Meijer, E. W. *Curr. Opin. Colloid Interface. Sci.* **1999**, *4*, 24.
[18]   Fredericks, J. R.; Hamilton, A. D. *Comprehensive Supramolecular Chemistry*, chapter 16, Lehn, J.-M (ed.),
        Pergamon, New York, **1996**.
[19]   Jorgenson, W. L.; Pranata, J. *J. Am. Chem. Soc.* **1990**, *112*, 2008.
[20]   Pranata, J.; Wierschke, S. G.; Jorgenson, W. L. *J. Am. Chem. Soc.* **1991**, *113*, 2810.
[21]   Sessler, J. L.; Wang, R. *Angew. Chem., Int. Ed. Engl.* **1998**, *37*, 1726.
[22]   Kolotuchin, S. V.; Zimmerman, S. C. *J. Am. Chem. Soc.* **1998**, *120*, 9092.
[23]   Corbin, P. S.; Zimmerman, S. C. *J. Am. Chem. Soc.* **1998**, *120*, 9710.
[24]   Beijer, F. H.; Sijbesma, R. P.; Kooijman, H.; Spek, A. L.; Meijer, E. W. *J. Am. Chem. Soc.* **1998**, *120*,
6761.
[25]   Beijer, F. H; Kooijman, H.; Spek, A. L.; Sijbesma, R. P.; Meijer, E. W. *Angew. Chem., Int. Ed. Engl.* **1998**,
        37, 75.
[26]   Söntjens, S. H. M.; Sijbesma, R. P.; van Genderen, M. H. P.; Meijer, E. W. *J. Am. Chem. Soc.* **2000**, *122*,
        7487.
[27]   Sijbesma, R. P.; Beijer, F. H.; Brunsveld, L.; Folmer, B. J. B.; Hirschberg, J. H. K. K.; Lange, R. F. M.;
        Lowe, J. K. L.; Meijer, E. W. *Science* **1997**, *278*, 1601.
[28]   Folmer, B. J. B.; Sijbesma, R. P.; Meijer, E. W. *Polym. Mater. Sci. Eng.* **1999**, *217*, 39.
[29]   Folmer, B. J. B.; Cavini, E.; Sijbesma, R. P.; Meijer, E. W. *Chem. Commun.* **1998**, 1847.
[30]   Folmer, B. J. B.; Sijbesma, R. P.; Meijer, E. W. *J. Am. Chem. Soc.* **2001**, *123*, 2093.
[31]   Hirschberg, J. H. K. K.; Beijer, F. H.; van Aert, H. A.; Magusin, P. C. M. M.; Sijbesma, R. P.; Meijer, E.
W.   *Macromolecules* **1999**, *32*, 2696.
[32]   Folmer, B. J. B.; Sijbesma, R. P.; Versteegen, R. M.; van der Rijt, J. A. J.; Meijer, E. W. *Adv. Mater.* **2000**,
        12, 874.
[33]   a) El-Ghayoury, A.; Peeters, E.; Schenning, A. P. H. J.; Meijer, E. W. *Chem. Commun.* **2000**, 1969. b) El-
        Ghayoury, A.; Schenning, A. P. H. J.; van Hal, P. A.; van Duren, J. K. J.; Janssen, R. A. J.; Meijer, E. W.
        Angew. Chem. Int. Ed. **2001**, *40*, 3660.
[34]   Rieth, L. R.; Eaton, R. F.; Coates, G. W. *Angew. Chem., Int. Ed.* **2001**, *40*, 2153.
[35]   Lange, R. F. M.; van Gurp, M; Meijer, E. W. *J. Polym. Sci. A* **1999**, *37*, 3657.

Status and Dreams of Photonics Polymer for IT

Yasuhiro Koike,* Takaaki Ishigure, Akihiro Tagaya

Faculty of Science and Technology, Keio University, 3-14-1, Hiyoshi, Kohoku-ku, Yokohama 223-8522, Japan
Email: koike@appi.keio.ac.jp
ERATO Koike Photonics Polymer Project, Japan Science and Technology Corporation, K² Town Campus, 144-8, Ogura, Saiwai-ku, Kawasaki 212-0054, Japan

Summary: We have proposed a low-loss, high-bandwidth and large-core graded-index plastic optical fiber (GI POF) in data-com. area. The GI POF enables us to eliminate the "modal noise" problem which is observed in medium-core silica fibers. Therefore, stable high-speed data transmission can be realized by the GI POF rather than medium-core silica fibers. Furthermore, advent of perfluorinated (PF) polymer based GI POF network can support higher transmission than silica fibers network because of the small material dispersion of PF polymer compared with silica. In addition, we proposed a "highly scattering optical transmission (HSOT) polymer" and applied it to a light guide plate of a liquid crystal display backlight. The HSOT polymer backlight that was designed using the HSOT designing simulator demonstrated twice the brightness of the conventional taransparent backlight with sufficient color uniformity. Furthermore, we proposed the two types of zero-birefringence polymers synthesized by the random copolymerization method and the anisotropic molecule dopant method. Both of the polymers exhibited no orientational birefringence for any orientation of polymer chains.

Keywords: GI plastic optical fibers, HSOT, injection molding, material dispersion, zero-birefringence polymers

Introduction

During the past several years, the IT industry was flourishing and leading the economic growth. However, in these days, it has gone into a temporary stall. The main cause of the stall should be a large difference between the current status of hardware and software. While the broadband Internet access has not penetrated into homes and offices yet, more and more expectations are placed on software including contents. In order to break through the situation surrounding the IT industry, a new development is required for the filed of information technology. And in order to make the new development, the biggest challenge will be how to install gigabit optical fiber to local area networks at homes and offices, and how to display the high quality motion pictures. In this article, a "high-speed graded-index plastic optical fiber (GI POF)", a "highly scattering

© 2003 WILEY-VCH Verlag GmbH & KGaA, Weinheim DOI: 10.1002/masy.200351118

optical transmission (HSOT) polymer", and "zero-birefringence polymers" for liquid-crystalline displays (LCDs) are described in detail.

1 Graded Index Plastic Optical Fiber (GI POF)

1.1 Modal Noise Elimination in GI POF Link

It was shown by previous works on the silica based multimode fiber (MMF) with 50 to 62.5 μm core diameter that the modal noise deteriorated the system performance.[1] The modal noise is observed in the MMF and a coherent light source such as laser diode or Vertical Cavity Surface Emitting Laser (VCSEL) systems. In such a system, "speckle" is formed on the output end face of the fiber by interference among the propagating modes. Therefore, if an offset connection between two fibers is included in the MMF link, the speckle power distribution is translated into the optical power fluctuation, which causes the modal noise. Therefore, the permissible misalignment in the connector of the silica based MMF is several micrometers, although MMF's larger core diameter than that of the single mode counterpart tolerates the misalignment in the fiber connection from the aspect of the connection loss.

On the other hand, in the case of the GI POF with much larger core than that of the silica based MMF, such a small misalignment in fiber connection is negligible in maintaining the low connection loss. In addition to the advantage in the connection loss, it was clarified that the modal noise was virtually eliminated in the large core (>120 μm) GI POF link, because of the huge number of propagating modes.

1.2 Bandwidth Achieved by GI POF

Despite the large modal noise problem mentioned above, the MMF networks have been expected to be a viable solution for the premises network, because it is still more cost effective than the use of single mode counterpart. Gigabit Ethernet and 10Gigabit Ethernet standards specify the use of MMF and an inexpensive VCSEL sources as a light source. However, the dispersion of the MMF is the serious problem particularly in the 10Gigabit transmission systems. When the refractive index profile of the MMF is optimized, a chromatic dispersion would be a dominant factor of the total dispersion.

For the premises network applications, we have proposed a low-loss perfluorinated (PF) polymer based GI POF.[2-3] The attenuation of the current PF polymer based GI POF is 10 dB/km in 0.8 − 1.3-μm wavelength range. In addition to the low attenuation, we have focused on the large advantage in the PF polymer based GI POF, that is lower material dispersion than that of silica.[3] This result means that the PF polymer based GI POF enables a higher data transmission rate than the conventional silica based MMF. Comparison of calculated bandwidth for 500-m length between the PF polymer based GI POF and silica based multimode fiber is shown in Figure 1 when their index profiles are optimised for 0.85-μm use. In this calculation, actually measured material dispersion was taken into account. It is noted that 10 GHz bandwidth for 500 m is achieved by the PF polymer based GI POF 1, which is approximately twice higher than that of silica based multimode fiber.

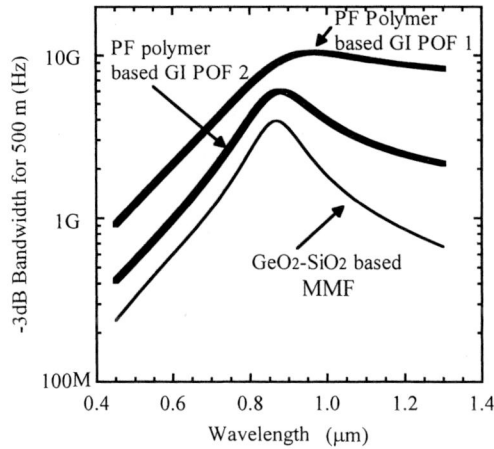

Fig. 1. Wavelength dependence of the bandwidth of PF polymer based GI POF compared to that of silica based multimode fiber (MMF). Spectral width of light source was assumed to be 1 nm.

In addition to the bandwidth advantage at 0.85-μm wavelength, it is noteworthy that the wavelength dependence of the bandwidth of PF polymer based GI POF is remarkably small. This indicates that if the index profile is optimized for 0.85-μm wavelength, the same fiber is also applicable at another wavelengths, such as 1.3-μm. Therefore, the same PF polymer based GI POF covers more than 10 Gb/s at almost any wavelength from visible to 1.3-μm regions.

2 Photonics Polymer for Display

2.1 Highly Scattering Optical Transmission (HSOT) Polymer

A display with high-visual quality and low energy consumption is one of the most important devices in the concept of the Ubiquitous Network. Liquid crystal displays (LCDs) are the mainstream flat panel display and have been widely used for a monitor of desktop computers and other portable devices, and a television. LCDs will be more important flat panel displays especially for portable devices in the coming highly-networked information society.

We proposed a novel photonics polymer, a highly scattering optical transmission (HSOT) polymer.[4-6] Light injected into the HSOT polymer is multiply scattered and homogenized, and then comes out as a directive illuminating light because of the microscopic heterogeneous structures formed by doping with spherical particles in the HSOT polymer. The scattering property of the HSOT polymer depends on the size and relative refractive-index of the heterogeneous structures. We applied the HSOT polymer to a light guide plate (LGP) in a backlight unit for LCDs and designed the heterogeneous structures by the multiple scattering simulator that we developed using the Monte Carlo method based on Mie scattering theory. Conventionally, before our proposal of the HSOT polymer, all LGPs in backlights were made of transparent polymers, and it was thought that polymers for the LGPs must be transparent without any contaminant, because the contaminant would absorb and scatter light. Even if the contaminant does not absorb light, it was supposed that the contaminant causes a decrease in brightness and degradation of color uniformity of the backlight, because scattered lights are distributed to all directions and the scattering efficiency depends on wavelength of the propagating light. On the other hand, the HSOT backlight demonstrated twice the brightness of the conventional one with sufficient color uniformity by optimizing the heterogeneous structures with using the multiple scattering simulator. This disproved the speculation that the LGPs must be transparent. Consequently, the HSOT backlight has become commercially used in some types of thin notebook computers because of its higher brightness.

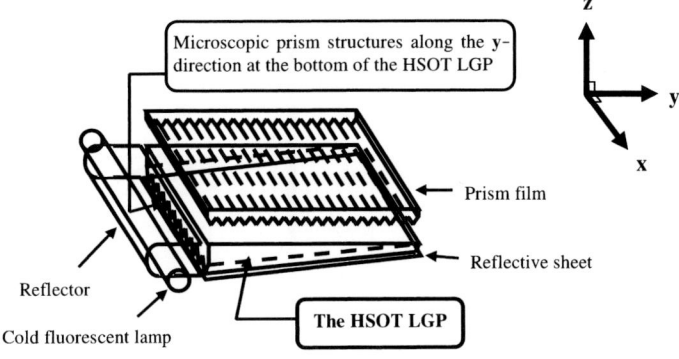

Fig. 2. Schematic diagram of the advanced HSOT backlight, in which the wedge-shaped HSOT LGP has the microscopic prism structures at the bottom in the y-direction.

Details of the optimization using the multiple scattering simulator are described in Ref. [5-6]. Here we describe the relation between scattering efficiency and wavelength. Scattering efficiency against the diameter of the spherical heterogeneous structure at 615, 545, 435nm corresponding to red, green and blue (RGB) lights are shown in Figure 3. Generally, we tend to think that blue light is always scattered stronger than red light based on Rayleigh scattering theory. However, it is not always true and depends on the size of heterogeneous structures. By injecting white light from a typical cold fluorescent lamp or a white LED into the scattering medium containing the heterogeneous structure (A), yellowish transmitted light with a lower color temperature was obtained because blue light was scattered stronger than red light. This is the same phenomenon as the red sunset. However, bluish transmitted light with a higher color temperature was obtained in the scattering medium containing the heterogeneous structure (B), because red light was scattered stronger than blue light. By controlling the diameter of the heterogeneous structure to give almost the same scattering efficiencies for RGB, output light having almost the same color temperature as that of injected light can be obtained. We realized sufficient color uniformity by using particles having almost the same scattering efficiencies.

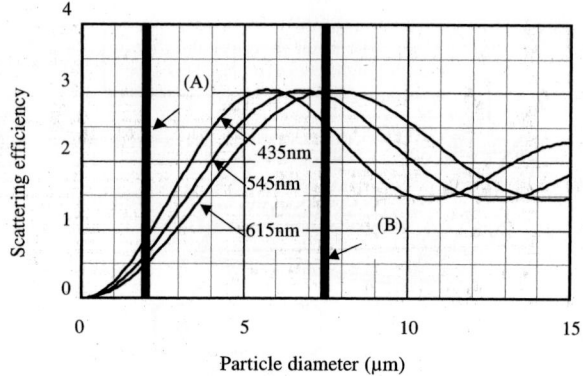

Fig. 3. Scattering efficiency curves of a single particle for 435, 545 and 615 nm-wavelengths, respectively. Typical cold fluorescent lamps have spectral peaks around these wavelengths. Relative refractive index: m = 0.965. (A) and (B) mean particle diameters of 2 μm and 7.5 μm as the heterogeneous structures, respectively.

2.2 Zero-Birefringence Polymer

Optical polymers have been widely used as key-materials for a variety of optical devices in recent optical technology, for example, polymer optical fibers, optical films for LCDs, optical disks and lenses because of their easy processing, easy handling, light weight, high transparency, and low cost. Although most polymers are composed of monomers exhibiting an anisotropic polarizability, they exhibit no birefringence in a perfectly amorphous state. However, the optical polymers tend to exhibit birefringence caused by the orientation of polymer chains in the process of injection-molding or extrusion, which restricts their application in optical devices that handle polarized light. Optical polymers that exhibit no birefringence for any orientation of polymer chains are desirable to realize high performance optical devices for handling polarized light. We define such polymers as "zero-birefringence polymers" in this article.

We developed the zero-birefringence polymers by the random copolymerization method[7] and by the anisotropic molecule dopant method.[8-9] In the random copolymerization method, negative and positive birefringence monomers are randomly copolymerized with the specified composition in order to compensate polarizability anisotropy in a polymer chain as

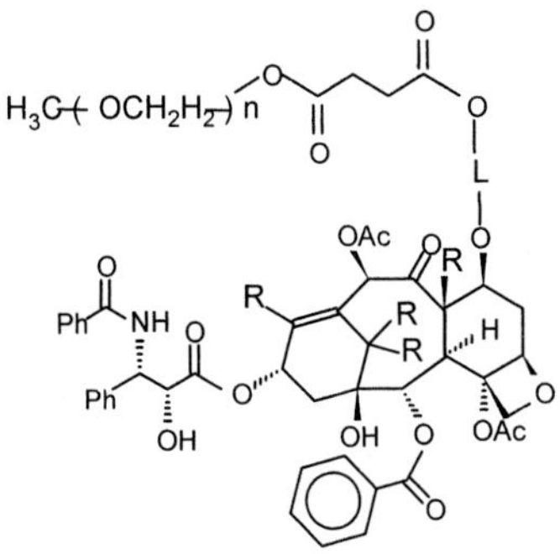

(R = CH$_3$)

(L = self immolating linker)

(n = 113)

Fig. 2. Paclitaxel coupled at C7 through a self-immolating linker with a functionalised PEO, in the following text termed PP7. The degree of polymerisation in this work is either 113 or 452 according to a molar mass of 5,000 g/mol or 20,000 g/mol, respectively, for the highly uniform PEO which was used here.

Coupling a hydrophilic molecule with a hydrophobic molecule might give rise to an amphiphilic molecule able to form colloidal associates in solution such as micelles. Therefore, we have studied the concentration dependent surface tension of PP7 (5,000 g/mol).

For spherical rigid molecules Einstein[3] has derived an equation which correlates the specific viscosity η_{spec} with the volume fraction of the solute by:

$$\eta_{spec} = \frac{\eta_1 - \eta_{01}}{\eta_{01}} = v \cdot \varphi \tag{1}$$

with the index 1 denoting the solution and 01 the pure solvent. v is Einsteins "rigidity-factor" and φ is the volume fraction of the solute which is defined through the concentration c and the overlap-concentration c* of the polymer coils. For an ideal rigid sphere $v = 2.5$ is valid. Simha [4] expanded the theory for ellipsoids and depending on the aspect ratio for oblate or prolate shapes, respectively.

$$\varphi = \frac{c}{c^*} = \frac{c \frac{4}{3} \pi R_H^3 N_A}{M} \tag{2}$$

The self-diffusion coefficient D of rigid spheres can be described by the Stokes-Einstein equation and a series expansion of D with respect to the volume fraction is given by:

$$D = \underbrace{\frac{k_b T}{6\pi \eta_{01} R_H^3}}_{D_0}\left(1 - \lambda\varphi + ...\right)$$

(3)

R_H is the apparent hydrodynamic radius of the particle, k_b is the Boltzmann-faktor, T denotes the thermodynamic temperature, the expansion factor λ is sometimes called the rigidity factor and depends on the hydrodynamic and the pair interactions between the particles, therefore on their shape and the rigidity. For an ideal, hard sphere $\lambda = 2$ is valid [5-7]. D_0 is the self-diffusion coefficient of a single particle, that is in an infinitely dilute solution where φ of the solute becomes zero. The denominator in the expression for D_0 is called friction factor f.

For rigid rods of aspect ratio a Dhont e. a. [8] have expanded the theory: λ increases with a. Also, λ becomes larger than the theoretical value when there are attractive interactions. On the (time) average two interacting particles share a longer time in an attractive field than in a neutral or a repulsive one so that the diffusion is slowed down.

Materials and Methods

Paclitaxel was modified with methyl terminated PEO (uniformity of 1.05 and molar mass averages of 5,000 g mol^{-1} and 20,000 g mol^{-1}) and attached to carbon 7 (see Fig. 2) with succinic acid through a self-immolating linker as described elsewhere, Jo [9]. The modified drug is termed PP7, the α-methyl- and ω-succinyl-terminated PEO is termed PEGS in the following text.

The measurements of the surface tension were performed with a Wilhelmy balance and also controlled by the contact angle. The solution viscosity was determined with an Ubbelohde viscosimeter. All experiments were carried out temperature-controlled at 25°C.

The ("long-time", that means $> 10^{-4}$s) self-diffusion coefficient was calculated from pulsed-gradient-field NMR spectroscopic measurements (Hahn-echo and stimulated echo) on a Bruker DRX 500 with a water cooled Diff30 z-gradient, gradient strength 0.3 T m^{-1} A^{-1} in a standard bore Bruker 11.74 T magnet with a pulsed field gradient magnitude g up to 6 T m^{-1}. In order to avoid undesired convection in the sample, the temperature of the sample was controlled by the water jacket of the z-gradient coil. The samples reached equilibrium after around 30 min.

Results and Discussion

The surface tension of the prodrug PP7 (5,000 g mol^{-1}) behaves over a wide concentration range within experimental error almost like the corresponding PEO chain, as it is shown in Fig. 3.

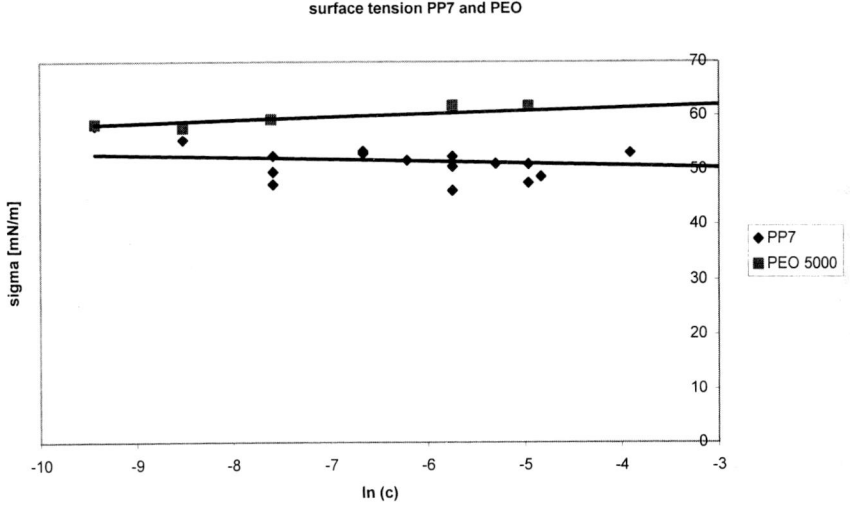

Fig. 3. surface tension vs. The concentration which is normalised by the unit of the concentration to give the argument of the logarithm the unit 1. Both substances show nearly the same behaviour only on a different level.

Assuming that the prodrug, as schematically shown in Fig. 2, consists of a hydrophobic "head" and a hydrophilic "tail", one might expect a detergent-like behaviour with the formation of aggregates of the prodrug like micelles. Apparently, however, there is no such aggregation. Concluding from the concentration dependence of the surface tension the prodrug does not show a detergent-like anisotropy of shape, even at a chain length of PEO as short as 113 repetition units.

The specific viscosity vs. the volume fraction of the solute shows the linear behaviour predicted by eq. 1, see Fig. 4. However, the slope of the function is steeper than the theoretical value of 2.5. According to Simha [4] this can be interpreted as a (prolate or oblate) deviation from a spherical shape of the molecules in solution.

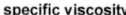

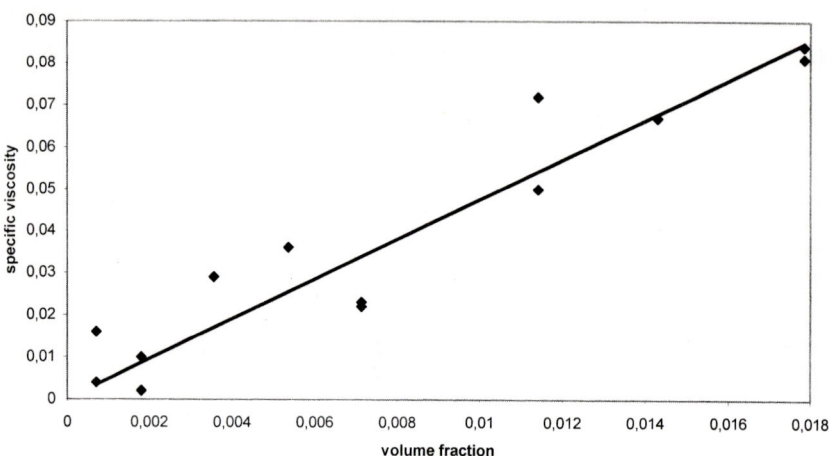

Fig. 4. specific viscosity of PP7 vs. the volume fraction of the solute. Within experimental error the behaviour is linear, although the slope is significantly larger than the theoretical value of 2.5.

In the present case this would mean an axial ratio of about 4.5 (prolate) or 6.2 (oblate), respectively. Which of these alternative molecule shapes is formed cannot yet be decided without further investigations. However, the results of the linear behaviour of the solution-viscosity with the volume fraction supports the interpretation of the surface tension measurements in that there is no discontinuity in the viscosity with the composition of the solution that would indicate a formation of larger aggregations.

Further information can be drawn from the temperature-, time- and concentration dependence of the self-diffusion coefficient of the prodrug in aqueous solution. The interpretation of the NMR pulsed gradient field experiments in terms of the self-diffusion coefficient are shown in Fig. 5. The Self-diffusion coefficient D can be determined from the attenuation of spin-echo experiments in a gradient of the magnetic field along the z-axis of the sample. No time-dependence of the self-diffusion coefficient could be found. The energy of activation does not change between 15°C and 33°C and hence indicating no change of the transport mechanism in this temperature range.

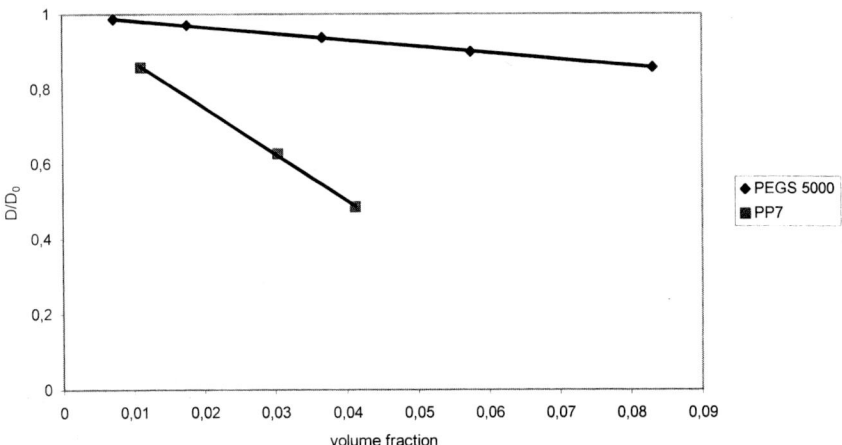

Fig. 5. the ratio of the concentration-dependent self-diffusion coefficient D and the self-diffusion coefficient in an infinitely dilute solution D_0 is plotted vs. the volume fraction of the solute PEGS (5,000) and PP7 (5,000), respectively.

Analysis of the data according to eq. 3 reveals that the effective hydrodynamic radius R_H of PP7 (5,000 g mol^{-1}) and PEGS (5,000) at infinite solution is very similar. The difference in their self-diffusion coefficient is mainly caused by differences in the λ-term. The λ-term is much larger for PP7. As already concluded from the results of the viscosity measurements this indicates a deviation from the behaviour of a rigid sphere as it is shown by PEGS or PEO. There is definitely a significant change of the interaction potential after the PEO chain is coupled with the drug. This becomes clear by comparison with the behaviour of PEO [10]. According to Faraone e. a. [11] pure PEO shows no interaction in solution and behaves like a rigid sphere. PP7, on the other hand, does not behave like a rigid sphere. But there are apparently also no a strong interaction between the PP7 molecules leading to larger aggregations in solution, see the behaviour of the surface tension. Also, the self-diffusion coefficient of an isolated PP7 molecule is not too much different from that of PEO. That leads to the conclusion that the surface of the molecule prodrug still is rather PEO-like. However, the shape of PP7 seems to be different from PEO, at least a deformed sphere. This deformation appears not to be so strong that the anisotropy of the shape causes lyotropic behaviour. The present explanation of the experimental results is that the PEO-chain coils-up around the lipophilic core of the paclitaxel in a way comparable to the tertiary structure of proteins where the hydrophilic parts of the polymer chain form a conformation which makes the surface hydrophilic while the inner part of the shell is lipophilic and covers the hydrophobic core. The hydrophobic interactions between core and shell could also explain the small decrease in the apparent hydrodynamic radius which is observed comparing PP7 with PEOS. Further investigations, in particular the influence PEO-chains on the hydrodynamic properties and a more detailed analysis of the hydrodynamic radius are planned to determine structure and interaction of the polymer modified drug in solution. In particular this will be of importance concerning possible interactions with components of the body fluid like proteins and polysaccharides and with respect of a possible targeting effect.

Conclusion

Molecular engineering on the paclitaxel molecule not only changes its solubility, a number of other properties such as activity, targeting and the transport properties in solution are affected by the modification. Examination of the details of these structure property relations can improve the applicability and the efficiency of the drug. Conformation of the polymeric solubiliser and the stability of the polymer-drug linkage play a crucial role in these interactions.

Acknowledgement

Financial support of the Deutsche Forschungsgemeinschaft (DFG), Bonn, and the Korean Science Foundation (KOSEF), Seoul, is greatly acknowledged.

[1] N. P. Desai, P. Soon-Shiong, P. A. Sandford, *U.S. Pat. 2,648,506* (1997)
[2] Byung-Wook Jo, *Korean Pat. 2000-0019873* (2000)
[3] A. Einstein, Ann. Physik **19** (4), 289 (1905)
[4] R. Simha, J. Chem. Phys. **44**, 25 (1905)
[5] J. K. G. Dhont, An Introduction to Dynamics of Colloids, Elsevier, Amsterdam (1996)
[6] Hanna, S., Hess, W., Klein, R. Physica **111A**, 181 (1982)
[7] Russel, W. B., Saville, D. A. , Schowalter, W. R., Colloidal Dispersions, Cambridge University Press (1989)
[8] Dhont, J. K. G., van Bruggen, M. P. B., Briels, W. F., Macromolecules **32**, 3809 (1999)
[9] Jo, B. W. and Kolon Inc., PCT/Kr01/00168
[10] Venkatesan, C.S., Hirzel, R., Rajagopalan, J. Chem. Phys. **82**, 5685 (1985)
[11] Faraone, Magazù, A., Maisano, P., Migliardo, E., Tettamanti, E, Villari, V, J. Chem. Phys. **110**, 1801 (1999)

Macromol. Symp. **2003**, *201*, 171—178

Quantifying Supercritical CO$_2$ Dilation of Poly(vinylidene fluoride) and Poly(vinylidene fluoride-*co*-hexafluoropropylene) Utilizing a Linear Variable Differential Transducer: Plasticization and Melting Behavior

Suresh L. Shenoy, Tomoko Fujiwara, Kenneth J. Wynne*

Chemical Engineering Department, School of Engineering, Virginia Commonwealth University, 601 West Main St., Richmond, VA 23284-3028, USA

Summary: Melting behavior of semicrystalline poly(vinylidene fluoride) (PVDF) and poly(vinylidene fluoride-*co*-hexafluoropropylene) is investigated as a function of supercritical CO$_2$ pressure using a Linear Variable Displacement Transformer (LVDT). The melting temperature (T_m) of both polymers is lowered due to supercritical CO$_2$ plasticization. For PVDF, the maximum lowering of T_m (ΔT_m=23°C) occurs between 483 and 552 bar. The corresponding value for the copolymer is ΔT_m = 26°C at 552 bar. At higher pressures, hydrostatic effects override plasticization and T_m increases for both polymers. By comparing T_m in N$_2$, a noninteracting gas, the opposing effects of plasticization and hydrostatic pressure on T_m are explored.

Keywords: LVDT, melting point, PVDF, supercritical CO$_2$, swelling

Introduction

Polymer plasticization by high-pressure carbon dioxide has been investigated by obtaining sorption isotherms,[1-3] measuring polymer dilation,[1,4-6] and by depression of glass transition temperatures (T_g).[7-10] A vast majority of these studies have focused on amorphous polymers.[1,4,5,7,8,11,12] The behavior of semicrystalline polymers in high pressure CO$_2$ is more complex.[2,10,13-23] In semicrystalline polymers, plasticization of strained amorphous regions adjacent to crystalline lamellae, can result in CO$_2$-induced crystallinity or "anti-plasticization".[13-16] For example, plasticization and anti-plasticization of poly(ethylene terephthalate) (PET)[13,15,16] and polycarbonate (PC)[14,24] in high pressure CO$_2$ have been extensively studied. The effect of CO$_2$ on other semicrystalline polymers such as low density polyethylene (LDPE),[17] polytetrafluoroethylene (PTFE),[19] poly(vinylidene fluoride)[2] (PVDF) and polyurethanes[18] has also been studied at relatively low temperatures.

Recently we developed a method that utilizes a Linear Variable Differential Transformer (LVDT) to quantify plasticization of polymers. The versatility and robustness of this technique permit in situ measurements at high temperatures and pressures. Using this method

© 2003 WILEY-VCH Verlag GmbH & KGaA, Weinheim

DOI: 10.1002/masy.200351120

we reported swelling behavior in $SCCO_2$ of styrene-butadiene-styrene triblock copolymer (elastomer)[25] and PVDF (semicrystalline).[26]

To employ supercritical CO_2 as a processing aid for semicrystalline polymers information about melting behavior as a function of pressure is desirable. The melting temperature (T_m) is defined as the temperature at which the chemical potentials of amorphous and crystalline phases are equal ($\Delta\mu_a = \Delta\mu_{cry}$). The chemical potential of the amorphous regions decreases due to polymer-CO_2 interactions. As a result, T_m decreases. For example, at 77 bar in supercritical CO_2 the T_m of syndiotactic PS (270°C) decreases by about 12°C.[27]

In addition to quantifying plasticization, the LVDT technique can be utilized to monitor the effect of supercritical CO_2 on T_m as a function of pressure. The lowering of PVDF homopolymer T_m in supercritical CO_2 was previously examined.[26] In this paper we compare and contrast the melting behavior of PVDF homopolymer and a PVDF copolymer containing 4.9 mol% hexafluoropropylene (HFP) comonomer units.

Experimental

Materials. Liquid CO_2 (bone dry) was obtained from Roberts Oxygen Company, Inc. Poly(vinylidene fluoride), [Catalog #102, M_w=530,000 and T_m =158°C], was obtained from Scientific Polymer Products, Inc. and was used as received. PVDF powder was melt pressed in a Carver Laboratory Press at 230°C for 20 minutes followed by fast cooling to room temperature. The PVDF-copolymer containing 4.9 mol% hexafluoropropylene (PVDF-HFP) [M_w=290,000 and T_m =138.9°C] was supplied in the form of 4.7 mm OD x 3.2 mm ID tubing by Daikin America Technical Center. For PVDF T_m measurements, rectangular bars ($\approx$1 cm x 0.35 cm x 0.2 cm) were cut from melt pressed discs with a small saw and sanded to regular dimensions. For PVDF-copolymer, tubing sample (1cm length) was cut using a razor.

Equipment. Differential scanning calorimetry (DSC) measurements were carried out using a Perkin Elmer DSC Pyris-1. Details of the experimental setup have been described in a previous paper. [25]

Measurements. Calibration, sample setting, and data acquisition were performed as previously reported.[25] Dilation is reported as percent change in the sample length ($\Delta L/L_o$)x100%, where $\Delta L = L_t - L_o$, L_t is the length of the sample at time t, and L_o is the initial sample length.

To measure T_m, supercritical CO_2 pressure was increased to the desired value at a temperature below T_m. LVDT readings were noted at regular intervals. When the sample attained maximum swelling, the LVDT readings remained constant for 20-30 minutes. The temperature was then increased by 3-4°C incrementally, while employing the manual pressure generator to maintain constant $SCCO_2$ pressure during temperature increase. The dilation includes thermal expansion as well as swelling due to CO_2. At the melting temperature T_m, $\Delta L/L_o$ decreased markedly due to sample softening and the experiment was terminated.

Results and Discussion

Thermal analyses of melt pressed PVDF and as received PVDF-copolymer were performed by DSC (data not shown). For PVDF homopolymer, the maximum in melting peak occurs at 158.0°C. The corresponding maximum in the melting endotherm for the copolymer appears at 138.9°C. The lower T_m reflects presence of more imperfect crystallites due to the bulky hexafluoropropylene comonomer.

PVDF homopolymer. Figure 1 shows a representative plot of $\Delta L/L_o$ as a function of time at 276 bar for PVDF homopolymer. With each increase of temperature, dilation increases with time, reaching a maximum value in about 1 hr. As the temperature increases from 127.2°C to 144.1°C, $\Delta L/L_o$ increases by 11.2%. Further increase to 145.9°C, results in a decrease of $\Delta L/L_o$. This trend continues till 149.2°C where a decrease in $\Delta L/L_o$ characteristic of melting is observed.

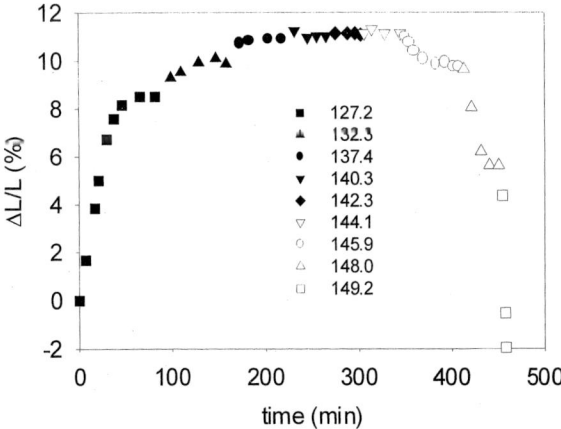

Fig. 1. $\Delta L/L_o$ for PVDF as a function of time at 276 bar from 127.2°C to 149.2°C.

Figure 2 shows a plot of PVDF dilation as a function of temperature in N_2 at ambient pressure and in supercritical CO_2 at 276, 476, 674 bar. At ambient pressure, the decrease in linear swelling occurs between 157°C and 159°C. We have designated T_m as the temperature halfway between the temperature of maximum swelling and the temperature at which the first loss of linear dimension occurs. Using this convention, LVDT measured T_m is 158°C. Alternatively, T_m may be described as the temperature at which $d(\Delta L/L_0)/dT$ becomes negative. This corresponds well to the maximum in the DSC melting endotherm (158°C).

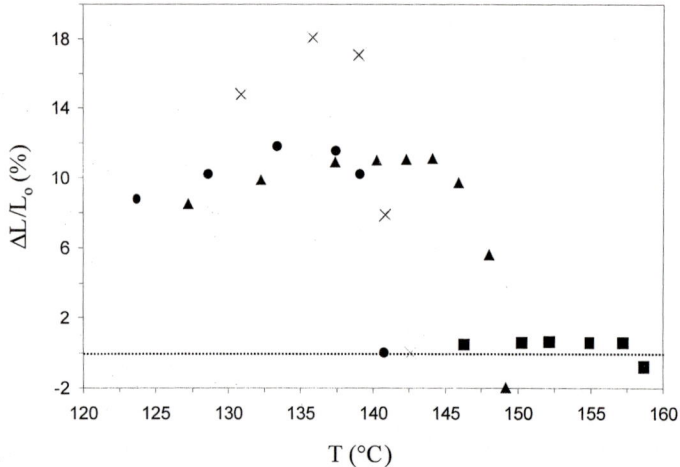

Fig. 2. Dilation of PVDF homopolymer as a function of temperature in nitrogen (N_2) at ambient pressure (■) and in supercritical CO_2 at 276 (▲), 476 (●), and 674 (x) bar.

In supercritical CO_2, initial dilation increases with temperature for all pressures (276, 476 and 674 bar) indicating homopolymer plasticization. The maximum for $\Delta L/L_0$ is ≈18% at 674 bar and 136°C. The temperature at which $d(\Delta L/L_0)/dT$ becomes negative is a function of the supercritical CO_2 pressure. The melting temperature, T_m decreases from 158°C at ambient pressure to ≈ 136°C in supercritical CO_2 at 476 bar, a decrease of 23°C. Further increase in CO_2 pressure results in a small increase in T_m to 137°C.

Figure 3 compares $\Delta L/L_0$ in nitrogen (N_2) at 276, 476, 674 bar with the dilation at ambient pressure. At lower temperatures, PVDF undergoes compression due to hydrostatic pressure effects. This is in contrast to the linear dilation in supercritical CO_2. Above 145°C, N_2 plasticization is minimal but a function of temperature with a 2.5% maximum dilation at 674 bar. In addition, $d(\Delta L/L_0)/dT$ becomes negative at temperatures greater than 158°C indicating a slight increase in T_m with N_2 pressure. At 476 bar, T_m increases to 170°C and at 674 bar T_m increases to 173°C, an increase of 15°C.

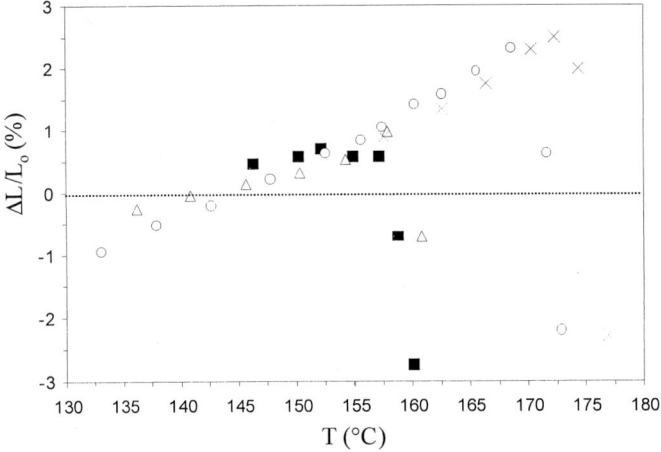

Fig. 3. $\Delta L/L_o$ of PVDF homopolymer as a function of temperature in nitrogen (N_2) at ambient pressure (■), 276 (Δ), 476 (O), and 674 (x) bar respectively.

The melting point of a polymer in high-pressure gas is a balance between plasticization due to polymer-gas interactions and hydrostatic pressure of the gas. The lowering of PVDF T_m in supercritical CO_2 is a result of reduction in the chemical potential of the amorphous phase due to favorable interactions between the CO_2-philic F and carbon dioxide[28]. Thus as pressure increases, T_m decreases to a minimum of 136°C at 476 bar due to increased CO_2 plasticization. However at 674 bar, our experimental limit, T_m increases to 137°C. At higher pressures, the decrease in T_m due to plasticization is offset by a modest increase in T_m due to higher hydrostatic pressure corresponding. Previously it has been documented that PVDF T_m increases ≈1°C for a 31 bar increase in hydrostatic pressure.[29,30] Thus above 476 bar, hydrostatic pressure effects offsets plasticization resulting in a small increase in T_m. In contrast to supercritical CO_2, N_2 is a "noninteracting gas".[31] Nitrogen plasticization is almost an order of magnitude smaller and the increase in T_m primarily reflects the effect of hydrostatic pressure.

PVDF copolymer. Figure 4 shows $\Delta L/L_o$ as a function of temperature for PVDF copolymer in N_2 at ambient pressure and in supercritical CO_2 at 276, 552 and 640 bar. At ambient pressure, the LVDT T_m using our convention (d($\Delta L/L_o$)/dT) = negative) is 132°C. Copolymer plasticization is observed for all pressures (276, 476 and 674 bar) with a maximum of $\approx 18\%$ at 552 bar and 104°C. Plasticization and melting point depression are different for the copolymer due to lower crystallinity from disorder introduced by bulky CF_3 groups ($\Delta H_{copolymer} = 32$ J/g versus $\Delta H_{homopolymer} = 51$ J/g).

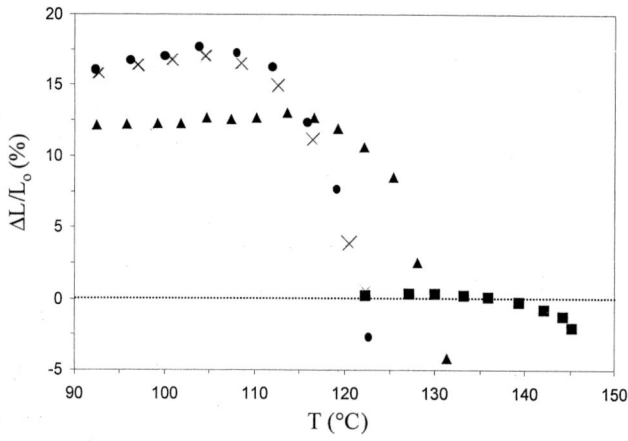

Fig. 4. PVDF copolymer dilation as a function of temperature in $SCCO_2$ in nitrogen (N_2) at ambient pressure ($\blacksquare$) and in $SCCO_2$ at 276 ($\blacktriangle$), 552 ($\bullet$), and 640 (x) bar respectively.

For the copolymer, lowering of T_m in supercritical CO_2 is similar to the homopolymer. At 276 bar, T_m is ≈ 115°C, about 17°C lower than at ambient pressure (132°C). Increasing pressure to 552 bar, decreases copolymer T_m to 106°C, a decrease of 26°C. Further increase in pressure (640 bar) results in a small increase in T_m to 107°C. Though close to our experimental limits, it appears that PVDF-copolymer T_m is minimized at 552 bar. In summary, PVDF-copolymer plasticization (18%) is somewhat greater than homopolymer (14%) and T_m depression (27°C) is larger than PVDF homopolymer (23°C).

Figure 5 plots copolymer dilation in N_2 at 1, 276, 552, and 640 bar. At lower temperatures, N_2 plasticization is observed in contrast to the observed homopolymer compression. However, copolymer plasticization by N_2 is minimal (maximum $\Delta L/L_o = 1.3\%$) as expected. The LVDT measured T_m increases with increasing N_2 pressure. At 276 bar, copolymer T_m is 137°C an increase of 4°C. Copolymer T_m increases by 13°C at 552 bar and by 15°C at 630 bar.

A balance between plasticization and hydrostatic pressure determines copolymer T_m. At lower pressures (up to 552 bar) in supercritical CO_2 plasticization dominates (max $\Delta L/L_o = 18\%$) and hence T_m decreases. At higher pressures, this decrease is offset by an increase in T_m due to plasticization is observed for all pressures (276, 476 and 674 bar) with a maximum of $\approx 18\%$ at 552 bar and 104°C. Plasticization and melting point depression are different for the copolymer due to lower crystallinity from disorder introduced by bulky CF_3 groups ($\Delta H_{copolymer} = 32$ J/g versus $\Delta H_{homopolymer} = 51$ J/g).

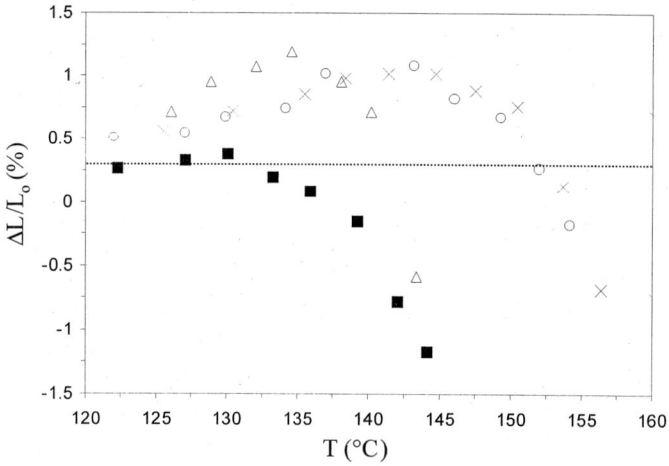

Fig. 4. PVDF copolymer dilation as a function of temperature in $SCCO_2$ in nitrogen (N_2) at ambient pressure ($\blacksquare$) and in $SCCO_2$ at 276 ($\blacktriangle$), 552 ($\bullet$), and 640 (x) bar respectively.

For the copolymer, lowering of T_m in supercritical CO_2 is similar to the homopolymer. At 276 bar, T_m is ≈ 115°C, about 17°C lower than at ambient pressure (132°C). Increasing pressure to 552 bar, decreases copolymer T_m to 106°C, a decrease of 26°C. Further increase in pressure (640 bar) results in a small increase in T_m to 107°C. Though close to our experimental limits, it appears that PVDF-copolymer T_m is minimized at 552 bar. In summary, PVDF copolymer plasticization (18%) is somewhat greater than homopolymer (14%) and T_m depression (27°C) is larger than PVDF homopolymer (23°C).

Figure 5 plots copolymer dilation in N_2 at 1, 276, 552, and 640 bar. At lower temperatures, N_2 plasticization is observed in contrast to the observed homopolymer compression. However, copolymer plasticization by N_2 is minimal (maximum $\Delta L/L_o = 1.3\%$) as expected. The LVDT measured T_m increases with increasing N_2 pressure. At 276 bar, copolymer T_m is 137°C an increase of 4°C. Copolymer T_m increases by 13°C at 552 bar and by 15°C at 630 bar.

A balance between plasticization and hydrostatic pressure determines copolymer T_m. At lower pressures (up to 552 bar) in supercritical CO_2 plasticization dominates (max $\Delta L/L_o = 18\%$) and hence T_m decreases. At higher pressures, this decrease is offset by an increase in T_m due to

178

Acknowledgments

Financial support from Daikin Institute of Advanced Chemistry and Technology (DAI-ACT) and from the VCU School of Engineering Foundation is gratefully acknowledged.

[1] R. G. Wissinger; M. E. Paulaitis *J. Polym. Sci., Part B: Polymer Physics* **1987**, 25, 2497.
[2] B. J. Briscoe, Lorge, O., Wajs, A., Dang, P. *J. Polym. Sci., Part B: Polymer Physics* **1998**, 36, 2435.
[3] J. S. Wang; Y. Kamiya; Y. Naito *J. Polym. Sci., Part B: Polymer Physics* **1998**, 36, 1695-1702.
[4] Y. M. Kamiya, Keishin; Terada, Katsuhiko; Fujiwara, Yukihiko; Wang, Jin-Sheng. *Macromolecules* **1998**, 31, 472.
[5] *High Pressure Solid Polymer-Supercritical Fluid Phase behvaior*; I. S. M. Liau, M. A., Ed.; Elsevier Science Publishers: Amsterdam, netherlands, 1985.
[6] Y. G. Zhang, K. K.; Lemert, R. M. *J. Supercritical Fluids* **1997**, 11, 115.
[7] J. S. Chiou; J. W. Barlow; D. R. Paul *J. Appl. Polym. Sci.* **1985**, 30, 2633.
[8] P. D. J. Condo, K. P. *J. Polym. Sci., Part B: Polym. Phys.* **1994**, 32, 523.
[9] Y. P. Handa; P. Kruus; M. Oneill *J. Polym. Sci., Part B: Polymer Physics* **1996**, 34, 2635-2639.
[10] Z. K. Zhong; S. X. Zheng; Y. L. Mi *Polymer* **1999**, 40, 3829-3834.
[11] W. V. Wang; E. Kramer; W. H. Sachse *J. Polym. Sci., Part B: Polymer Physics* **1982**, 20, 1371.
[12] J. R. Royer; J. M. DeSimone; S. A. Khan *Macromolecules* **1999**, 32, 8965-8973.
[13] J. S. Chiou; J. W. Barlow; D. R. Paul *J. Appl. Polym. Sci.* **1985**, 30, 3911.
[14] S. M. R. Gross, G. W.; Kiserow, D. J.; DeSimone, J. M. *Macromolecules* **1999**, 32, 8965.
[15] W. J. P. Koros, D. R. *J. Poly. Sci.: Part-B: Polymer Physics* **1978**, 16, 1947.
[16] K. H. Mizoguchi, Takuji; Naito, Yasutoshi; Kamiya, Yoshinori. *Polymer* **1987**, 28, 1298.
[17] Y. T. Shieh; J. H. Su; G. Manivannan; P. H. C. Lee; S. P. Sawan; W. D. Spall *J. Appl. Polym. Sci.* **1996**, 59, 707-717.
[18] B. J. Briscoe; C. T. Kelly *Polymer* **1995**, 36, 3099-3102.
[19] B. J. Briscoe; S. Zakaria *J. Polym. Sci., Part B: Polymer Physics* **1991**, 29, 989.
[20] Y. P. Handa; J. Roovers; F. Wang *Macromolecules* **1994**, 27, 5511.
[21] Y. P. Handa; Z. Y. Zhang; J. Roovers *J. Polym. Sci., Part B: Polymer Physics* **2001**, 39, 1505-1512.
[22] K. Hatada; T. Kitayama; K. Ute; N. Fujimoto; N. Miyatake *Macromol. Symp.* **1994**, 84, 113-126.
[23] M. A. Singh; R. Hutanu; M. Shea; R. Fraser; T. Plivelic; Y. P. Handa *J. Polym.Sci., Part B: Polymer Physics* **2000**, 38, 2457-2467.
[24] Y. L. Mi; S. X. Zheng *Polymer* **1998**, 39, 3709-3712.
[25] Shenoy. S.; D. Woerdeman; R. Sebra; A. Garach-Domech; K. J. Wynne *Macromol. Rapid Commun.* **2002**, 23, 1130-1133.
[26] S. L. Shenoy; T. Fujiwara; K. J. Wynne *Macromolecules* **2003**, accepted for publication.
[27] Z. Zhang; Y. P. Handa *Macromolecules* **1997**, 30, 8505.
[28] S. G. V. Kazarian, M. F.; Bright, F. V.; Liotta, C. L.; Eckert, C. A. *J. Am. Chem. Soc.* **1996**, 118, 1729.
[29] T. H. Hattori, M.; Ohigashi, H. *Polymer* **1996**, 37, 85.
[30] N. Mekhilef *J. Appl. Polym. Sci.* **2001**, 80, 230-241.
[31] M. Mulder *Basic Principles of Membrane Technology*; Second ed.; Kluwer Academic Publishers, Boston, 1997.

Macromol. Symp. **2003**, *201*, 179—186

New Biodegradable Polymers for Delivery of Bioactive Agents

Young Min Kwon, Sung Wan Kim*

Department of Pharmaceutics and Pharmaceutical Chemistry, Center for Controlled Chemical Delivery, University of Utah, 30 S. 2000 E. Rm 205, Salt Lake City, UT 84112, USA
Email: rburns@pharm.utah.edu

Summary: Biodegradable, thermosensitive triblock copolymer, PLGA-PEG-PLGA, can be easily fabricated into drug-loaded microspheres or injectable in situ hydrogel system for protein or water-insoluble drugs without use of organic solvent. Aqueous-based microsphere exhibited continuous release of intact insulin in vitro for 3 weeks while the microspheres prepared using dichloromethane showed initial burst and incomplete release. Confocal miscoscopy images of microspheres corroborated the release pattern. Next study with an injectable in situ hydrogel (ReGel[TM]) exhibited zero-order insulin release in vitro and sustained plasma insulin level for 2 weeks in vivo upon single subcutaneous injection in SD rats.

Keywords: biodegradable polymers, drug delivery, hydrogels, microspheres, thermosensitive polymers

Introduction

Over the past two decades extensive research has been performed in the design area of polymeric drug delivery systems. Among them, the use of biodegradable polymers have been successfully carried out. They include polyesters, poly(orthoesters), polyanhydrides, polyamino acid, poly(alkyl-cyanoacrylates), polyphosphazenes, copolymers of (PLA/PGA) and aspartate or PEO.

Although these biodegradable polymers were used for drug delivery and some are successfully for human application, there remains fabrication problems, such as difficult processability and limited organic solvent and irreproducible drug release kinetics. In this presentation, new design of biodegradable polymers and their application for drug delivery will be discussed.

A series of thermoplastic biodegradable hydrogels (TBH) based on star-shaped poly(ether-ester) block copolymers have been synthesized in this laboratory.[1-4] Physically crosslinked TBH may

          DOI: 10.1002/masy.200351121

present improved biocompatibility, mass transport, biodegradability, and processability, and thus can provide a better way of parenteral injectable drug delivery.

New star-shaped block copolymers, of which the typical molecular architecture is presented, results from their distinct solution properties, thermal properties and morphology. Their unique physical properties are due to the three-dimensional, hyperbranched molecular architecture and influence microsphere fabrication, drug release and degradation profiles.[1]

We recently synthesized thermosensitive biodegradable hydrogel consisting of polyethylene oxide and poly (L-lactic acid). Aqueous solution of these copolymers with proper combination of molecular weights exhibit temperature dependent reversible sol-gel transition. HPL-HPB-HPL (HPL=hydrophilic; HPB=hydrophobic) triblock copolymers (Hygel™) shows sol formation at an elevated temperature (~45 °C) and drug loaded sol forms gel upon rapid cooling to body temperature which acts as a sustained drug release matrix.[2-4] On the contrary, HPB-HPL-HPB molecular arrangements (ReGel™) provide unique behavior that sol (at low temperature) form gel (at body temperature). [5-6] The use of these two biodegradable polymers have great advantages for sustained injectable drug delivery systems. The formulation is simple, which is totally free of organic solvent and thus suitable for hydrophobic drug or protein drug loading.

Due to its amphiphilic nature, this polymer in sol or aqueous state can solubilize poorly water soluble drugs prior to forming gel matrix. ReGel™ forms a controlled release drug depot with delivery times ranging from 1 to 6 weeks. ReGel™ 's inherent ability to solubilize (400 to >2000-fold) and stabilize poorly soluble and sensitive drugs, inculding proteins is a substantial benefit. The gel provided excellent control of the release of paclitaxel for approximately 50 days. Direct intratumoral injection of ReGel /paclitaxel (Oncogel™) results in a slow clearance of paclitaxel from the injection site with minimal distribution into any organ. Efficacies equivalent to maximum tolerated systemic dosing were observed at Oncogel™ doses that were 10-fold lower.[6]

Protein drugs can easily be loaded into the triblock copolymer system to yield an injectable depot system via microspheres as well as in situ gelling systems. We will give examples of insulin release from using both microsphere system and in situ gel forming device.

Preparation of Triblock Copolymer Microspheres

Fabrication of protein-loaded microspheres, either a water-in-oil-in-water (w/o/w) double-emulsion-solvent evaporation method or spray-dry method, usually involves the use of water-immiscible, volatile organic solvent such as dichloromethane (CH_2Cl_2). However, there are unresolved drawbacks associated with protein-loaded microspheres in the context of protein release and stability. Initial burst release followed by slow and incomplete release of proteins has often been observed. Also, residual level of organic solvent that is difficult to remove completely may bring about toxicity issues. The use of organic solvent as well as the harsh preparation condition are thought to be main reasons for these observed drawbacks that hamper controlled delivery of protein drugs due to the physical degradation of proteins at the interface of aqueous phase and the organic phase.

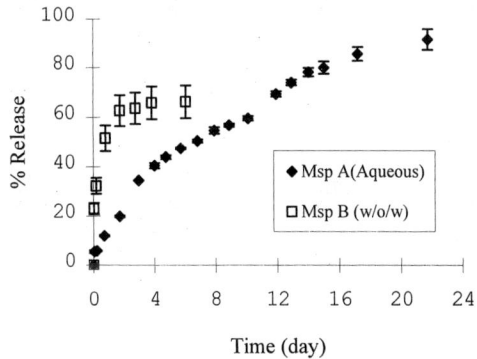

Fig. 1. In vitro release of human insulin from msp A (aqueous-based) and msp B (dichloromethane-based). (n = 3, mean ± S. D.).[7]

Microspheres of the biodegradable, triblock copolymer (PLGA-PEG-PLGA, Mw=4000, 1500-1000-1500 by NMR) was prepared in two methods: microspheres A (aqueous-based) and B (dichloromathane).[7] For both microspheres, same amount of Zn-insulin was loaded (~4% of polymer mass). In vitro release studies were carried out with both. As shown in Figure 1, microsphere A exhibited continuous and nearly complete release of insulin over 3 weeks. The first phase of insulin release (first 10 days) from msp A seems to be dependent more upon diffusion since release rate was slightly decreasing. Then after day 10 the insulin release rate turned to an increasing mode and this is probably the degradation of the matrix, at this time point, begins to play more significant role in release than in the earlier phase. However, microsphere B showed initial burst release (~50 % in 1 day) and release was discontinued at ~ 60% afterwards. In preparing microsphere B, during the formation of primary emulsion where it involves high shear and heat generation to create a large water/organic solvent interfacial area, proteins can undergo rapid aggregation under this environment and thus the incomplete release of proteins from microspheres may be due to this trapped aggregates formed during microsphere fabrication. This accounts for slow and incomplete release after initial release phase with burst effect.

In the case of microsphere A, microsphere was prepared in a mild environment in that organic solvent and high shear was absent. As shown in Figure 2, circular dichroism (CD) spectrum of insulin released from microsphere at day 12 is virtually identical as that of freshly prepared native insulin solution. This means that released insulin preserved its secondary structure. In contrast, the CD spectrum for microsphere B indicates loss of secondary structure integrity due to the use of dichloromethane and the harsh preparation condition employed.

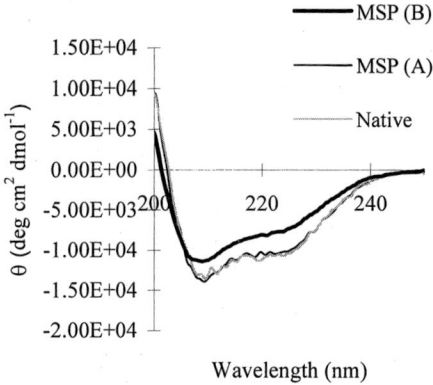

Fig. 2. Circular dichroism (CD) spectra of released insulin from msp A and msp B with respect to the native insulin solution.[7]

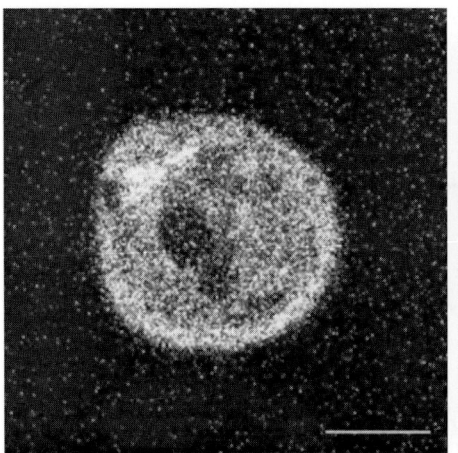 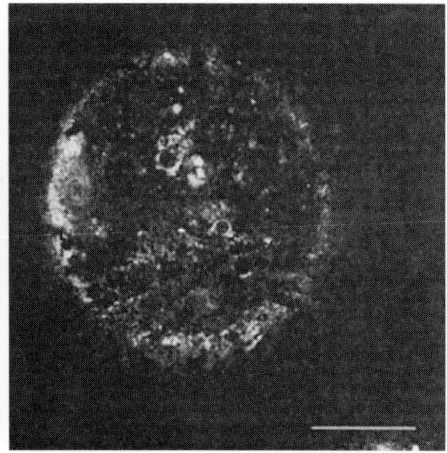

Fig. 3. Confocal microscopy images of FITC-insulin loaded microspheres. (Left) msp A. (right) msp B. Bar indicates 50 μm.

The observed release pattern from both types of microspheres lies in the distribution of the protein inside a microsphere, which is associated with the preparation method. In order to see this, FITC (fluorescein isothiocyanate)-insulin incorporated microspheres were observed under a confocal microscope. The fluorescence distribution is shown in Figure 3. For msp A, homogeneous distribution of fluorescence was observed while msp B exhibited rather heterogeneous distribution of FITC-insulin. In addition, msp B shows significant surface fluorescence. Hence, this is consistent with the observed initial burst from msp B and from constant insulin release from msp A over prolonged period of time. It is reported that the constant release of insulin from triblock copolymer hydrogel may be attributed to the hydrophilic/hydrophobic domain structure of the gel.[5] Low-molecular weight triblock copolymer of PEG and PLGA is known to form micelles at low concentrations and at higher concentrations, gel forms via packing of the micelles and interaction between hydrophobic phases of the micelles by partial overlap.[8] Hence the matrix possesses these microdomains throughout. Thus, significant fraction of insulin is incorporated in the hydrophobic domain that allows sustained release of insulin.

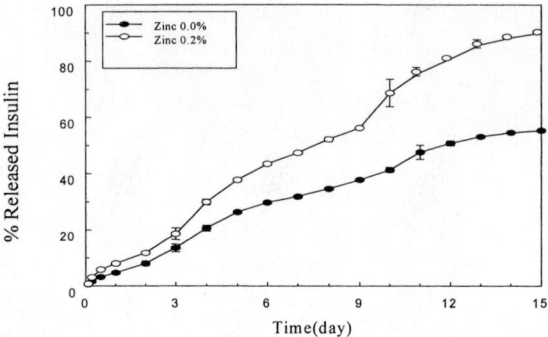

Fig. 4. In vitro release of human insulin from ReGel formulation (n=5).[5]

Insulin Release from Injectable in situ Hydrogel

Insulin release from triblock copolymer hydrogel both in vitro and in vivo was carried out. Aqueous solution of triblock copolymer (23% by mass) was prepared at 5 °C and mixed with insulin. This solution was transferred to 37 °C to form a hydrogel. In vitro release study was carried out at 37 °C by measuring released insulin by HPLC. Figure 4 shows the result of insulin release from the hydrogel (ReGel[TM]) in vitro.[5] There was no initial burst effect of the insulin release from the ReGel[TM] formulations. This hydrogel system is thought to have a core-shell structure in an aqueous environment. The hydrophilic PEG occupies the shell region and hydrophobic PLGA hides into the core in order to decrease surface free energy. Assuming a domain structure of the hydrogel, the partitioning of drug between the hydrophilic domain and the hydrophobic domain was considered. Insulin can be hydrophobic (around isoelectric point) and may mostly be located inside the hydrogel network. Drug release from the hydrophilic domain can be described by diffusion and this is represented by the release profile till day 7. After day 7 the hydrogel network, especially hydrophobic PLGA started to degrade so that the diffusion and degradation governed the release profile of day 7 to day 15. The release profile of the insulin with zinc showed a constant (zero-order) release rate and almost 90% of the initial amount was released over 15 days.[5]

Animal studies using male Sprague-Dawley rats were performed with ReGel[TM]/insulin. As shown in the in vitro release study, aqueous triblock copolymer solution containing insulin (10 IU/ml) at 5 °C was injected subcutaneously (s.c.). Upon injection, the polymer/insulin mixture formed a gel at body temperature. Figure 5 shows plasma insulin level at designated time points. There have been steady amount of insulin release from ReGel[TM] formulation up to day 15 after an s.c. injection. Current protocol of insulin supplementation relies on daily or continous subcutaneous injection of insulin to meet the basal and postprandial requirements In this study, ReGel formulation maintains insulin injection twice per month for basal insulin requirements.[5]

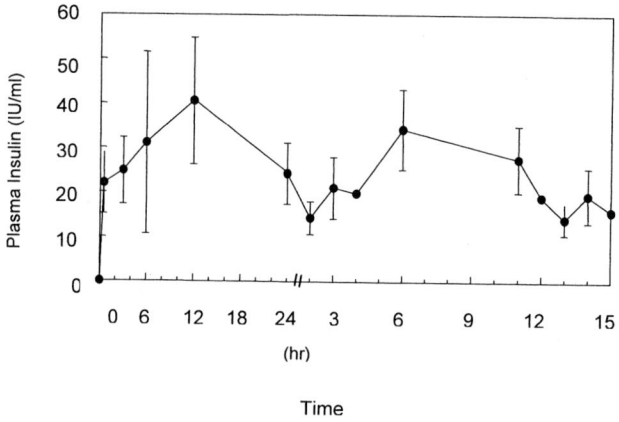

Fig. 5. Plasma insulin level in SD rats in vivo.[5]

Conclusion

Biodegradable, thermosensitive triblock copolymers are potentially useful drug delivery system for therapeutic protein drugs or hydrophobic drugs. As a model study, continuous insulin release for 2-3 weeks can be achieved using either microspheres of PLGA-PEG-PLGA or injectable in situ hydrogel device. While the insulin release from the typically prepared microspheres using CH_2Cl_2 exhibited initial burst release and incomplete release, continuous and nearly complete release (>90%) of insulin in vitro was achieved by aqueous-based microspheres and ReGel[TM] system. In addition to in vitro data, constant insulin release for two weeks in vivo after single s.c. injection of ReGel^{TM} loaded with insulin in rats makes the polymer a promising.candidtate for protein delivery.

Acknowlegments

The authors wish to thank S. Choi, K. S. Ko, Y. J. Kim, S. J. Oh, M. Baudys and G. Zentner for their contribution. This work was supported by MacroMed, Inc.

[1] Y. K. Choi, Y. H. Bae, S. W. Kim, *Macromolecules* **1998**, *31*, 450.
[2] B. Jeong, Y. H. Bae, D. S. Lee, S. W. Kim, *Nature* **1997**, *388*, 860.
[3] B. Jeong, D. S. Lee, J. Shon, Y. H. Bae, S. W. Kim, *J. Polym. Sci. Part. A* **1999**, *37*, 751.
[4] B. Jeong, Y. H. Bae, S. W. Kim, *J. Controlled Release* **2000**, *63*, 155.
[5] Y. J. Kim, S. Choi, J. J. Koh, M. Lee, K. S. Ko, S. W. Kim, *Pharm. Res.* **2001**, *18*, 548.
[6] G. M. Zentner, R. Rathi, C. Shih, J. C. McRea, M. H. Seo, H. Oh, B. G. Rhee, J. Mestecky, Z. Moldoveanu, M. Morgan, S. Weitman, *J. Controlled Release* **2001**, *72*, 203.
[7] Y. Kwon, S. W. Kim *Proc. Int'l Symp. Control. Rel. Bioact. Mater.* **2002**, 683.
[8] D. S. Lee, M. S. Shim, S. W. Kim, H. Lee, I. Park, T. Chang, *Macromol. Rapid Commun.* **2001**, *22*, 587.

Macromol. Symp. **2003**, *201*, 187—201

Impact of Microcellular Plastics on Industrial Practice and Academic Research

Nam P. Suh

The Ralph E. & Eloise F. Cross Professor, Massachusetts Institute of Technology, Cambridge, MA 02139 USA

Summary: Microcellular plastics (MCP) refer to any plastic with tiny bubbles of less than about 50 microns. It is made by subjecting polymers with a large amount of dissolved gas to a thermodynamic instability so as to nucleate a large number of cells instantaneously. MCP has been extruded and injection molded to make various industrial products. The fundamental theory for design of MCP and the processing methods are reviewed. It also discusses the design of equipment, including the die or mold, for MCP processing. The performance of MCP and the advantages of using MCP are presented. Some of industrial applications are also highlighted.

Keywords: cell density, microcellular plastics, MuCell, nucleation of cells, polymer processing

Introduction

Microcellular plastics (MCPs), which were first invented at the Massachusetts Institute of Technology in 1979,[1] refer to any polymeric materials that have closed cells of very small diameters, typically smaller than 50 microns. The cell density can be made to vary a great deal depending on the final application of a given microcellular plastic. MCPs can have as high as 10^{15} bubbles/cm^3 when the bubble diameter is 0.1 microns, 10^{12} bubbles/cm^3 for 1 micron, and 10^9 for 10 micron diameter cells. They can be created in thermoplastics, thermosetting plastics, and elastomers.

The original impetus for the invention of MCPs was to create a plastic consuming less material without sacrificing mechanical properties, especially toughness. The saving of material was achieved by creating voids and toughness was a result of making the diameter of the bubble smaller than a critical size. The central idea was to replace some of the polymers with a large number of very small bubbles that are smaller than the pre-existing flaws in polymers. Small bubbles can blunt the crack-tips and act as crazing initiation sites, this making the material tougher.

     DOI: 10.1002/masy.200351122

The basic processing method for all microcellular plastics is the use of thermodynamic instability phenomena. A large amount of gas, typically CO_2 or N_2, is dissolved the in plastics under high pressure at the processing temperature so as to create a driving force for phase separation when the pressure is suddenly lowered.

Depending on the magnitude of the driving force, various nucleation sites are activated. The number of nucleation sites increases nearly exponentially with the amount of gas dissolved, when the polymer is super-saturated with the dissolved gas – relative to its equilibrium concentration at the pressure of one bar and the operating temperature. Micro-cells form for the following reasons: the amount of gas dissolved must be shared equally by an extremely large number of nucleated sites, since the cells nucleate nearly simultaneously, preventing the preferential diffusion of the gas to the sites that have nucleated first. Because the driving force is so large, homogeneous nucleation dominates even when there are second-phase particles that would be the preferred heterogeneous nucleation sites because of its low activation energy.

Microcellular plastics have unique processing characteristics. The processing temperature is substantially less than the conventional processes because the viscosity of plastics is substantially reduced due to the presence of gas between polymeric molecules. The throughput rate of a given extruder can be also greater because of the low viscosity. The cycle time of injection molding machines is also reduced because the processing temperature is lower and the phase separation of gas from polymer instantaneously increases the rigidity of plastics. Furthermore, there is no shrinkage of the injection-molded part because it is compensated by the internal expansion in the microcells, creating parts with minimal residual stress and warpage. Sometimes, depending on the color of the plastic and the smoothness of the molded surface, swirl marks may appear, which can be hidden through painting or texturing.

Certain properties of microcellular plastics, such as modulus and strength, follow the rule of mixture, whereas such properties as toughness and coefficient of thermal expansion do not. When the cell size is less than a few microns, the toughness of certain microcellular plastics should be equal to or better than the plastic without the cells. Small cells also lower the thermal conductivity when they are smaller than a critical size.

Many industrial firms worldwide are now making microcellular products through extrusion and injection molding.[2] It is very likely that the number of new applications that use the microcellular technology will continue to increase at a rapid rate in the years to come.

The field of microcellular plastic technology is in some ways in the early stages of research and development, notwithstanding its relatively long history. It has raised many interesting scientific and technological issues that can be the basis for thought provoking ideas and research. Many academic institutions worldwide are conducting their research in the field of microcellular plastics, which should further generate new ideas and applications. Many industrial firms are developing new applications for injection molding and extrusion processes.

Design of Microcellular Plastics and Processing Techniques

Review of the Design of Microcellular Plastics

Microcellular plastics were designed to satisfy the following three functional requirements (FRs) based on axiomatic design:[1,3] FR1 = Reduce plastics consumption, FR2 = Maintain the toughness of plastics, and FR3 = Make three dimensional parts.

To satisfy these FRs, the concept of microcellular plastics was created by envisioning plastics with tiny bubbles.[1] Then the design parameters (DPs) of microcellular plastics are the following: DP1 = Total volume of cells (i.e., bubbles), V; DP2 = Diameter of cells, d; DP3 = Die or Mold.

The design equation that relates the FRs to the DPs of microcellular plastics may be written as

$$\begin{Bmatrix} FR1 \\ FR2 \\ FR3 \end{Bmatrix} = \begin{bmatrix} X & X & 0 \\ 0 & X & 0 \\ 0 & 0 & X \end{bmatrix} \begin{Bmatrix} DP1 \\ DP2 \\ DP3 \end{Bmatrix} \quad (1)$$

Equation (1) indicates that the design of an MCP is a decoupled design. It indicates that the bubble size must be determined first before setting the total volume of the bubbles.

In an ideal microcellular plastic, where spherical bubbles are packed in body- centered cubic structure, the bubble size can be directly related to the bubble density. In a 1-cm cube of foamed material, the number of cells is inversely proportional to the cube of the bubble diameter. Therefore, a microcellular plastic with 10-mm bubbles has approximately 10^9 bubbles per cm^3 of unfoamed material, whereas microcellular plastics with 1-μm and 0.1-μm size

190

bubbles have approximately 10^{12} and 10^{15} bubbles per cm^3 of unfoamed material, respectively. Since the volume taken by spherical bubbles in an ideal, closely packed hexagonal or cubic structure is approximately 74%, the plastic occupying the interstitial space is 26%. Therefore, the cell density of an ideal closely packed spherical microcellular plastic is equal to (1/cell size)^3 times (1/0.26). For a microcellular plastic with 1-μm cell diameter, the bubble density is 3.85×10^{12} cells/cm^3 of the solid plastic. The overall density of foam can decrease further when these cells expand, thinning the wall diameter and reducing the interstitial materials between the cells.

Dissolution of Gases in Polymers

The basic physics involved are as follows:[4]

(1)   The plastic must be supersaturated with sufficient gas such as N_2 and CO_2 to nucleate a large number of cells simultaneously.

(2)   The temperature of the plastic must be set so as to control the flow of plastics during processing.

(3)   A gas with a suitable solubility and diffusivity for the plastic must be selected.

(4)   Homogeneous nucleation must dominate the nucleation process to create a large number of microcells even when heterogeneous nucleation sites are available by providing sufficient driving force with a sufficient amount of dissolved gas.

The processing technique consists of forming a polymer/gas solution and then suddenly inducing a thermodynamic instability by either lowering the pressure or raising the temperature to change the solubility S. The solubility of gas in polymers decreases with an increase in temperature. The solubility of N_2 is considerably less than that of CO_2. Since the amount of gas that can be dissolved is a function of the saturation pressure and since the gas diffusion rate is the rate-limiting process, we can use supercritical CO_2 to enhance the solubility and diffusion rate. CO_2 is supercritical at pressures and temperatures greater than 7.4 MPa and 31.1°C.

With dissolution of a large number of gas molecules in polymers, the glass transition temperature and viscosity decrease with the increase in gas concentration. The change in the glass transition temperature is quite substantial at high gas concentrations. These changes affect the processibility of polymers.

To decrease the solubility and induce the thermodynamic instability, either the pressure must be decreased (i.e., $\Delta p<0$) or the temperature must be increased (i.e., $\Delta T>0$). Furthermore, regardless of whether the process is continuous or batch type, the thermodynamic instability must be induced quickly so that the cells will nucleate simultaneously before significant diffusion of gas has taken place. Therefore, the higher the temperature of the polymer, the quicker nucleation has to occur since the diffusion of the gas occurs faster at higher temperatures. Such simultaneous cell nucleation will assure a uniform cell size distribution. The following two dimensionless numbers must be less than one for this to happen:

$$\frac{\text{Characteristic nucleation time}}{\text{Characteristic diffusion time}} \propto \frac{\alpha}{\dfrac{dN}{dt}d_c} \ll 1 \tag{2}$$

$$\frac{\text{Characteristic gas diffusion distance}}{\text{Characteristic spacing between stable nuclei}} \propto 2\rho_c^{1/3}(\alpha\, t_D)^{1/2} \ll 1 \tag{3}$$

The number of cells nucleated is a function of the supersaturation level relative to the equilibrium concentration at ambient pressure at the processing temperature. The higher the supersaturation level, the greater is the number of cells nucleated. Furthermore, since the amount of dissolved gas that fills the nucleated cells is finite, and since all the cells are nucleated almost simultaneously, the gas distributes more or less evenly among all these cells -- a condition for making microcellular plastics. The final bubble size is then determined by the total gas per bubble, and by the flow characteristics of the polymer at the nucleation temperature.

To create a continuous process, we designed processes and associated equipment to perform the following functions in extrusion and injection molding: (1) Rapid dissolution of gas into molten flowing polymer to form a polymer/gas solution; (2) Nucleation of a large number of cells; (3) Control of the cell size; and (4) Control of the geometry of the final product.

To produce the microcellular plastics at an acceptable production rate through a continuous process, we must dissolve the gas in polymers quickly despite the slow diffusion rate. The diffusivity increases with temperature by an Arrhenius relationship. The time for gas diffusion is proportional to the thickness of the plastic ℓ as $t \propto \dfrac{\ell^2}{\alpha}$. The diffusivity of CO_2 and N_2 are nearly the same and it takes a long time to diffuse gas into a polymer at room temperature. For example, the diffusivity of CO_2 in most thermoplastics at room temperature is in the range of 5×10^{-8} cm^2/s and the diffusion time is approximately 14 hours when ℓ is 0.5 mm. The diffusivity at 200°C is 3 to 4 orders of magnitude greater than that at room temperature. Even at high temperatures, the diffusion rate is still the rate-limiting step in continuous processes.

To accelerate the diffusion rate and shorten the time for the formation of gas/polymer solutions, we must raise the temperature and shorten the diffusion distance. This is done by deforming the two-phase mixture of polymer and gas through shear distortion to decrease the diffusion path. This type of deformation occurs in an extruder under laminar flow conditions. The bubbles are stretched by the shear field of the two-phase mixture and eventually break up to minimize the surface energy when a critical Weber number is reached.[5] The disintegrated bubble size is calculated to be about 1 mm and the initial striation thickness after bubble disintegration is calculated to be about twice the bubble diameter.[6] This striation thickness decreases with further shear, and the gas diffusion occurs faster as a result of the increase in the surface area and the decrease in striation thickness. The striation thickness in an extruder is estimated to decrease to about 100 μm. At this thickness, the diffusion time is in the range of 1 minute in PET, from 10 to 20 seconds in polystyrene (PS), polyvinylchloride (PVC), and high density polyethylene (HDPE), and in the range of a few seconds in low density polyethylene (LDPE).

Nucleation

The key idea in the formation of an MCP is the nucleation of an extremely large number of bubbles (cells). Although cells can nucleate either homogeneously or heterogeneously, the driving force is so high due to such a large amount of supersaturation of the gas in the polymer that the both homogeneous and heterogeneous nucleation sites are expected to be nucleated. This can be seen from micrographs, which show that cells are nucleated both at and away from the heterogeneous sites.

For nucleation to occur, a finite energy barrier has to be overcome. The energy barrier depends on two competing factors: (a) the energy available in the gas diffused into the embryo of the cell and (b) the surface energy that must be supplied to form the surface of the cell. There is a critical cell size beyond which the cell becomes stable and grows, and below which the cell embryo collapses. Typically the cell nucleation rate is expressed as:

$$\frac{dN}{dt} = N_0 f e^{\frac{-\Delta G}{kT}}$$
(4)

where N = the number of cells; N_0 = the number of available sites for nucleation; f = the frequency of atomic or molecular lattice vibration; ΔG = the activation energy barrier; k = the Boltzmann constant; T = the absolute temperature. A variety of different nucleation sites may be nucleated when the driving force is very large, the most prominent of which are the free volume sites. Table 1 shows the potential activation sites and the expected cell density when these sites become activated.

Table 1. Potential activation sites for cells and rough estimates of potential cell density.

Solid/polymer interface	10^5 to 10^6 cells/cc
Non-polar polymer/polar polymer interface	--
High strain region	10^9 cells/cc
Free volume	10^9 cells/cc
Crystalline/amorphous interface in a polymer	10^{12} cells/cc
Interface between crystallites	10^{18} cells/cc
Morphological defects in a polymer	--
Polar groups of polymers	10^{22} cells/cc

The activation energy associated with each one of these potential activation sites is expected to be substantially different, probably increasing with the available sites. The activation energy may be represented in terms of its probability density function as shown in Figure 1. The activation energy also changes when the gas is dissolved.

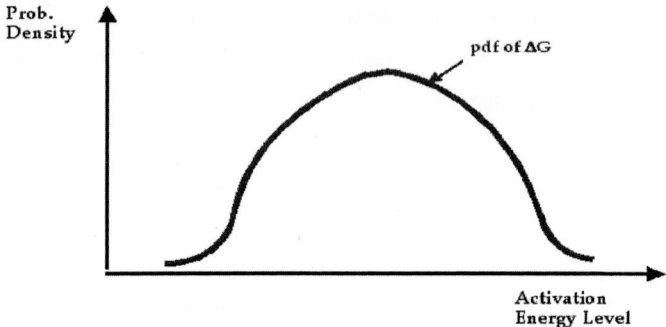

Fig. 1. Probability distribution of ΔG. Note as the amount of dissolved gas increases, the sites with high activation energy are expected to be activated.

The number of the available sites N_0 is also affected by the gas dissolved since the gas changes the intermolecular forces as indirectly evidenced by the change in the viscosity and melting point of the polymer/gas solution (see Figure 2). N_0 is a function of both the original activation energy ΔG and the amount of the gas dissolved. Although there is no data available, the N_0 is expected to increase with higher activation energy since it appears that there are more activation sites at these higher-level activation energies, which is represented schematically as shown in Figure 3.

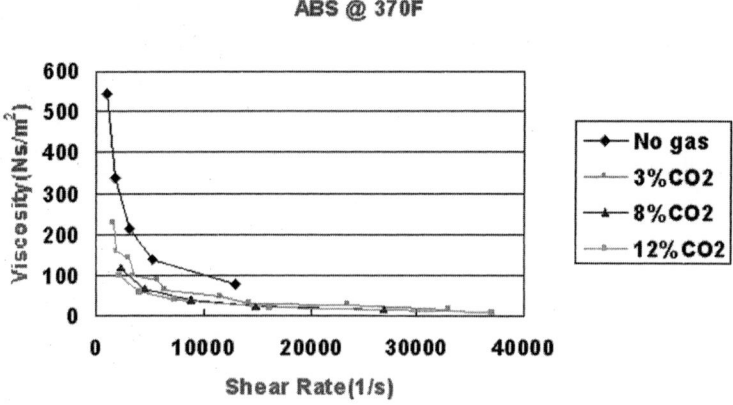

Fig. 2. Viscosity of ABS as a function of CO_2 concentration and shear rate at 370F (Courtesy of Trexel, Inc.). Note that the relative viscosity change is most pronounced at low shear rates.

Fig. 3. Number of available sites for cell nucleation as a function of the gas dissolved. It is conjectured that $(N_0)_{max}$ is greater as the activation energy increases.

Cell Growth

Immediately after the cells are nucleated, the pressure in the bubble is equal to the saturation pressure. Therefore, the cells expand if the polymer matrix is soft enough to undergo viscoelastic-plastic deformation. The cell expands until the final pressure inside the cell is equal to the pressure required to be in equilibrium with the surface forces and the stress in the viscoelastic cell wall. Unlike in conventional foaming, in the case of microcellular plastics, there are so many cells nucleated and the diffusion length is so short that the diffusion of the gas to the cell growth stops relatively quickly. In practice, the temperature of the surface of the extrudate changes as a result of heat transfer, and thus, the expansion of the cell is constrained by the outer stiff layer. Also some of the gases from the cells near the surface escape, reducing the tendency to expand.

Cell Density and Cell Size

The cell density is a function of both the pressure drop and the pressure drop rate. During the cell nucleation stage, there is a competition for gas between cell nucleation and cell growth if the cells do not nucleate simultaneously. When some cells nucleate before others, the gas in the solution will preferentially diffuse to the nucleated cells to lower the free energy of the system. As the gas diffuses to these cells, low gas concentration regions where nucleation cannot occur are generated adjacent to the stable nuclei. As the solution pressure drops further, the system will either both nucleate additional microcells and expand the existing cells by gas diffusion or only expand the existing cells. Therefore, when the pressure drop occurs rapidly, the gas-depleted region where nucleation cannot occur will be smaller and a more uniform cell distribution will result. It has been determined experimentally that a drop rate of 2 GPa/second is the minimum pressure drop rate required for microcellular plastics processing. Figure 4 and show examples of the cell nucleation density as a function of the gas pressure.

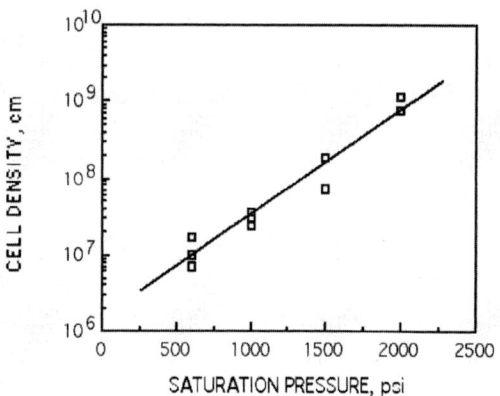

Fig. 4. Cell nucleation density as a function of N_2 pressure in polystyrene.[7]

Equipment and Die Design

The role of the extruder (or the plasticating unit of an injection-molding machine) is to melt the plastic, create a single phase polymer/gas solution, and pump the solution through a die or inject it into a mold. To achieve these functions, high-pressure CO_2 or N_2 gas is introduced into the extruder barrel by metering the exact amount of CO_2 or N_2 at pressures greater than 2,000 psi. The flow rate of CO_2 into the extruder can be controlled using a special metering pump. The gas forms a large bubble in the extruder since the flow of the gas is briefly interrupted whenever the screw flight wipes over the barrel. Then to diffuse the gas in the bubble quickly in the molten plastic, the polymer/gas interfacial area is increased and the striation thickness of polymers between the gas bubbles decreased. This is done by elongating the bubble in the barrel through the shear deformation of the two-phase mixture of the polymer and gas. The approximate residency time required for diffusion and solution formation in the extruder is estimated to be as follows: less than 100 seconds for PET and less than 10 seconds for polystyrene at typical operating temperatures.

To design the process, the FRs are selected as: FR1 = Reduce the amount of plastic used; FR2 = Increase the toughness of the plastic product; FR3 = Make three-dimensional geometrical shape. The DPs are chosen as: DP1 = Microcellular plastics (uniform cell distribution in large numbers); DP2 = Diameter of microcells; DP3 = Die shape. The process variables (PVs) for the process described that can satisfy the DPs given are: PV1 = Supersaturation of the plastic with a large amount of gas and sudden pressure change (dp/dt); PV2 = Temperature of the molten polymer to control the expansion of cells at the die; PV3 = Cross-sectional dimensions.

The design equation for the extrusion process may be written as

$$\begin{Bmatrix} DP1 \\ DP2 \\ DP3 \end{Bmatrix} = \begin{bmatrix} X & 0 & 0 \\ X & X & 0 \\ 0 & 0 & X \end{bmatrix} \begin{Bmatrix} PV1 \\ PV2 \\ PV3 \end{Bmatrix} \qquad (5)$$

Equation (5) shows that the process design is also a decoupled design. Therefore, each design satisfies the Independence Axiom.

DP1 and PV1 can be further decomposed as: DP11 = Large number of nucleated cells; DP12 = Uniform-size cells; PV11 = The level of supersaturation of CO_2; PV12 = Rapid pressure drop dp/dt. The design matrix for this design may be represented as

$$\begin{Bmatrix} DP11 \\ DP12 \end{Bmatrix} = \begin{bmatrix} X & x \\ X & X \end{bmatrix} \begin{Bmatrix} PV11 \\ PV12 \end{Bmatrix} \qquad (6)$$

Equation (6) states that DP11 and DP12 are coupled slightly in that if the pressure drop rate is really slow, we cannot get a large number of cells and uniform-sized cells. In most cases, the effect of dp/dt on the number of cells is negligible.

The role of the plasticating section of the injection molding machine or the extruder is to melt the plastic and dissolve the gas in the polymer. The extruder must be under high pressure to maintain a single-phase solution. The cell density is primarily controlled by the amount of the dissolved gas and also partly by the pressure drop rate.

The die must be designed to control the pressure drop rate, which controls the uniformity of cell size. The desired pressure drop rate is greater than 1 GP/s. It also removes the thermal energy from the molded part. The die also creates 3D shapes in the case of injection molding or the profile in the case of extrusion. It can be seen that the die design is as important as the extruder or the injection molding design.

The highest-level FRs and DPs are given as; FR1 = Control cell size; FR2 = Control the number of cells; FR3 = Control the geometry of the extrudate. DP1 = P_i*; DP2 = dp/dt; DP3 = Die shape & Accessories.

The design equation is given by

$$\begin{Bmatrix} \text{Cell size} \\ \text{Cell density} \\ \text{Geometry} \end{Bmatrix} = \begin{bmatrix} X & x & 0 \\ X & X & 0 \\ x & 0 & X \end{bmatrix} \begin{Bmatrix} P_i \\ dp/dt \\ Die\,\&\,Acc. \end{Bmatrix} \tag{7}$$

The corresponding PVs are chosen as: PV1 = Extruder RPM; PV2 = Die lip length; PV3 = Means of controlling the profile.

The design equation for the process is given by

$$\begin{Bmatrix} P_i \\ dp/dt \\ Die\,\&\,Acc. \end{Bmatrix} = \begin{bmatrix} X & x & 0 \\ X & X & 0 \\ 0 & 0 & X \end{bmatrix} \begin{Bmatrix} \Omega \\ L \\ Profile \end{Bmatrix} \tag{8}$$

Advantages of Injection Molding with Microcellular Plastics

The injection pressure of injection-molding process decreases due to the presence of dissolved gas, which lowers the viscosity. The cycle time is also reduced due to the elimination of the "hold and pack" time and also due to about 25% reduction in cooling time. Table 2 presents a comparison of the injection molding process with and without the dissolved gas.

Table 2. Comparison of injection-molding process for various products with and without microcellular structure (Courtesy of Mar Lee Companies).

Air Bag Canister		Conventional	MCP	%
(33% glass filled	**Part Weight**	365 g	252 g	30.9
Nylon)	*Cycle Time*	45 sec	35 sec	22.2
	Clamp.Tonnage	150 tons	15 tons	90
Connector	*Part Weight*	48.8 g	42.9 g	10.6
(polycarbonate)	*Cycle Time*	17.5 sec	15.9 sec	9.1
	Clamp. Tonnage	140 tons	20 tons	85.7
Battery cover	*Part Weight*	201 g	159 g	20.8
(poypropylene)	*Cycle Time*	60 sec	37 sec	38.3
	Clamp. Tonnage	200 tons	15 tons	92.5

Figure 5 shows an injection-molded printer chassis for inkjet printers made of glass filled engineering plastic (PPO/HIPS). The chassis made of microcellular plastics has 50 % less warpage, 25% reduction in cycle time, and 8% weight reduction. The microcellular plastic also had higher toughness -- 9.0 ft-lb vs 6.7 ft-lb by drop weight test, and 9.7 kJ/m^2 vs 7.3 kJ/m^2 by notched Izod impact test.

Fig. 5. Injection-molded Printer Chassis with Microcellular Plastics (Courtesy of Trexel, Inc.).

There are a large number of advantages of using the microcellular plastics: reduction of material consumption (between 5% and 30 %); faster cycle time; higher productivity; greater toughness in some plastics; low residual stress; dimensional accuracy; minimal residual stress, dimensional stability; reduction in warping of injection molded parts; appearance (no visible cells); thin sections; no sink marks; low temperature process; low pressure process; large number of cavities or smaller machines; most polymers; use of non-hydrocarbon solvents; no additives for nucleation; and no special equipment other than gas supply system.

Performance and Applications of Microcellular Plastics

Since the cell size is extremely small, the cells cannot be seen by the naked eye. Therefore, the foamed plastic resembles a solid plastic, having a good physical appearance. At conventional cell sizes, they are opaque without the need to introduce pigments such as titanium dioxide.

MCPs save money for manufacturers due to the use of less material and faster cycle time. Since about 70% of the cost of foamed plastic goods is the material cost and since up to 50% weight reduction is possible for some applications, the cost of plastic parts can be reduced by as much as 35%.

MCPs are environmentally acceptable since they are processed using carbon dioxide (CO_2) or nitrogen (N_2), instead of hydrocarbons or fluorinated materials. Since smaller amounts of plastics are used in a given product, there is less material to recycle or dispose. Furthermore, less raw material and energy are used to make the same plastic article.

MCPs find many applications in housing and construction, sporting goods, vehicles, electrical and electronic products, chemical and biochemical applications, and the textile and apparel industry. They can be used in siding, pipes, electrical wire, automotive seats and other parts, airplane parts, filters, shoe soles, office equipment housing, artificial paper, food containers, polishing cloth, thermal insulation around pipes, and other uses as well.

MCPs are processed at lower processing temperatures, since the glass-transition and melting temperatures and the viscosity of plastics decrease with the increase in dissolved gas. As the gas is formed in the bubble during the nucleation and cell growth phase, the viscosity and the melting temperature of plastics increases, reverting back to the original state. Therefore, MCPs "solidify" much more quickly – by almost a factor of two – and therefore, injection-molded MCP parts can be taken out of the mold quickly.

Conclusions

Because of their advantages in performance, cost, and processing microcellular plastics show great promise in changing the polymer processing industry. A great deal is known about MCPs and their processing techniques, but the field is still full of research and development opportunities to satisfy the diverse requirements of the polymer processing field.

[1] See for detailed historical account and a review, Suh, N. P. "Microcellular Plastics," *Innovation in Polymer Processing: Molding*, (Ed. by J. Stevenson), SPE Books of Hanser Publishers, New York, **1996**. (ISBN 3-446-17433-8)
[2] Under the license from Trexel, Inc. Trexel, as the sole licensee of M.I.T., has developed the MIT technology further for commercial applications. The trade name is MuCell.
[3] Suh, N. P., "*Axiomatic Design: Advances and Applications*", Oxford University Press, New York, 2001.
[4] Martini, J., Waldman, F. A., and Suh, N. P. "The Production and Analysis of Microcellular Thermoplastic Foam," *Society of Plastics Engineers Technical Papers*, Vol. 28, **1982**, pp. 674-676.
[5] Taylor, G. I. "The Formation of Emulsion in Definable Fields of Flow," *Proceedings of the Royal Society*, London, Vol. 146A, p. 501, **1934**.
[6] Park, C. B. "The Role of Polymer/Gas Solutions in Continuous Processing of Microcellular Polymers," Ph.D. Thesis in Mechanical Engineering, Massachusetts Institute of Technology, Cambridge, MA, May **1993**.
[7] Kumar, V., Ph.D. Thesis, Department of Mechanical Engineering, M.I.T., **1988**

The Processing of Starch as a Thermoplastic

R.F.T. Stepto

Polymer Science and Technology Group, Manchester Materials Science Centre, UMIST and University of Manchester, Grosvenor Street, Manchester, M1 7HS, UK.

Summary: The thermoplastics processing of native starch in the presence of water is a recent development with very wide possible applications. Eventually, oil-based polymer materials have to be replaced in many applications by sustainable, inexpensive, natural materials from renewable resources. As with conventional thermo-plastics, starch-water melts may be processed by injection moulding and extrusion. The present contribution focuses on injection moulding. The bases of the processing and the thermal and molecular changes occurring are described. In addition, the rheological behaviour of the starch-water melts during processing is analysed quantitatively to give apparent melt viscosities. The dimensional, thermal and mechanical properties of moulded thermoplastic starch polymer (TSP) materials and the products presently being produced from them are discussed.

Keywords: native starch, starch-water, thermoplastic starch polymer (TSP)

Introduction

One of the emerging major themes in polymer science for the 21st Century is the preparation of sustainable polymeric chemicals and materials from renewable resources, as distinct from petrochemicals. Eventually, oil-based polymer materials will be replaced in many applications by inexpensive, natural-based products. When designed properly, such products have a useful life and properties and are biodegradable with natural degradation products. Within this field, the thermoplastics processing of natural hydrophilic polymers, particularly starch, in the presence of water is a recent development with very wide possible applications.[1-5]

It has been found that by heating hydrophilic polymers in closed volumes in the presence of given amounts of water for given times, homogeneous melts may be formed. If such melts are produced in injection-moulding machines and extruders then they may be processed like thermoplastics. The processing of various starches and of gelatin and other hydrophilic

© 2003 WILEY-VCH Verlag GmbH & KGaA, Weinheim   DOI: 10.1002/masy.200351123

polymers and blends to useful thermoplastics materials has been achieved in this way.[1-5] Essential features are that limited amounts of water are used and a confined volume is maintained throughout the process if a solid rather than a foamed product is to be formed.

The present contribution is the latest in a series of papers describing the injection moulding of starch.[6-8] First, comparisons with the conventional processing of celluloses and starch are made and the thermal and molecular changes on heating starch-water mixtures are discussed. Second, the rheological behaviour of starch-water melts during the refill part of the injection-moulding cycle is analysed quantitatively to give apparent melt viscosities. Finally, the dimensional, thermal and mechanical properties of moulded thermoplastic starch polymer (TSP) materials and emerging TSP-based products are reviewed.

The Bases of Thermoplastic Starch Melt Formation

Comparison with the Processing of Cellulose and Cellulose Derivatives and Conventional Starch Processing. In polymeric terms, a main distinction between starch and cellulose is that the former contains highly branched molecules, whereas the latter contains linear molecules. The branching means that crystalline sequences are shorter in starch and fibres do not form. Accordingly, native starch is more readily destructured than native cellulose in the presence of water. Indeed, native cellulose cannot be processed as a thermoplastic and it has to be converted to derivatives, e.g., esters and ethers, to reduce the strength of intermolecular forces so that molecular flow can occur under the action of heat and shear. The thermoplastics processing of cellulose derivatives is well-established and well-understood and does not involve water as an integral part of the processes.

The destructuring of starch under the action of heat, water and shear is, of course, the basis of much food preparation and the processing of starch for food and adhesives dates back several millennia into human history. Such conventional processing of starch is in the presence of heat and *excess* water. Initially, a process occurs that is termed gelatinisation,[9,10] resulting in a breaking down of the structures in the starch granules to different extents, depending on the starch and the processing conditions. The structure of the starch granule is very complex and hierarchical, and it is partly crystalline. Importantly, the crystallinity and the supramolecular structure are based on the amylopectin component, and not on the amylose. The initial granules can be up to 100 μm in diameter. In excess water, the amylopectin crystallinity is lost, some hydrolytic degradation occurs, granules swell and eventually disappear, and the linear amylose molecules diffuse into solution. On aging, the starch solution or suspension undergoes so-called retrogradation to a swollen network material with a structure now based principally on associations between sections of amylose molecules.

From a structural polymer materials point of view, the preceding, conventional processing of starch uses too much plasticiser (water) and eventually lays emphasis on the wrong component, namely, the lower molar mass amylose, of native $M_n < 10^6$ g mol^{-1}. Native amylopectins, on the other hand, can have M_n and M_w in excess of 10^6 g mol^{-1} and 10^8 g mol^{-1}, respectively. Superior mechanical properties of amorphous materials will be obtained if the molecular, solid or network structure is formed at lower water contents and is based on the branched component of higher molar mass. In this respect, important break-throughs occurred in the 1980s,[1,2,11] culminating in the thermoplastics processing of starch at approximately its natural water content (≈ 15 %), in a closed volume at temperatures above 100 °C. Using conventional injection moulding, glassy, amorphous, thermoplastic starch polymers (TSPs) were obtained. An important characteristic of thermoplastic starch formation is the thermal and mechanical (shear) destructuring of the starch granules to form a homogeneous melt, unaccompanied by swelling.

Compatibility with Water. Starch is a hydrophilic polymer, that is, for present purposes, a polymer whose uptake of water in equilibrium with pure water is unlimited. Hydrophilic polymers are characterised by water-vapour adsorption isotherms of sigmoidal shape, indicating the presence of bound and unbound water, tending to an infinite amount of water adsorbed in the presence of pure water.[2] Such behaviour ensures no phase separation will occur during processing. Both gelatin and starch show the required form of adsorption isotherm and both can be successfully injection moulded in the presence of water.[1,2]

Thermal Changes during Processing. The thermal changes occurring on heating starch-water mixtures can be followed using differential scanning calorimetry (DSC) employing completely filled pans with seals designed to withstand the pressure generated by the sample (up to 30 bar). Fig. 1 shows examples of the endothermic changes in a potato starch at two water contents. The endotherm for the higher water content occurs at les than 100 °C and is characteristic of the gelatinisation of conventional starch processing, in which the starch granules become swollen and destructured and lose amylose by diffusion. The endotherm for the lower water content is characteristic of melt formation, namely, a thermal and aqueous destructuring of the amylopectin crystallites and molecular order in the granule without the mass diffusion of water.[1-3,12,13] At 12% water, there are only about 1.2 molecules of water per anhydroglucose unit. A similar variation in the temperature of the destructurisation endotherm with water content also occurs in gelatin-water mixtures.[2]

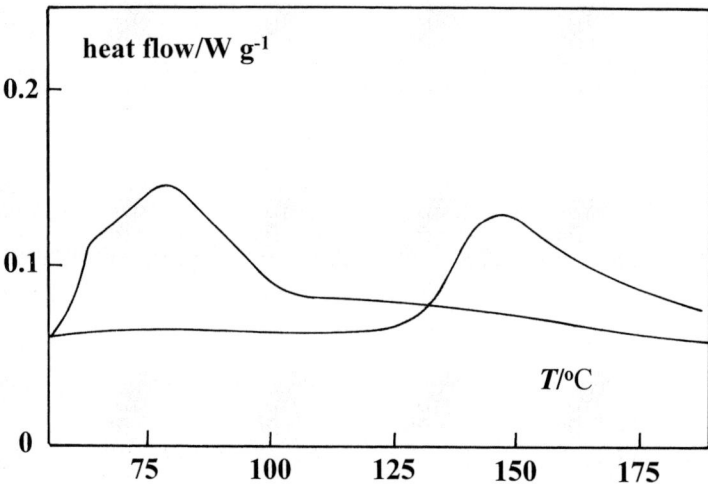

Fig. 1. Examples of DSC endotherms for a potato starch at 42% and 12% water content $(= 100 W_{H2O} / (W_{H2O} + W_{starch}))$.

The temperature range and size of the melt-formation endotherm for starch and water mixtures depend on the type of starch and also on the particular batch of starch. For example, different trace amounts of metallic ions in potato starches can affect the temperature range and, hence, the processing conditions.[14] In general, the temperature range of the endotherm has to be exceeded before destructuring is complete and a homogeneous melt can be achieved.[1]

Molecular Changes on Heating Starch-Water Mixtures. Figs. 2 and 3 illustrate the molecular changes that occur on heating potato starch-water mixtures in closed glass ampoules for various lengths of time at 140-160 °C. Generally, due to hydrolysis, molar mass reduces as the length of time of heating increases. Fig. 2 shows the logarithmic (Mark-Houwink) plot of intrinsic viscosity ($[\eta]$) versus mass-average molar mass (M_w), as determined by Rayleigh light scattering. The low Mark-Houwink exponent of 0.39 is due to the hydrodynamic dominance of the highly branched amylopectin species. The intrinsic viscosity of the native starch was found to be about 280 cm^3 g^{-1} consistent with $M_w >$ 10^8 g mol^{-1}.

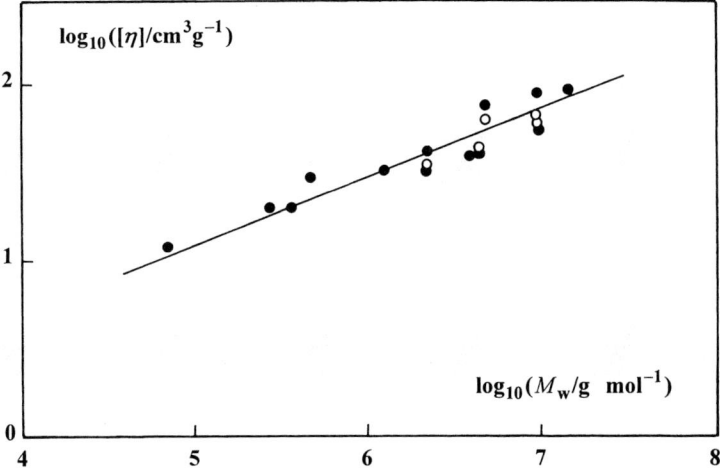

Fig. 2. $\log_{10}[\eta]$ versus $\log_{10}M_w$ (Mark-Houwink plot) for hydrolysed potato starches.

● $[\eta]$ measured in dimethyl sulfoxide at 40 °C; ○ $[\eta]$ measured in dimethyl sulfoxide / 6M urea at 40 °C; M_w measured in dimethyl sulfoxide. Least-squares line gives $[\eta]/cm^3\ g^{-1} = 0.23(M_w/g\ mol^{-1})^{0.39}$

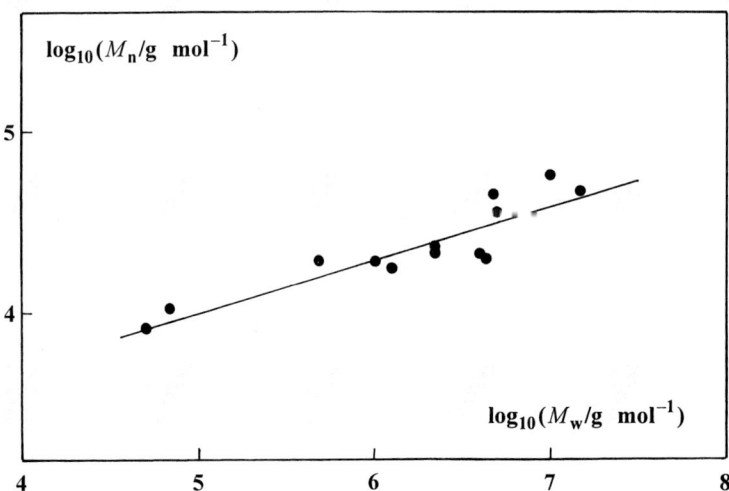

Fig. 3. $\log_{10}M_n$ versus $\log_{10}M_w$ for for hydrolysed potato starches. Least-squares line gives $M_n/g\ mol^{-1} = 405(M_w/g\ mol^{-1})^{0.28}$.

Fig. 3 gives the logarithmic plot of number-average molar mass (M_n) versus M_w. The values of M_n were determined from assays of the numbers of reducing end-groups per unit mass of sample. There is only one reducing end-group per molecule, hence, the number of reducing end-groups present in a sample is equal to the number of molecules present and M_n/g mol^{-1} = (dry mass of sample/ moles of end-groups). It can be seen that $M_n \propto M_w^{0.28}$, showing that M_w/M_n decreases rapidly as more hydrolysis occurs. At the highest molar mass shown in Fig. 3, $M_w \approx 1.5 \times 10^7$ g mol^{-1}, $M_w/ M_n \approx 300$, whilst for the lowest molar mass shown, $M_w \approx 5 \times 10^4$ g mol^{-1}, $M_w/ M_n \approx 6.2$. This large decrease in the polydispersity ratio is consistent with random chain scission through hydrolysis, with the larger molecules being the more likely to be attacked, causing M_w to be reduced more quickly than M_n. To obtain injection-moulded TSPs with satisfactory mechanical properties it is found that M_w should lie in the approximate range 10^6-10^7 g mol^{-1}, with Fig. 3 indicating that M_w/ M_n then lies in the approximate range 100-300.

Injection-Moulding Behaviour and Apparent Melt Viscosity

By carrying out injection moulding with a given mould (shot volume), given screw and given temperature profile, it is possible to measure the variation of refill times for material to feed in front of the screw at different screw-rotation speeds and under different applied back-pressures. The shear rate ($\dot{\gamma}$) is determined by the rotational speed of the screw and the back-pressure defines the reverse pressure drop along the metering zone of the screw.[15,16] As back-pressure is increased for a given screw speed, refill time increases as the backward (viscous) flow rate increases, detracting more from the forward, drag flow due to screw rotation, to give a lower net flow rate. In addition, assuming Newtonian behaviour, the backward flow rate is inversely proportional to the viscosity of the melt (η). Hence, η can be determined from the change of refill time (net flow rate) with back-pressure at a given $\dot{\gamma}$ (screw speed).

Fig. 4 shows apparent values of η for starch-water melts, determined from back-pressure experiments, plotted versus $\dot{\gamma}$ and compared with values of η for polyethylene melts, taken from a manufacturer's literature. From the similarity of the values of η obtained for a given shear rate, it can be seen that, once the other processing parameters, such as water content, temperature profile and screw characteristics, are properly defined, starch processes like polyethylene. The similarity of the viscosity-shear rate behaviour of starch-water melts to that of polyethylene shown in Fig. 4 is corroborated by the measurements of viscosities of pre-processed corn starch-water melts by Willet, Jasberg and Swanson.[5]

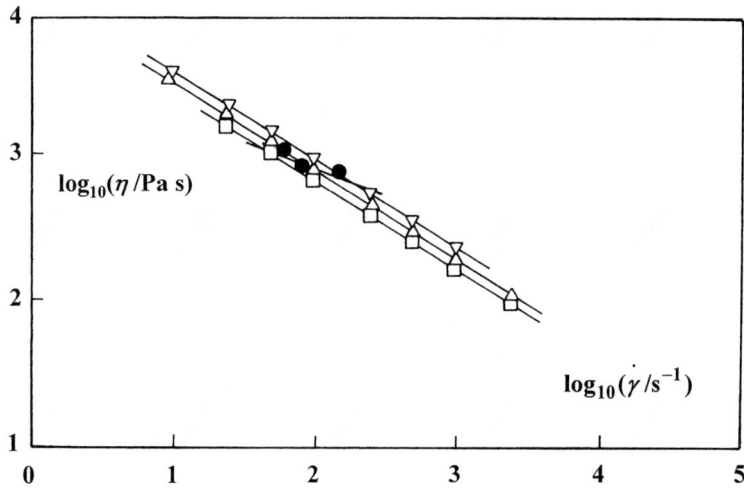

Fig. 4. Comparison of melt viscosities of a medium density polyethylene and a starch-water mixture. Polyethylene: □ 230°C, △ 210°C, ▽ 190°C. Starch-water, 17% water, • 175°C (metering zone); apparent melt viscosities from back-pressure experiments using a standard Arburg injection moulding machine.

Properties of Injection-Moulded TSPs

Dimensional Stability. Obviously, the dimensions of moulded objects from hydrophilic polymers depend on their water contents. If precise dimensions are required, processing should be carried out so that products are formed at approximately the equilibrium in-use water content. For potato starch, for example, this means water contents of around 14% for use under ambient conditions (50% relative humidity, 20-25 °C). If higher water contents are used in processing, distortion and shrinkage will occur as the equilibrium water content is naturally achieved after processing. In addition, higher water contents can induce more hydrolytic degradation of the starch chains during processing and also gelatinisation rather than melt formation. If lower water contents are used, thermal degradation can occur during processing, as well as swelling after processing.

Thermal Properties. Fig. 5 shows the glass-transition temperatures of injection-moulded starch-water and gelation-water mixtures. Similar values of T_g (60 °C to 80 °C) are observed for the two materials under normal ambient conditions, when values of w_{H_2O} are in the range 0.12 to 0.14. The materials are then in their glassy states.

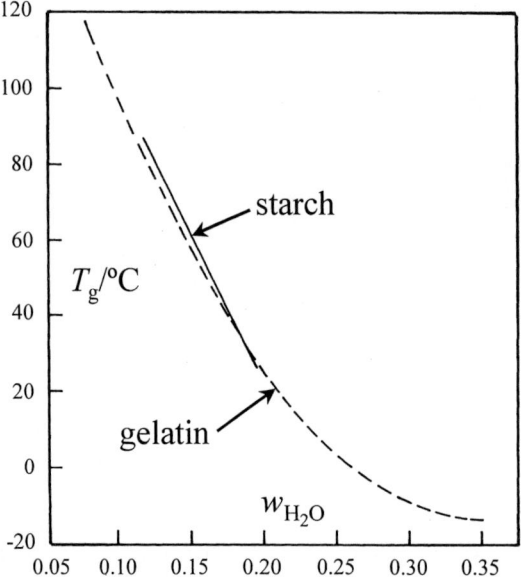

Fig. 5. Glass-transition temperature (T_g) versus weight fraction of water (w_{H_2O}) for injection-moulded gelatin-water and starch-water mixtures.

Stress-Strain Behaviour. Fig. 6 illustrates the stress-strain behaviour at ambient temperature of tensile test-pieces moulded from potato starch at 17% water and conditioned to the water contents shown. Accordingly, the glass-transition temperatures (see Fig. 5) vary from about 60°C to 100°C and the behaviour shown in Fig. 6 is typical of that of glassy thermoplastics. The initial moduli are about 1.5 GPa, similar in value to those of glassy polyolefins, polypropylene and high-density polyethylene, and the materials show yield points at between 5 to 10% extension. The changes in properties with decrease in water content are consistent with the loss of free water, which has a plasticising action on the materials.

It is important that further systematic and fundamental studies on TSP processing and TSP materials are pursued in the future. In comparison with that of synthetic polymers, knowledge of the processing-structure-property relationships for TSP-based materials is still in its infancy. However, it is an exciting field with many possibilities for discoveries and developments.

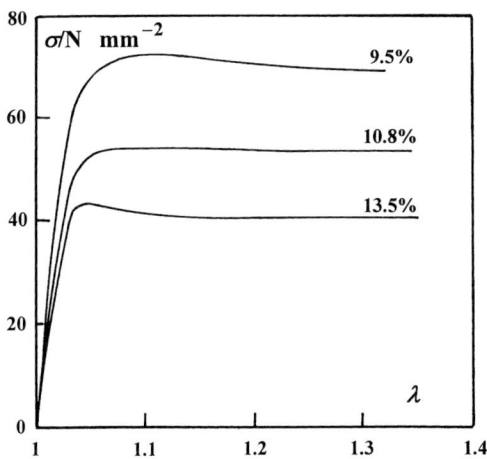

Fig. 6. Tensile stress (σ) versus deformation ratio (λ) of injection-moulded potato starch at ambient temperature and the water contents specified with the curves.

TSP-Based Products

The first commercial product made of injection-moulded TSP was the drug-delivery capsule, Capill[1,12,17,18] and further products are gradually appearing, e.g., golf tees, cutlery, plates, food containers. In addition, extrusion has been applied to produce rigid foams, suitable for loose-fill packaging. Generally, the polymers are dimensionally and mechanically stable under ambient, indoor conditions and are completely biodegradable and compostable. They break-down in water. Hence, TSPs can be considered as a new class of inexpensive, green polymers that can be returned to the natural cycle, with no pollution, after use.

Further developments include the thermoplastic processing of blends of starch with hydrophilic synthetic polymers,[4,19] e.g., poly(ε-caprolactone) and poly(ethylene-vinyl alcohol), to give the possibility of flexible films and materials with improved mechanical properties, lower water sensitivity, but also lower biodegradability. Table 1 summarises the TSP-based materials and types of products currently available.

Table 1. Examples of TSP-based materials and products.

PRODUCT OR MATERIAL NAME	PRODUCT TYPE	MANUFACTURER
CAPILL	pharmaceutical capsules	Capsugel/Warner Lambert; West Pharmaceutical Services
ECO-FOAM	foam packaging pellets	NationalStarch
Mater-Bi (Starch and Starch Blends with Hydrophilic Polymers)	moulding & extrusion powders for flexible films, rigid packaging, moulded articles, etc.	Novamont, Italy

Acknowledgements

My sincere thanks to the research group of Capsugel AG at Riehen, Switzerland for making this research possible and to UMIST for an extended leave of absence during the period of the work.

[1] L. Eith, R.F.T. Stepto, I. Tomka, F. Wittwer, *Drug Dev. & Ind. Pharm.* **12** 2113 (1986)

[2] R.F.T. Stepto, I. Tomka, *Chimia* **41** 76 (1987)

[3] I. Tomka, in *Water Relationships in Food*, eds. H. Levine, L. Slade, Plenum Press, New York, 1991, p.627

[4] G. Lay, J. Rehm, R.F.T. Stepto, M. Thoma, J-P. Sachetto, D.J. Lentz, J. Silbiger, US Patent 5,095,054; 1992

[5] J.L. Willet, B.K. Jasberg, C.L. Swanson, in *ACS Symposium Series 575, Polymers from Agricultural Coproducts*, eds. M.L. Fishman, R.B. Friedman and S.J. Haag, Amer.Chem.Soc., Washington D.C., 1994, Chapter 3.

[6] R.F.T. Stepto, *Polymer International*, **43**, 155 (1997)

[7] R.F.T. Stepto, *Macromol. Symp.*, **252**, 73 (2000)

[8] S.B. Ross-Murphy and R.F.T. Stepto, in *Emerging Themes in Polymer Science*, ed. A.J. Ryan, Special Publication No. 263, The Royal Society of Chemistry, Cambridge, 2001, Chapter 13

[9] *Starch Chemistry and Technology*, eds. R.L. Whistler, J.N. BeMiller, E.F. Paschall, Academic Press, New York, 1984

[10] R. Lapasin, S. Pricl, *Rheology of Industrial Polysaccharides: Theory and Applications*, Blackie Academic and Professional, Glasgow, 1995

[11] R.F.T. Stepto, B. Dobler, UK Patent 88 01562; 1988

[12] L Eith, R.F.T. Stepto, I. Tomka, F. Wittwer, *Proc. Interphex '86 Conference, Cahners Exhibitions Ltd., Brighton*, 1986, p.2-22

[13] J.W. Donovan, *Biopolymers*, **18** 263 (1979)

[14] J.-P. Sachetto, R.F.T. Stepto, H. Zeller, UK Patent 87 15941; 1987

[15] *Extrusion and Other Plastics Operations*, ed. N.M. Bikales, Wiley-Interscience, New York, 1971

[16] J.-F. Agassant, P. Avenas, J.-Ph. Sergent, P.J. Carreau, *Polymer Processing*, Hanser Publishers, Munich, 1991

[17] H. Augart, A. Borgmann, R.F.T. Stepto, *Proc. 6th Pharmaceutical Technolog Conference*, Canterbury, 1987, p.257

[18] V.D. Vilivalam, L. Illum, K. Iqbal, *Pharmaceutical Sci. & Tech. Today*, **3**, 64 (2000)

[19] *Mater-Bi Brochure*, Novamont S.p.A., Novara, 1999

Macromol. Symp. **2003**, *201*, 213—221

Lignin: Its Functions and Successive Flow

Masamitsu Funaoka

School of Bioresources, Mie University, 1515 Kamihama, Tsu, Mie 514-8507, Japan Email: funaoka@bio.mie-u.ac.jp

Summary: Phenolic lignin-based polymers with structure-variable function have originally been designed, and a process has been developed for synthesizing them directly from native lignins. The key point of the process is to set up two different solvents, which are immiscible each other, for selective modification and separation of lignin and carbohydrates: hydrophobic solvent for hydrophobic lignin and hydrophilic solvent for hydrophilic carbohydrates. The native lignins, three dimensional network polymers, are subjected to selective modifications at the interface between both phases to give linear type polymers (lignophenols) composed mainly of 1,1-bis(aryl)propane type units, almost quantitatively. The process provides a new system for successively utilizing lignocellulosics in the molecular level.

Keywords: lignin, lignocellulosics, lignophenols, phase-separation, recycling

Introduction

Lignin is an amorphous, aromatic network polymer, second to cellulose in natural abundance. The biosphere is estimated to contain 3×10^{11} tons of lignin with an annual biosynthesis rate of about 2×10^{10} tons.[1] However, in contrast to the importance and potential of lignin in nature, lignin-based products have scarcely been in human life. This strange situation is due to the fact that lignin molecules lack stereoregularity, and repeating units in its molecule are too heterogeneous and complex. In addition, non-selective modifications during isolation from the cell wall make lignin molecules much more heterogeneous.

Lignin-based functional polymers (lignophenols) have originally been designed and their synthesis process from native lignins has been developed.[2-5] This process includes the phase-separation reaction system composed of phenol derivatives and concentrated acid. In the present paper are described the design of lignin-based polymers (lignophenols) with the structure variable function, the selective conversion process for lignocellulosics and the features of the resulting lignophenols.

 DOI: 10.1002/masy.200351124

218

devices) within the molecules. The switching devices are divided into two types: one is the reactive device with a reactive point on the nucleus, and the other is the stable device without any reactive point. The phenols linked to lignin side chains through *p*-position to phenolic hydroxyl group do not work as switching devices, due to the steric factor and can be used as control devices for controlling the frequency of switching devices within the molecule. There is a good correlation between the controlled molecular weights of lignophenols and the frequency of switching devices within the molecule (Figure 5). Using reactive- and stable devices, network type- and linear type polymers can be prepared, respectively. By mixing both devices, the polymer network structures from lignophenols can be controlled (Figure 6). The resulting lignophenol polymers can be readily released to subunits using the switching function, producing new raw materials for the next industrial system.

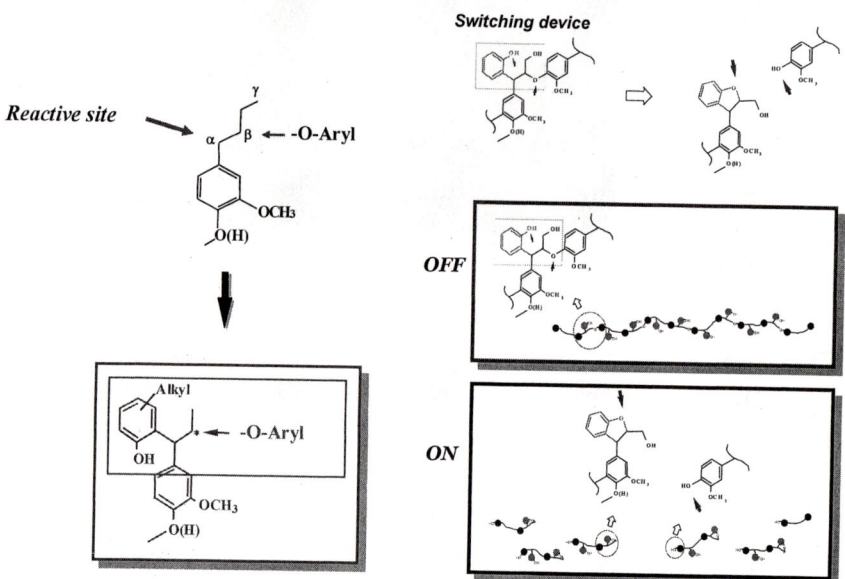

Fig. 4. Design of intramolecular switching units and the second structural control of lignophenols.

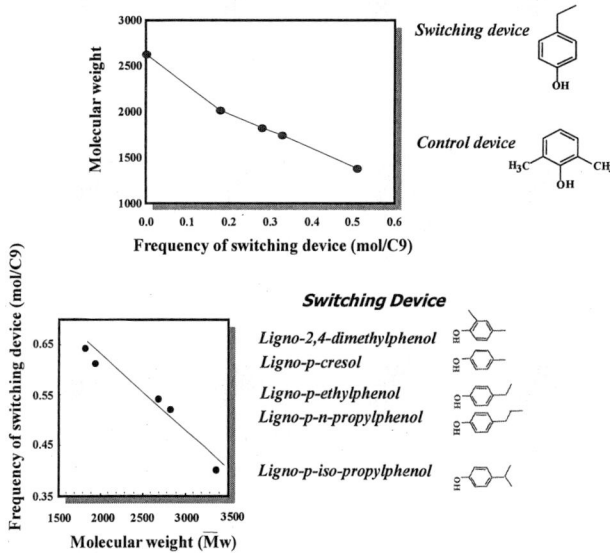

Fig. 5. Relationship between the frequency of switching units and controlled molecular weights.

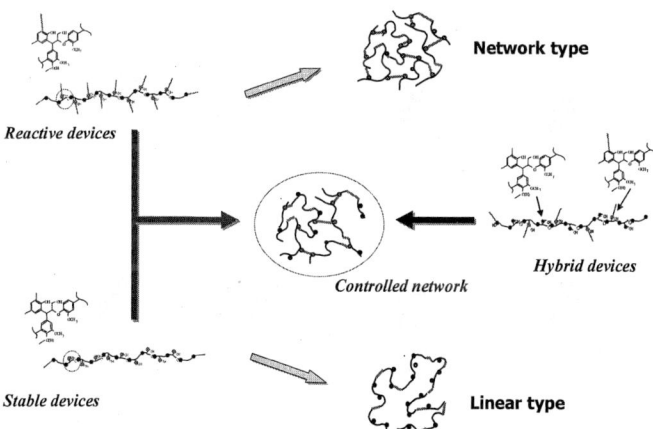

Fig. 6. Second structural control of lignophenols using the intramolecular switching devices.

New Application Fields of Lignin

Lignin is a natural polymer with very complicated network structures. The complexity is due to the random linking between building units, phenylpropanes. The side chain carbon (C1) combined with aromatic unit has a reactive substituent (hydroxyl group, carbonyl group or ether linkage). Therefore, the selective grafting of monomeric phenol derivatives to C1-positions of propane units leads to the formation of new phenylpropane units between grafted phenolic units and lignin propane units. This results in a dramatic change of the original lignin functions, forming a new type of lignin-based polymers composed mainly of 1,1-bis(aryl)propane type units.

New application fields of lignophenols are shown below.

- Recyclable composites with cellulose, biopolyesters, and inorganic materials (glasses, metals)[12,14-17]
- Raw materials for recyclable polymers[7]
- Detachable adhesives[8,13,15-17]
- Switching devices for material recycling[8,13,15-17]
- Electromagnetic shielding materials[18-20]
- Carbon molecular sieving membranes[21-24]
- Enzyme supports for bioreactors and affinity chromatography[25]
- Adsorbents for proteins and metals[25]
- Performance control agents for lead-acid battery and enzymes[25,26]
- Photoresists[27]
- Antioxidants[28]
- UV barriers[14]

Although lignophenols can be applied in various fields as a new type of phenolic polymers with structure-variable function, their application fields would be greatly extended through its secondary functionality control, leading to a new type of lignocellulosic industry.

[1] H. Sandermann, D. Scheel, T. Van der Trenck, *J. Appl. polym. Sci.*, **1983**, *37*, 407.

[2] M. Funaoka, I. Abe, *Tappi .J.* **1989**, *72*, 145.

[3] M. Funaoka, M. Matsubara, N. Seki, S. *Fukatsu, Biotechnol. Bioeng*, **1995**, *46*, 545 .

[4] M. Funaoka, S. Fukatsu, *Holzforschung*, **1996**, *50*, 245.

[5] M. Funaoka, *PolymerInternational*, **1998**, *47*, 277.

[6] M. Funaoka, H. Ioka, T. Hosho, Y. Tanaka, *J. Network Polymer, J*, **1996**, *17*, 121.

[7] M. Funaoka, *Polymer Processing*, **1999**, *48*, 66.

[8] Y. Nagamatsu, M.Funaoka, *Trans. Materials Res. Soc.J*, **2001**, *26*, 821.

[9] Y. Nagamatsu, M. Funaoka, *Sen'I Gakkaishi*, **2001**, *57*, 54.

[10] Y. Nagamatsu, M. Funaoka, *Sen'I Gakkaishi*, **2001**, *57*, 75.

[11] Y. Nagamatsu, M. Funaoka, *Sen'I Gakkaishi*, **2001**, *57*, 82.

[12] Y. Nagamatsu, M. Funaoka, *J. Adhesion Soc. J*, **2001**, *37*, 479.

[13] Y. Nagamatsu, M. Funaoka, *J. Advan. Sci.*, **2002**, *13*, 402.

[14] E. Ohmae, M. Funaoka, S.Fujita : Functions of Biopolyester, *Trans.Material. Res.Soc.J*, **2001**,*26*, 829.

[15] Y. Nagamatsu, M. Funaoka, *Material Sci. Res.International*, **2002**, *9, 108.*

[16] M. Funaoka, M. Maeda, M. Matsubara, *Trans.Material. Res.Soc.J*, **1996**, *20*, 167.

[17] M. Uehara, Y.Nagamatsu, M.Funaoka, *Trans.Material. Res.Soc.J*, **2001**, *26*, 825.

[18] X.-S.Wang, T. Suzuki, M. Funaoka, *Material Sci. Res. International*, **2003**, in press.

[19] Xiao-S. Wang, T. Suzuki, M. Funaoka, Y. Mitsuoka, T.Yamada, S.Hosoya, *Material Sci. Res. International*, **2002**, *8, 249.*

[20] Xiao-S.Wang, N.Okazaki, T.Suzuki, M.Funaoka, *Chemistry Letters*, **2003**, *32*, 42.

[21] H. Kita, K. Nanbu, M. Yoshino, K. Okamoto, M. Funaok*a, Prep. Polym. Materials: Sci. and Eng. (ACS)*, **2001**, *85*, 296.

[22] H. Kita, M. Hamano, M. Yoshino, K. Okamoto. M. Funaoka, *Trans.Materials Res. Soc.J.*, **2002**, *27*, 423.

[23] H. Kita, K. Nanbu, T. Hamano, M. Yoshino, K. Okamoto, M. Funaoka, J. Polymers. Environment, **2002**, *10, 69.*

[24] H. Kita, M. Hamano, M. Yoshino, K. Okamoto, M. Funaoka, *Prep. Polym. Materials: Sci. and Eng.* (ACS), **2002**, *86*, 376.

[25] M. Funaoka, H. Ioka, N. Seki, *Trans. Material Res. Soc. J*, **1996**, *20*, 163.

[26] M. Funaoka, H. Tamura, CREST symposium (Mie Univ.), **2001**.

[27] J. Kadota, K. Hasegawa, M. Funaoka, T. Uchida, K. Kitashima, *Network polymer*, **2002**, *23*(3), 15.

[28] S. Fujita, M. Funaoka, *J.Jap. Assoc.Dietary Fiber Res.*, **2003**, in press.

Macromol. Symp. **2003,** *201,* 223—236

The New Science of Protein Mimetics

Murray Goodman, * *Weibo Cai, Garth A. Kinberger*

Department of Chemistry and Biochemistry, University of California - San Diego, La Jolla, California 92093, USA
Email: mgoodman@ucsd.edu

Summary: New chemistries have been developed for *de novo* protein design. Protein mimetics of different structural and functional properties such as synthetic peptide ligases and Dn symmetrical helical bundles have been reported. The Template-Assembled Synthetic Protein (TASP) method (as well as the "Molecular Kit" approach) has also been utilized to prepare protein-like molecules. Here we report the synthesis of single chain, scaffold (TRIS)- and dendrimer-assembled collagen mimetics composed of the Gly-Nleu-Pro sequence where Nleu denotes N-isobutyl glycine. From the CD spectra and the thermal denaturation studies it can be seen that the collagen mimetics prepared form stable triple helices except the single chain structure. Furthermore, the 162-residue collagen mimetic dendrimer exhibits enhanced triple helical stability compared to the equivalent scaffold-terminated structure by an increase in the melting temperature in both H_2O and 2:1 ethylene glycol/H_2O (4 °C and 12 °C respectively). The concentration dependence for the melting transition of the collagen mimetic dendrimer was measured from which it was determined that the stabilization effect arises from the intramolecular clustering of the triple helical arrays about the core structure. This ensemble excludes solvent from the interior portion of the array which stabilizes the triple helix cluster.

Keywords: collagen mimetics, circular dichroism, dendrimers, peptides; triple helix

Introduction

Proteins fold into specific three-dimensional structures to carry out their biological functions. Although there are vast numbers of possible conformations for each protein, the native state is achieved rapidly. Much research has been undertaken in *de novo* protein design to prepare protein mimetics and to understand the protein folding problem. In the past decade, many designed protein mimetics with specific functions have been reported.[1-3] Ghadiri and co-workers have synthesized a 33-residue synthetic peptide ligase.[1] Based on coiled-coil structural motif, the two short negatively charged electrophilic and nucleophilic peptide

© 2003 WILEY-VCH Verlag GmbH & KGaA, Weinheim   DOI: 10.1002/masy.200351125

substrates (17 and 16 residues respectively) are pre-organized on a longer complementary, positively charged catalyst, forming the ternary complex which facilitates the ligation reaction between a C-terminal thioester and an N-terminal cysteine (Figure 1).

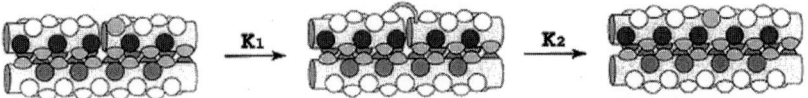

Fig. 1. Ghadiri's synthetic peptide ligase catalyzes the ligation of two short peptides.[1]

Anti-parallel helical bundles are found in a wide variety of proteins. DeGrado and co-workers synthesized Dn symmetrical helical bundles[2] where n equals 2, 3 and 6 (4, 6 and 12 helix bundles respectively), and the coiled helix backbone models generated of the synthesized compound superimpose with the experimental protein crystal structures (Figure 2).

Synthetic Helical Bundles   **Full Protein Crystal**

Structures

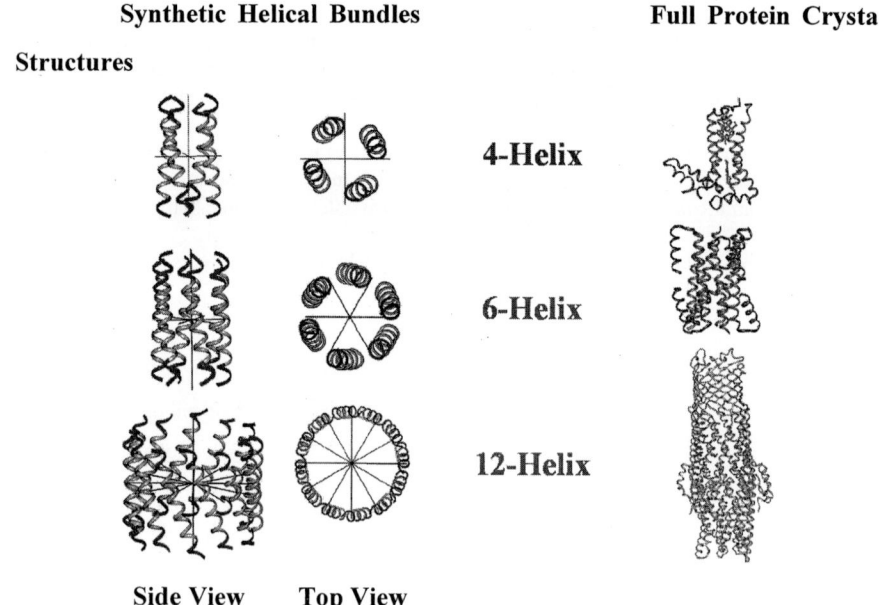

Fig. 2. DeGrado's Dn symmetrical anti-parallel synthetic helical bundles (the helices are represented as ribbons). Only the central helical portion of the proteins were synthesized.[2]

Mutter and co-workers extended his concept of the Template-Assembled Synthetic Protein (TASP) to a "Molecular Kit" approach (Figure 3).[3] Individual secondary structural elements such as α-helices, β-sheets, turns, and loops are covalently attached via both chain ends to appropriately functionalized templates. A series of protein-like molecules are formed exhibiting interesting structural and functional properties such as locked-in-4-helix bundle (α_4), β-sheet bundle (β_4), or more complex arrangements ($\beta_2\alpha$ and $\beta_3\alpha_2$).

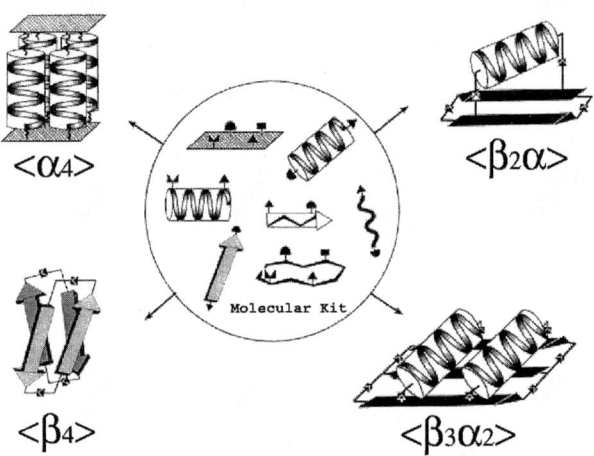

Fig. 3. Mutter's "Molecular Kit" and his synthetic protein mimetics.[3]

Collagen

Collagen is the most abundant extracellular protein in vertebrates. It has a unique triple helical motif in which three polypeptide chains form left-handed helices and are supercoiled into a right-handed triple helix. The primary sequence of triple helical collagen is composed of Gly-Xaa-Yaa trimer repeats. The imino acid proline (Pro) is usually found in the Xaa and Yaa positions whereas 4-hydroxyproline (Hyp) is normally located in the Yaa position.[4] There are two structural models for collagen. One is the Rich & Crick model where three polypeptide chains fold into a 10/3 helix with an axial 28.6 Å repeat.[5] The other model was proposed by Okuyama in which three strands form a 7/2 helix with an axial repeat of 20.0 Å.[6]

Collagen is an important biomaterial because of its low immunogenicity and high tensile strength. However, it is difficult to purify natural collagen without degrading its structural

© 2003 WILEY-VCH Verlag GmbH & KGaA, Weinheim

integrity. Furthermore, natural collagen is subjected to enzymatic degradation under biological conditions. Synthetic collagen structures including unnatural amino acid residues offer an alternative to natural collagens and have been shown to exhibit better enzymatic stability.[7]

Triple-helical conformations of collagen can be detected by a variety of experimental techniques such as X-ray diffraction,[5,6] electron microscopy, circular dichroism (CD) spectroscopy,[8] NMR spectroscopy and molecular modeling.[9] The most frequently used technique to study triple helical conformations is CD spectroscopy. The natural collagen triple helix has a characteristic CD spectrum with a small positive peak around 220 nm, a crossover at about 213 nm and a trough near 197 nm.[8]

Early studies of the preparation of synthetic collagens were conducted using polymerization methods. However, this approach lacks sequence and molecular weight control. Solid phase peptide synthesis (SPPS) has facilitated the synthesis of peptides with specific lengths and sequences. Sakakibara and co-workers have successfully synthesized (Pro-Pro-Gly)$_n$ (n = 10, 15, 20) and (Pro-Hyp-Gly)$_n$ (n = 5, 10) using the SPPS method.[10-11] (Pro-Pro-Gly)$_{20}$ and (Pro-Hyp-Gly)$_{10}$ formed stable triple helices in water. Many other peptides with different trimer sequences have also been prepared using the SPPS method.

Pioneering research by Prockop and co-workers revealed that Hyp greatly increases the thermal stability of collagen.[12] Later they also found that only 4-(R)-Hyp stabilized the triple helix while 4-(S)-Hyp destabilized triple helix.[13]

Raines and co-workers synthesized collagen mimetic peptides incorporating 4(R)-fluoro-L-proline (Flp) and found that (Pro-Flp-Gly)$_{10}$ forms a highly stabilized triple helix which melts at 91 °C.[14] They concluded that Flp (similar to Hyp) in the Yaa position stabilizes collagen-like triple helical structures by a combination of two effects: stereo-electronic (the gauche-effect) which fixes the pyrrolidine ring pucker and an increase of the trans/cis ratio of the peptide bond. Through these effects Flp as well as Hyp in the Yaa position preorganizes all three main chain torsion angles: ω, φ and ψ which stabilizes the collagen triple helix.

Scaffold-Assembled Collagen Mimetics

The folding of peptides into their secondary or tertiary structures is the essential requirement to induce the proper biological response or activity. To facilitate protein folding, Mutter and co-workers introduced the TASP approach for the design of a four α-helix bundle.[3]

The concept of scaffold-assembled structures has been applied to the design of collagen-like triple helices. The terms template and scaffold are often used interchangeably. We prefer to define the structures which hold the three peptide chains together as scaffolds. Fields and co-workers have used a lysine-lysine dimer as a scaffold to assemble peptide chains into collagen mimetics.[15] Later they also employed aliphatic fatty acid chains as a scaffold to self-assemble peptide chains into collagen mimetics (Figure 4).[16]

Lys-Lys Scaffold

Fatty Acid Termini

IV-H1 ══ Gly-Val-Lys-Gly-Asp-Lys-Gly-Asn-Pro-Gly-Trp-Pro-Gly-Ala-Pro

Fatty Acid Scaffold

Fig. 4. The scaffold (Lys-Lys dimer) assembled collagen mimetics[15] and self-assembling peptide amphiphiles[16] used by Fields and co-workers to study triple helicity.

Tanaka and co-workers also used the lysine-lysine dimer derived scaffolds to assemble collagen-like peptide chains. They extended this approach by synthesizing collagen-like peptides with both ends tethered by scaffolds and found enhanced thermal stability of the triple helix (Figure 5).[17]

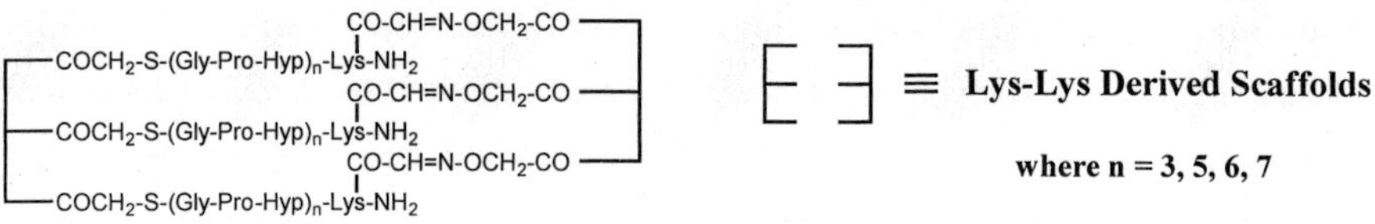

228

CO-CH=N-OCH₂-CO ─┐
─COCH₂-S-(Gly-Pro-Hyp)ₙ-Lys-NH₂
CO-CH=N-OCH₂-CO ─┤
─COCH₂-S-(Gly-Pro-Hyp)ₙ-Lys-NH₂
CO-CH=N-OCH₂-CO ─┘
─COCH₂-S-(Gly-Pro-Hyp)ₙ-Lys-NH₂

⊟ ⊟ ≡ **Lys-Lys Derived Scaffolds**

where n = 3, 5, 6, 7

Fig. 5. Tanaka's di-scaffold-assembled collagen mimetics.[17]

Moroder and co-workers synthesized heterotrimeric bioactive collagen mimetic peptides using a "Cystine knot" strategy (Figure 6).[18] They incorporated a collagenase cleavage site of collagen type I of the two α1-chains and one α2-chain which was stabilized by the N-terminal extension of (Pro-Hyp-Gly)₅ repeats. The three chains were tethered by a "Cystine knot". A tryptophan and a danzyl group were incorporated into the peptide for fluorescence resonance energy transfer. They discovered that cleavage of the peptide by collagenase led to a 6-fold increase in the Trp fluorescence intensity, thus making it a useful fluorogenic substrate for interstitial collagenases.

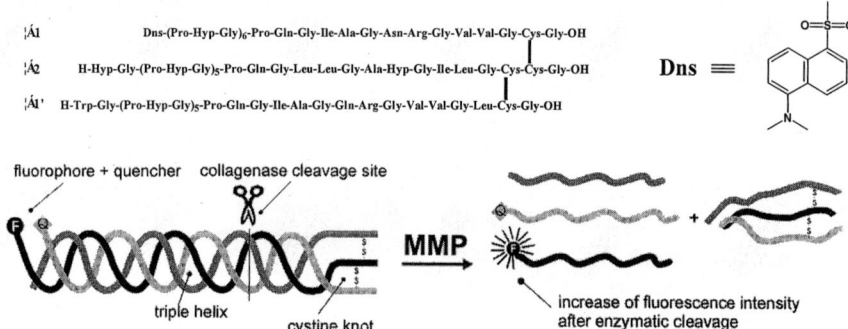

Fig. 6. Moroder's heterotrimeric bioactive collagen mimetic. Dns = dansyl group.[18]

In our design of scaffold-assembled collagen mimetics, we have incorporated both cis-1,3,5-trimethyl cyclohexane-1,3,5-tricarboxylic acid (also known as the Kemp triacid, KTA) and tris(2-aminoethyl)amine (TREN) into collagen mimetic peptides. Spacers were used for both scaffolds, glycine for KTA and succinic acid for TREN respectively (Figure 7).[19,20]

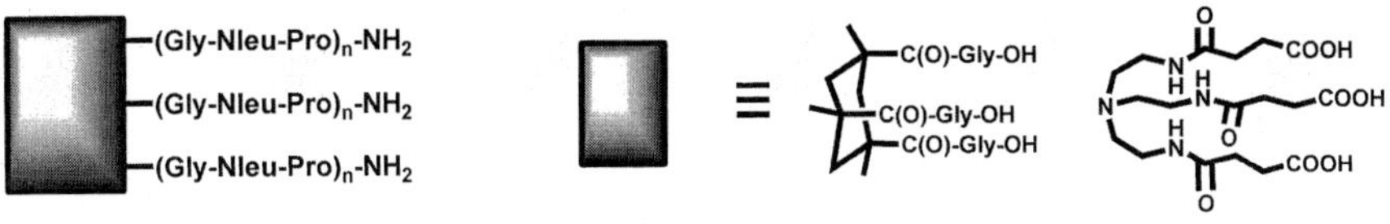

Fig. 7. The KTA-(Gly-OH)₃ and the TREN-(suc-OH)₃ scaffolds.[19,20]

We have also incorporated an unnatural amino acid N-isobutyl glycine (Nleu) as a proline surrogate (Figure 8).[21] Ring opening between the α and β carbons leads to N-propyl glycine which was synthesized and found to be too hydrophilic. Since hydrophobic interactions are important in triple-helix stabilization, a methyl group was added to achieve the desired hydrophobicity. Nleu was incorporated into both the Gly-Nleu-Pro and the Gly-Pro-Nleu sequences. It was found that scaffold-assembled collagen mimetic peptides composed of both sequences form stable triple helices when the peptide chains attain certain lengths (5 or 6 trimer repeats).

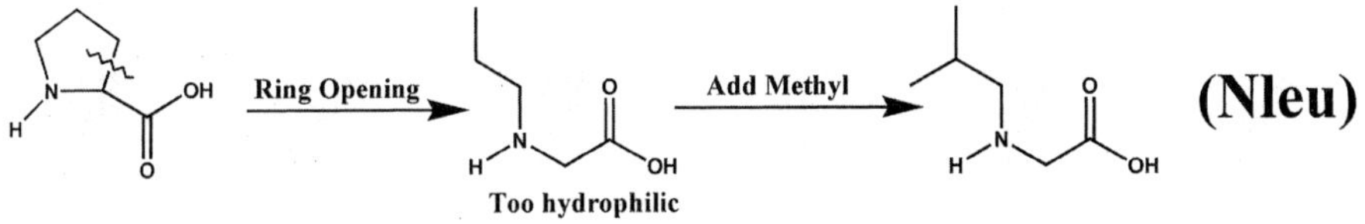

Fig. 8. Conceptual approach to the design of Nleu as a proline mimetic.[21]

Some of the results from biophysical studies of the KTA and TREN assembled collagen mimetic peptides composed of both sequences are shown below (Figure 9,10).[22,20]

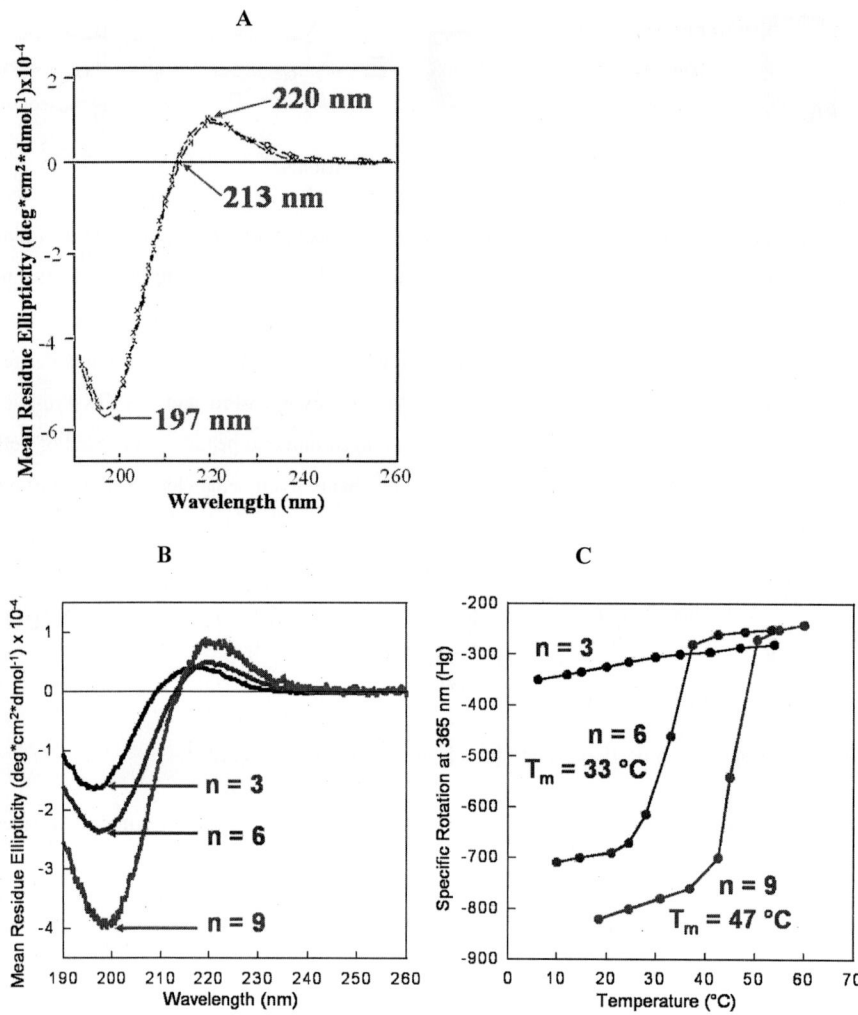

Fig. 9. (A) CD spectra of natural collagen. (B) CD spectra of KTA-[Gly-(Gly-Pro-Nleu)$_n$-NH$_2$]$_3$ (n = 3, 6 and 9). (C) Melting transitions of KTA-[Gly-(Gly-Pro-Nleu)$_n$-NH$_2$]$_3$ (n = 3, 6 and 9).[22]

Collagen Mimetic Dendrimers

Dendrimers are highly branched globular macromolecules with functional groups located at the periphery of the globules. We have incorporated another scaffold N-(t-butyloxycarbonyl)-β-alanyl-tris(carboxyethoxymethyl) aminomethane (Boc-β-Ala-TRIS[OH]$_3$) into our collagen mimetic research (Figure 11).[23] The TRIS scaffold not only has three carboxylic acid groups for attachment of peptide chains, it also has an amino group where it can be attached to a core structure to form collagen mimetic dendrimers.

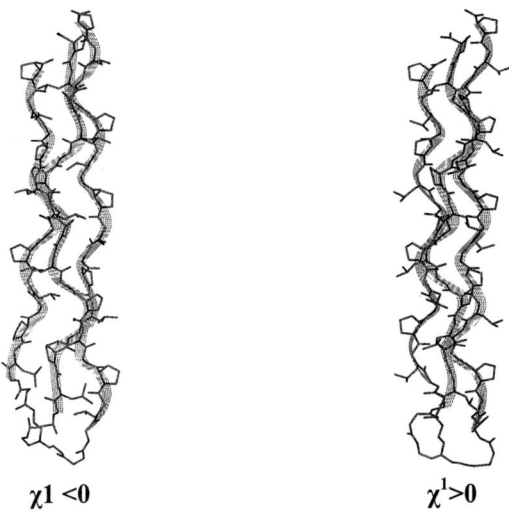

$\chi 1 < 0$ $\chi^1 > 0$

Fig. 10. Lowest energy structures of TREN-[suc-(Gly-Nleu-Pro)$_5$-NH$_2$]$_3$ from each of the two families of clusters generated from molecular modeling. The two families are based on the orientation of the isobutyl side chains of the Nleu residue: $\chi^1 < 0$ (pointing in) and $\chi^1 > 0$ (pointing out).[20]

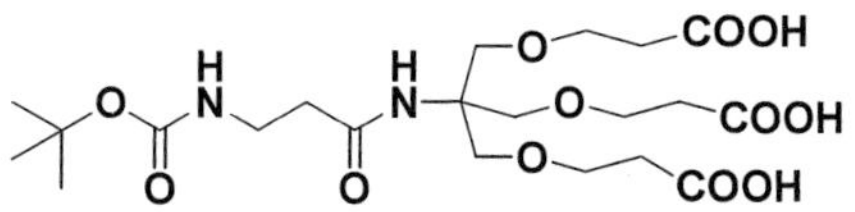

Fig. 11. The Boc-β-Ala-TRIS[OH]$_3$ scaffold for collagen mimetic assembly.[23]

Recently we reported the synthesis and characterization of collagen-like dendrimers.[23] We have synthesized the single chain structure Boc-(Gly-Nleu-Pro)$_6$-OMe, scaffold-assembled

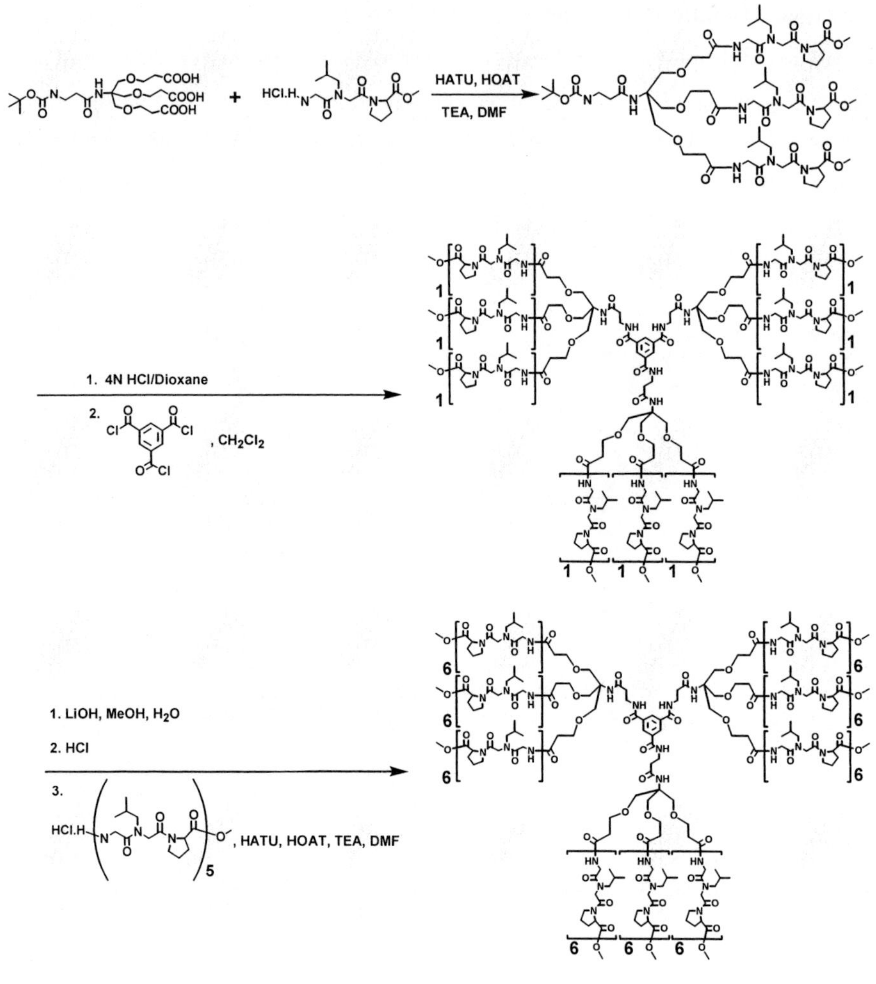

Fig. 12. The synthesis of the dendrimer TMA[β-Ala-TRIS-[(Gly-Nleu-Pro)₆-OMe]₃]₃.

structure Boc-β-Ala-TRIS-[(Gly-Nleu-Pro)₆-OMe]₃ and the collagen mimetic dendrimer TMA[β-Ala-TRIS-[(Gly-Nleu-Pro)₆-OMe]₃]₃ where TMA denotes the trimesic acid core. Isomeric structures composed of the Gly-Pro-Nleu sequence were also prepared and characterized in similar fashion. In this paper, only the results of the collagen mimetics composed of the Gly-Nleu-Pro sequence are discussed. The single chain structure and the scaffold-assembled structure were synthesized by a series of stepwise peptide condensations. A combination of divergent and convergent approaches was employed to synthesize the collagen mimetic dendrimer (Figure 12).

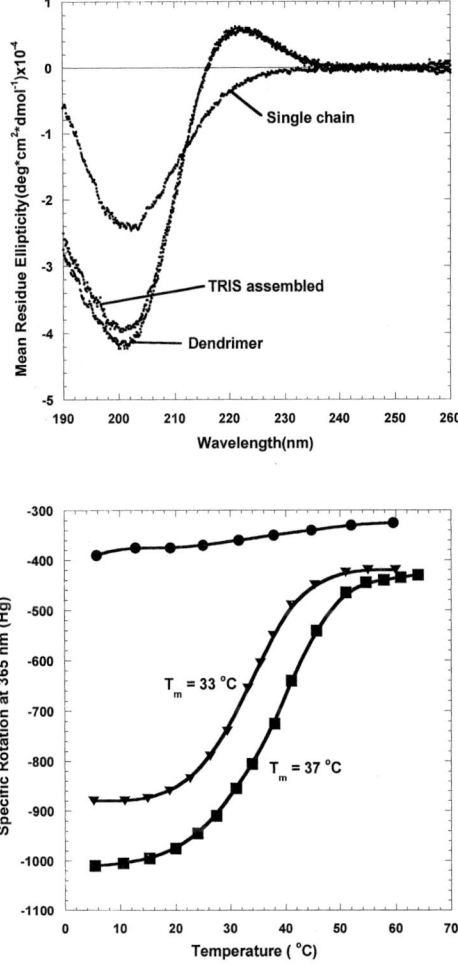

Fig. 13. CD spectra (carried out at 8 °C) and thermal denaturations measured by changes in optical rotations at 365 nm carried out in H_2O (0.2 mg/ml): Single chain structure (Boc-(Gly-Nleu-Pro)$_6$-OMe, ●), TRIS assembled structure (Boc-β-Ala-TRIS-[(Gly-Nleu-Pro)$_6$-OMe]$_3$, ▼), and Dendrimer (TMA[β-Ala-TRIS-[(Gly-Nleu-Pro)$_6$-OMe]$_3$]$_3$, ■).[23]

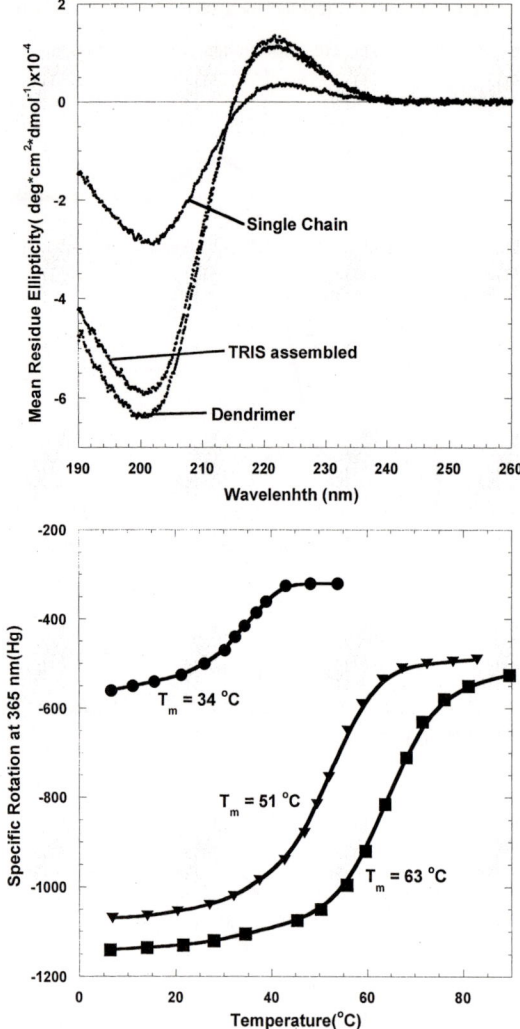

Fig. 14. CD spectra (carried out at 8 °C) and thermal denaturations measured by changes in optical rotations at 365 nm carried out in ethylene glycol/H$_2$O (2:1, v/v) (0.2 mg/ml): Single chain structure (Boc-(Gly-Nleu-Pro)$_6$-OMe, ●), TRIS assembled structure (Boc-β-Ala-TRIS-[(Gly-Nleu-Pro)$_6$-OMe]$_3$, ▼), and Dendrimer (TMA[β-Ala-TRIS-[(Gly-Nleu-Pro)$_6$-OMe]$_3$]$_3$, ■).[23]

The triple helicity of all structures were determined by CD measurements and thermal denaturation studies monitored by optical rotation (Figure 13, 14).[23] These studies were carried out in both H_2O and the triple helicity-enhancing solvent ethylene glycol:H_2O (EG:H_2O, 2/1, v/v).

The optical rotation data shown in Figure 13 and 14 indicates that the molecules studied exhibit cooperative melting transitions except the single-chain compound in water shown in Figure 13. A broad shallow transition for the single chain molecule can be seen in 2:1 EG/H_2O (Figure 14). This indicates that the single chain structure may form triple helices to a small extent in 2:1 EG/H_2O. The CD data shown in Figure 13 and 14 are also indicative of triple helical conformations for the scaffold-assembled and dendritic Gly-Nleu-Pro containing molecules at low temperature.

It is clear from Figure 13 and 14 that the dendritic collagen mimetic forms more thermally stable triple helices than the corresponding scaffold-assembled structure, the melting temperature (T_m) of the dendrimer is 4 °C (in H_2O) and 12 °C (in 2:1 EG/H_2O) higher than those of the scafold-assembled structure. In order to determine whether the stabilizing effect of the dendrimer arises from intermolecular or intramolecular interactions, we measured the melting transition of the collagen mimetic dendrimer in H_2O at different concentrations (0.05 mg/ml, 0.2 mg/ml and 2.0 mg/ml). No significant change in the T_m was observed over the concentration range.[23] We therefore believe that the stabilizing effect arises from an intramolecular clustering of the triple helical arrays about the core structure. Figure 15 shows a possible schematic representation for such a cluster. This cluster excludes solvent from the interior portion of the array which leads to stabilization of the triple helix bundle.

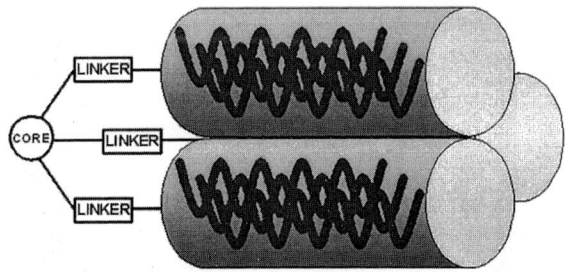

Fig. 15. A schematic representation of the triple helix cluster.[23]

Conclusion

New chemistries have been developed for the preparation of protein mimetics. As part of this new science, we have designed and synthesized single chain, scaffold (TRIS)- and dendrimer-assembled collagen mimetics composed of the Gly-Nleu-Pro sequence. From the CD spectra and the thermal denaturation studies it can be seen that the 162-residue collagen mimetic dendrimer exhibits enhanced triple helical stability compared to the corresponding scaffold-terminated structure by an increase in the melting temperature in both H_2O and 2:1 EG/H_2O (4 °C and 12 °C respectively). The concentration dependence for the melting transition of the collagen mimetic dendrimer was measured from which it was determined that the stabilization effect arises from the intramolecular clustering of the triple helical arrays about the core structure. This ensemble excludes solvent from the interior portion of the array which leads to the stabilization of the bundle of triple helices.

It is our intention to create other collagen-like dendrimers with functional groups attached including metal binding sites, integrin sequences, drugs and others. These ensembles will represent a novel class of structures for biomaterial applications.

[1] K. Severin, D. H. Lee, A. J. Kennan, M. R. Ghadiri, *Nature* **1997**, *389*, 706.
[2] B. North, C. M. Summa, G. Ghirlanda, W. F. DeGrado, *J. Mol. Biol.* **2001**, *311*, 1081.
[3] G. Tuchscherer, D. Grell, M. Mathieu, M. Mutter, *J. Peptide Res.* **1999**, *54*, 185.
[4] F. R. Brown, J. P. Carver, E. R. Blout, *J. Mol. Biol.* **1969**, *39*, 307.
[5] A. Rich, F. H. C. Crick, *J. Mol. Biol.* **1961**, *3*, 483.
[6] K. Okuyama, K. Okuyama, S. Arnott, M. Takayanagi, M. Kakudo, *J. Mol. Biol.* **1981**, *152*, 427.
[7] R. J. Simon, R. S. Kania, R. N. Zuckermann, V. D. Huebner, D. A. Jewell, S. Banville, S. Ng, L. Wang, S. Rosenberg, C. K. Marlowe, D. C. Spellmeyer, R. Tan, A. D. Frankel, D. V. Santi, F. E. Cohen, P. A. Bartlett, *Proc. Natl. Acad. Sci. U.S.A.* **1992**, *89*, 9367.
[8] F. R. III Brown, A. DiCorato, G. P. Lorenzi, E. R. Blout, *J. Mol. Biol.* **1972**, *63*, 85.
[9] M. Li, P. Fan, B. Brodsky, J. Baum, *Biochemistry* **1993**, *32*, 7377.
[10] S. Sakakibara, Y. Kishida, Y. Kikuchi, R. Sakai, K. Kakiuchi, *Bull. Chem. Soc. Jpn.* **1968**, *41*, 1273.
[11] S. Sakakibara, K. Nouye, K. Shudo, Y. Kishida, D. J. Prockop, *Biochim. Biophys. Acta* **1973**, *303*, 198.
[12] R. A. Berg, D. J. Prockop, *Biochem. Biophys. Res. Commun.* **1973**, *52*, 115.
[13] K. Inouye, S. Sakakibara, D. J. Prockop, *Biochimica. et Biophys. Acta* **1976**, *420*, 133.
[14] C. L. Jenkins, R. T. Raines, *Nat. Prod. Rep.* **2002**, *19*, 49.
[15] C. G. Fields, B. Grab, J. L. Lauer, G. B. Fields, *Anal. Biochem.* **1995**, *231*, 57.
[16] G. B. Fields, *Bioorg. Med. Chem.* **1999**, *7*, 75.
[17] Y. Tanaka, K. Suzuki, T. Tanaka, *J. Peptide.Res.* **1998**, *51*, 413.
[18] J. C. D. Müller, J. Ottl, L. Moroder, *Biochemistry* **2000**, *39*, 5111.
[19] M. Goodman, Y. Feng, G. Melacini, J. P. Taulane, *J. Am. Chem. Soc.* **1996**, *118*, 5156.
[20] J. Kwak, A. De Capua, E. Locardi, M. Goodman, *J. Am. Chem. Soc.* **2002**, *124*, 14085.
[21] M. Goodman, G. Melacini, Y. Feng, *J. Am. Chem. Soc.* **1996**, *118*, 10928.
[22] Y. Feng, G. Melacini, J. P. Taulane, M. Goodman, *Biopolymers* **1996**, *39*, 859.
[23] G. A. Kinberger, W. Cai, M. Goodman, *J. Am. Chem. Soc.* **2002**, *124*, 15162.

Macromol. Symp. **2003**, *201*, 237—244   237

DNA: Beyond the Double Helix

Nadrian C. Seeman

Department of Chemistry, New York University, New York, NY 10003, USA
Email: ned.seeman@nyu.edu

Summary: Reciprocal exchange can be used to produce DNA motifs based on branching at the level of secondary structure. These motifs can be combined by sticky-ended cohesion to produce a variety of structures. Stick polyhedra and nanomechanical devices have been produced by self-assembly from motifs based on branched DNA. Periodic arrays with tunable surface features has also been produced; aperiodic arrangements have been used for DNA-based computation.

Keywords: branched DNA, DNA-based computation, periodic arrays, self-assembly, stick polyhedra, surface features, tunable nanomechanical devices

Introduction

We are all aware that DNA is the genetic material of living organisms. It is able to function in that capacity because of its chemical nature. The key features of its chemistry are the specificity of one strand for its complement, the stiffness of the double helix, and the predictability of the DNA structure that two antiparallel strands will form. For over 20 years, our group has been engaged in using these features to exploit the architectural properties of DNA.

"Architectural properties!" What architectural properties? After all, the DNA double helix axis is just a line, topologically speaking, so concatenation of sticky-ended fragments of DNA double helices ought not be very interesting from an architectural standpoint. Not exactly. Biological DNA indeed is a molecule with a linear helix axis, a property necessary for the complement of any strand to be well-defined.[1] Nevertheless, branched molecules arise naturally as Holliday junction[2] intermediates in the process of genetic recombination; these intermediates are necessarily ephemeral. However, chemical DNA synthesis[3] can be used to produce stable branched species. It is straightforward to define sequences of DNA that will associate to form more complex topologies.[4] An example of a stable branched junction[5] is shown in Figure 1a. Similarly, the biological operation of reciprocal exchange provides a direct method to design a variety of motifs that can serve as basic units of DNA nanoconstruction.[6]

            DOI: 10.1002/masy.200351126

238

This operation is illustrated in Figure 1b: Two strands, one gray and one black are combined to make two hybrid strands, one gray-black and one black-gray. This operation is shown in a larger context in Figures 1c and 1d; the difference between these two figures is that in Figure 1c the operation takes place between strands of the same polarity, whereas in Figure 1d the strands are of opposite polarity. For a single exchange event, the two products are simply conformers of each other, but for two or more operations distinct species result.

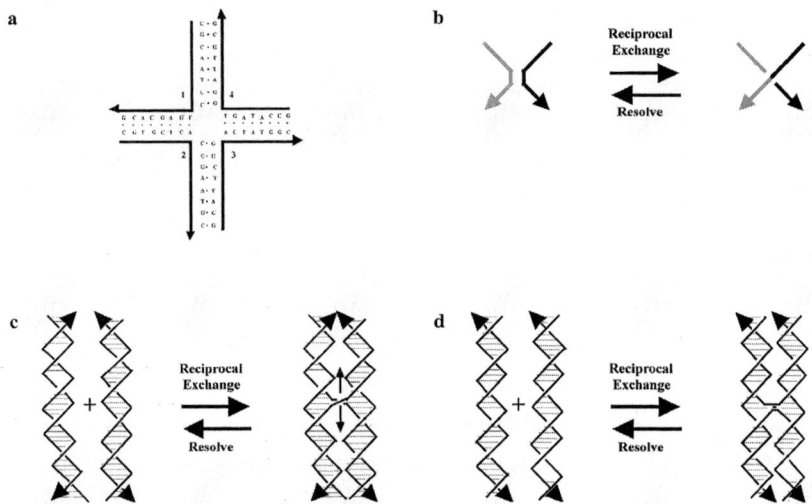

Fig. 1. (a) A four-arm DNA branched junction made from synthetic strands whose sequences have been designed. (b) The reciprocal exchange operation. Two juxtaposed strands, gray and black are exchanged to produce two hybrids, a gray-black strand and a black-gray strand. (c) Reciprocal exchange between two strands of the same polarity. (d) Reciprocal exchange between two strands of opposite polarity.

DNA 'sticky ends' have a central role in genetic engineering: These are short single-stranded overhangs can be used to direct the associations of two different DNA molecules.[7] They produce cohesive hydrogen bonded interactions, interactions that can be made covalent with the use of a DNA ligase. We combine synthetic versions of branched molecules with sticky ends. This enables us to generate new shapes of DNA, that lead to objects, arrays and devices. This concept is illustrated in Figure 2, where four branched junctions are combined to produce a quadrilateral that could be extended by the sticky ends on the outside. An extended complex would result in a 2-dimensional lattice. However, this system is not limited to two dimensions. The angle between the double helices shown in Figure 2 is a function of their separation. Thus, the drawing showing the DNA backbones as parallel lines is valid only for to an exact number of double helical half-turns. The relative orientations are a function of the separation, just like two wing-nuts on a screw.

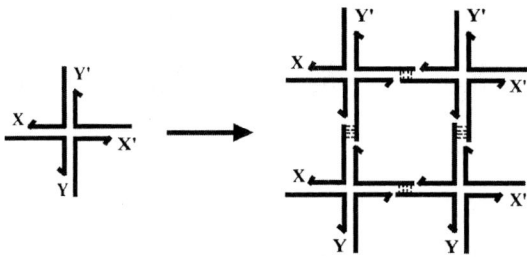

Fig. 2. A four-arm DNA branched junction with two sets of complementary sticky ends (X and X', Y and Y') forms a quadrilateral by self-assembly. The open valences on the outside can lead to the self-assembly of periodic arrays.

Constructions

In the course of our research, we have used DNA topological schemes to generate a variety of branched molecules in both flexible and stiff motifs.[8] Combining flexible motifs with sticky ends has enabled us to produce DNA molecules connected like geometrical stick figures, a cube and a truncated octahedron, as well as a variety of knots and Borromean rings (Figure 3).

More recently, we have used stiff motifs to produce a variety of two-dimensional arrays with programmable patterns from these building blocks. These arrays include patterns formed from double crossover molecules and from triple crossover molecules. The double crossover (DX) molecules can be decorated by a DNA hairpin perpendicular to the plane of the helix axes, to generate a topographic marker that is visible in the atomic force microscope (AFM). Different patterns may be programmed and modified (Figure 4).[9,10]

Similar arrays can be made from triple crossover (TX) molecules, containing three double helical domains, rather than two.[11] A more intriguing application of TX molecules is that we have used them to perform a prototype cumulative XOR calculation by self-assembly.[12] The components of this assembly are shown in Figure 5, and the way that the answer is extracted is shown in Figure 6.

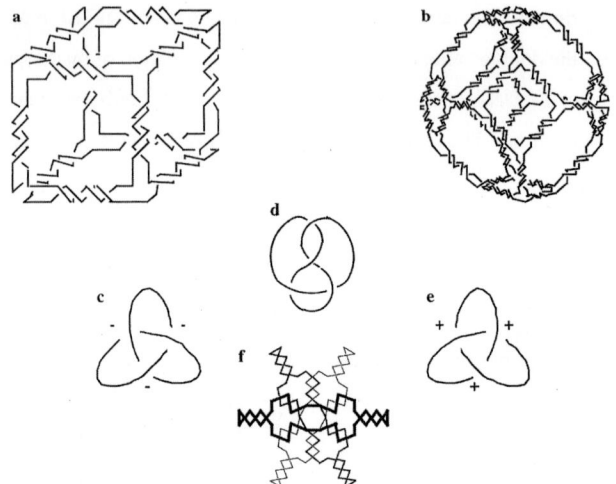

Fig. 3. Ligated products from flexible DNA components. (a) A stick cube and (b) a stick truncated octahedron. The drawings show that each edge of the two figures contains two turns of double helical DNA. The twisting is confined to the central portion of each edge for clarity, but it actually extends from vertex to vertex. Both molecules are drawn as though they were constructed from 3-arm junctions, but the truncated octahedron has been constructed from 4-arm junctions, which has been omitted for clarity. (c-e) Knots Constructed from DNA. The signs of the nodes are indicated. (c) A trefoil knot with negative nodes. (d) A figure-8 knot. (e) A trefoil knot with positive nodes. (f) Borromean rings. Scission of any of the three rings shown results in the unlinking of the other two rings.

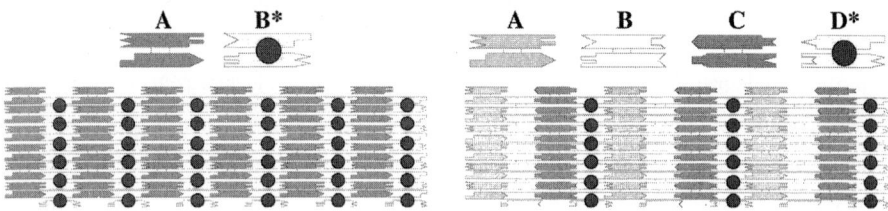

Fig. 4. Two dimensional DNA double crossover arrays. At the top of each panel are shown DX molecules, A and B* or A, B, C and D*. Their sticky ends are represented by geometrical shapes. It is clear from the patterns below these drawings that this set of molecules can tile the plane. Likewise, it is clear that different patterns can be programmed by changing the assignments of the sticky ends. The stripe-like pattern generated by the features in the B* and D* tiles can be visualized in the atomic force microscope. The stripes in the AB* array are separated by 32 nm, whereas those in the ABCD* array are separated by 64 nm. The patterns may be modified by restriction, ligation or hydrogen bonded association.

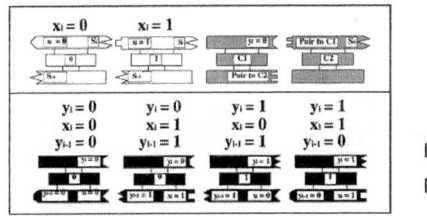

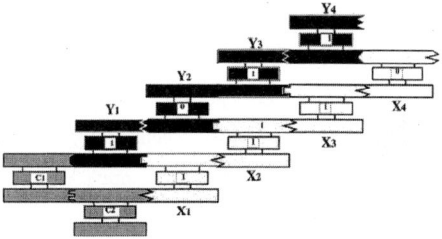

Fig. 5. The components of a DNA-based cumulative XOR calculation are shown on their left, and their self-assembly is shown on the right. The light gray TX tiles are the input to the calculation, the dark gray tiles are initializers, and the black tiles act as the four XOR gates. They contain the four possibilities on their bottom helical domains as sticky ends. The input and initializers have longer sticky ends, so they assemble first.

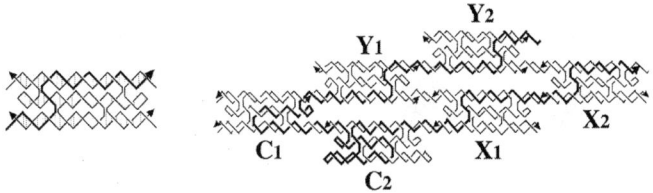

Fig. 6. Extracting the answer from the self-assembly. The strand structure of the TX tile is shown on the left. The value of the tile, 0 or 1, derives from a restriction site (one of two alternatives) on the dark strand of the tile. The strand structure of the assembly is shown on the right. The dark strands have been ligated, thereby connecting the input with the output. The long ligated dark strand is extracted and then subjected to restriction by the enzymes in separate experiments; the answer can be read off the resulting gel.

Stiff DNA motifs have also been used to produce two different DNA devices. One device is based on the transition between right-handed B-DNA and left-handed Z-DNA;[13] the second is based on controlling motif topology in a sequence-specific fashion (Figure 7).[14] The first device has a global trigger, the addition of a small molecule $Co(NH_3)_6^{3+}$, so all molecules will react in the same way. The sequence-dependent trigger of the second device allows us to address each species of device separately, creating a panoply of robust structural states.

Objectives

What are the objectives of this program? The first potential application of this system is to scaffold biomolecular crystallization, using a nucleic acid host lattice to organize a biological macromolecular guest into a crystal that can diffract X-rays and thereby enable structure determination.[4] Of course, if one can imagine organizing biological macromolecules into an array, one can imagine organizing other molecules as well. Prominent amongst these are the components of molecular electronics.[15] There are many species that are well-suited to nanoelectronics, e.g., metallic nanocrystals or carbon nanotubes, but it is difficult to organize them into functional arrays. We expect that structural DNA nanotechnology will be able to provide the organization needed for this purpose. Schematic diagrams of these goals are shown in Figures 8 and 9.

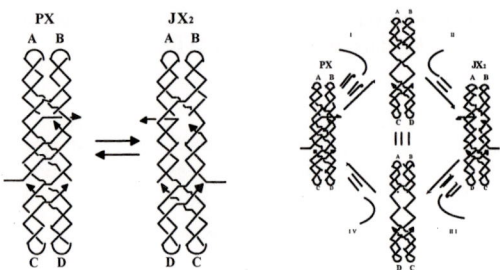

Fig. 7. A sequence-dependent DNA nanomechanical device. The left side shows the two end-state molecules of the device. The one on the left, a PX motif, contains crossovers at every point in the central region. The molecule on the right, known as JX_2, lacks two of those crossovers in the central portion of the molecule. The molecules differ by the relative positions of the C and D markers, although A and B are the same; thus JX_2 is a half-turn unwound, relative to PX. The strands containing the horizontal extensions (called set strands) set the conformation to one of the two states. The right panel illustrates the machine cycle. Removing the set strands by adding their full complements allows the opposite set strands to bind, changing the state of the molecule.

Fig. 8. Macromolecular guests in a DNA host lattice. The cube-like structure represents a polyhedron constructed from DNA, and the extensions represent sticky ends. The irregular blobs attached to it represent biological macromolecules arranged parallel to each other.

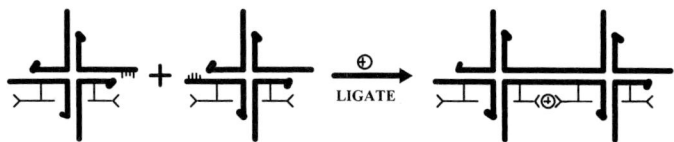

Fig. 9. DNA branched junctions organizing nanoelectronic components. Two branched junction structures are shown. They have complementary sticky ends, and a nano-wire is hanging from each of their horizontal arms. The idea illustrated here is that when the DNA structures cohere *via* their sticky ends, the nano-wires will be directed to make contact, as well. An ion is shown as an additional organizing element in this scheme.

A third species to organize into arrays is the sequence-dependent nano-device that we have already produced. If we can incorporate N different species of these 2-state devices, we should be able to generate 2^N different structural states. Short-range goals include molecular pegboards, and molecular assembly lines. Multiple structural states are a necessary concomitant of nanorobotics, so this system may lead to a future DNA-based nanorobotics.

The primary advantage of DNA for these goals is its outstanding molecular recognition properties, allowing precise structural alignment of diverse intermolecular species. DNA appears to be unique among biopolymers in this regard. Although there are other systems that lead to specific binding, only in the case of nucleic acids are the structures formed known in advance: Sticky ends form B-DNA.[16] If one considers a related cohesive system, antigen-antibody interactions, affinity can still be predicted, but the structure of every individual complex must be determined by experiment (e.g., crystallography) before it can be used in nanoconstruction.

Other advantages of DNA are the convenience of chemical synthesis,[3] and the presence of enzymes to manipulate it and trouble-shoot errors. For example, DNA ligases enable covalentbonding of complexes held together by sticky-ended cohesion.[17] Exonucleases can be used to purify cyclic target molecules from linear failure products. Restriction endonucleases can be used both to trouble-shoot syntheses and to create cohesive ends from topologically-closed species.[18,19] The persistence length of DNA is about 500 Å,[20] leading to a predictable overall structure for the short (70-100 Å) lengths we use. There is an external code on DNA that can be read, even when the double helix is intact.[21] The packing of nanoelectronics very

244

tightly will be aided by the high density of functional groups (every 3.4 Å or so) on DNA, regardless of tile size.

The enzymes noted above make it convenient for us to prototype our constructions with conventional DNA. However, we are not limited to the molecule evolved in nature for use as genetic material. A vast number of DNA analogs have been produced and analyzed for therapeutic purposes.[22,23] This means that systems prototyped by conventional DNA ultimately may be converted to other backbones and bases, as required by specific applications.

Acknowledgments

This research has been supported by grants GM-29554, from NIGMS, N00014-98-1-0093 from the ONR, DMI-0210844, EIA-0086015, DMR-01138790, and CTS-0103002, from the NSF and F30602-01-2-0561, from DARPA/AFOSR.

[1] Seeman, N.C. *Synlett.* **2000**, 1536-1548.
[2] Holliday, R. *Genet. Res.* **1964**, *5*, 282-304.
[3] Caruthers, M.H. *Science* **1985**, *230*, 281-285.
[4] Seeman, N.C. *J. Theor. Biol.* **1982**, 99, 237-47.
[5] Kallenbach, N.R., Ma, R.-I. & Seeman, N.C. *Nature* **1983**, *305*, 829-831.
[6] Seeman, N.C. *NanoLett.* **2001**, 1, 22-26.
[7] Cohen, S.N; Chang, A.C.Y.; Boyer, H.Y.; Helling, RB. *Proc. Nat. Acad. Sci. (USA)* **1973** 70, 3240-3244.
[8] Seeman, N.C. *Trends Biotechnol.* **1999** 17, 437-443.
[9] Winfree, E.; Liu, F.; Wenzler, L.A.; Seeman, N.C. *Nature* **1998** *394*, 539-544.
[10] Liu, F.; Sha, R.; Seeman, N.C. *J. Am. Chem. Soc.* **1999** *121*, 917-922.
[11] LaBean, T.; Yan, H.; Kopatsch, J.; Liu, F.; Winfree, E; Reif, J.H.; Seeman, N.C. *J. Am. Chem. Soc.* **2000** *122*, 1848-1860.
[12] Mao, C.; LaBean, T.H.; Reif, J.H.; Seeman, N.C. *Nature* **2000** 407, 493-496.
[13] Mao, C.; Sun, W.; Shen, Z.; Seeman, N.C. *Nature* **1999** *397*, 144-146.
[14] Yan, H.; Zhang, X.; Shen, Z.; Seeman, N.C. *Nature* **2002** 415, 62-65.
[15] Robinson, B.H.; Seeman, N.C. *Protein Eng.* **1987** *1*, 295-300.
[16] Qiu, H.; Dewan, J.C.; Seeman, N.C. *J. Mol. Biol.* **1997** *267*, 881-898.
[17] Chen, J; Seeman *Nature* **1991** *350*, 631-633.
[18] Zhang, Y; Seeman *J. Am. Chem. Soc.* **1992** *114*, 2656-2663.
[19] Zhang, Y; Seeman *J. Am. Chem. Soc.* **1994** *116*, 1661-1669.
[20] Hagerman, P. *Ann. Rev. Biophys. & Biophys. Chem.* **1988** *17*, 265-286.
[21] Seeman, N.C.; Rosenberg, J.M.; Rich, A. *Proc. Nat. Acad. Sci. (USA)* **1976** 73, 804-808.
[22] Freier, S.M.; Altmann, K.-H. *Nucl. Acids Res.* **1997** 25, 4429-4443.
[23] Nielsen, P.E.; Egholm, M.; Berg, R.H.; Buchardt, O. *Science* **1991** *254*, 1497-1500.

Novel Biomaterials Derived from Deoxyribozyme and NAPzyme

Yasuhide Okumoto,[1,3] Hiroyoshi Fujiki,[1] Junji Kawakami,[1,2] Shoji Nakashima,[2] Shu–ichi Nakano,[2] Tatsuo Ohmichi,[2] Daisuke Miyoshi,[1] Naoki Sugimoto,[1,2]*

[1]Department of Chemistry, Faculty of Science and Engineering and
[2]High Technology Research Center, Konan University, 8–9–1 Okamoto, Higashinada–ku, Kobe 658–8501, Japan
[3]Nippon Laser & Electronics Lab., 20-9, Sanbonmatsu-cho, Atsuta-ku, Nagoya 456-0032, Japan

Summary: We report the potential of a small Ca^{2+}–dependent deoxyribozyme as a novel biomaterial to distinguish RNA foldings. It is found that an immobilized deoxyribozyme using avidin–biotin interaction cleaves the target site within only single–stranded RNAs. The RNA cleavage reaction is also detected using the deoxyribozyme SPR sensor chip. Furthermore, we develop a novel NAPzyme (nucleic acid peptide deoxyribozyme) with its RNA cleavage function in the absence of divalent metal ions.

Keywords: biomaterials, deoxyribozyme, enzymes, metal ion, NAPzyme (nucleic acid peptide deoxyribozyme)

Introduction

High–density oligonucleotide captures for sequence–specific analysis has been developed as novel biotechnologies of DNA chips and microarrays.[1] These biotechnologies are very powerful tools for the genomic analysis of SNPs (single nucleotide polymorphisms) and mRNA/gene expressions.[2] However, it is difficult using these methods to detect the difference in the RNA folding. The RNA folding variation due to one or more mutations would lead to different biological functions as a result of differences in the primary or higher–ordered structures that interact with other cellular molecules, because the RNA folding influences the RNA splicing, RNA processing, and translational control.[3] Thus, to distinguish the RNA foldings is one of the guides to detect the gene functions related to diseases and drug responses. Recent studies on ribozymes and deoxyribozymes showed that they are also interesting tools in biotechnology and biomaterial.[4,5] For a cleavage reaction using ribozymes, the cleavage

© 2003 WILEY-VCH Verlag GmbH & KGaA, Weinheim   DOI: 10.1002/masy.200351127

activity is basically dependent on the sequence close to the cleavage site and the secondary structure of the target RNA. This fact indicates the possibility for the catalytic nucleic acid to distinguish or detect the target RNA foldings.

Ribozymes have a requirement for divalent metal ions,[6-8] very high concentration of monovalent metal ions,[9,10] or small molecules,[11] which act as base and acid catalysts. On the other hand, the hydrolysis of RNA by RNase A (ribonuclease A) is initiated by histidine residues at positions 12 and 119, which acts as acid and base catalysts in the absence of divalent metal ions.[12,13] Thus, a synthesis of nucleotide analogues containing covalently linked histidines has made it possible to investigate the catalytic activity of novel ribozymes in the absence of divalent metal ions.

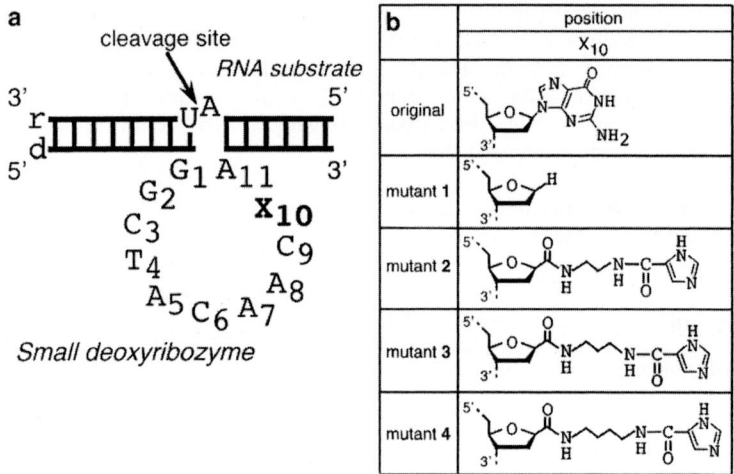

Fig. 1. (a) Secondary structure of the complex of the small deoxyribozyme and its RNA substrate. (b) Structures of the modified nucleotides at position 10 of its catalytic loop of the deoxyribozyme.

Here, we have now developed an immobilized small Ca^{2+}–dependent deoxyribozyme with 11 nts (nucleotides) catalytic loop as a novel and useful biomaterial to distinguish RNA foldings as shown in Figure 1a.[14-16] Furthermore, we have developed a novel NAPzyme (nucleic acid peptide deoxyribozyme) with its RNA cleavage function in the absence of divalent metal ions as shown in Figure 1b.

Materials and Methods

Materials. All DNA, DNA containing an abasic nucleotide or the histidine NAP, and RNA oligonucleotides were synthesized chemically on solid support and purified as described previously.[14,15] The final purity of the oligonucleotides was confirmed to be >99 %. The oligonucleotides were desalted with a Sep–Pak C18 cartridge column (Waters) before use. Single–strand concentrations of oligonucleotides were determined by measuring the absorbance at 260 or 280 nm. The single strand extinction coefficient was calculated from the mononucleotide and dinucleotide data using nearest–neighbor approximation.[17] The extinction coefficient of the single strand containing the abasic nucleotide or histidine NAP was obtained as a sum of two regions separated by the site.[18] The target was 5'–end labeled using 5–IAF (5–iodoacetamidofluorescein) at 37 °C as previously described.[16]

Synthesis of histidine NAP phosphoramidites. The synthesis of various histidine NAP phosphoramidites was synthesized using a 10–step synthesis from 2–deoxy–D–ribose as previously reported.[19,20] The phosphoramidites were identified by MS and ^{1}H NMR spectra.

RNA cleavage reactions. The target cleavage reactions by the immobilized deoxyribozymes were done in a buffer containing 50 mM Tris–HCl (pH 8.0) and 25 mM Ca^{2+} at 37 °C.[14,15] The RNA cleavage reactions by the mutant deoxyribozyme were done in a buffer containing 50 mM Tris–HCl (pH 8.0), 100 mM Na^+, and 25 mM Ca^{2+} at 37 °C. After cleavage reactions were stopped, the 5'–end fluorescein–labeled products and targets were separated by electrophoresis on 20 % polyacrylamide / 7 M urea denaturing gels. The cleavage yields were determined by quantifying the fluorescence intensity in the bands of the 5'–end labeled products and targets using a fluorescence imager. An observed rate constant (k_{ops}), an equilibrium constant (K_1), and a cleavage rate constant (k_2) of the original deoxyribozyme or NAPzyme with its RNA substrate were calculated from nonlinear least–square fitting analysis as previously described.[21-23]

Deoxyribozyme SPR sensor chip. A BIAcore (BIAcore 1000, Biacore AB, Uppsala, Sweden) was used for the SPR measurements.[15] To remove the background binding between the injected target and the immobilized streptavidin to the dextran matrix or the refractive index change in the injection, the SPR trace, after flowing a buffer containing the target over the sensor chip coated without the ligand, was deducted from those with the ligand. The target was injected in a buffer containing 50 mM Tris–HCl (pH 8.0) and 100 mM Na^+ at 37 °C at the slow flow rate of 5 µL min^{-1}. The concentration of the injected target was 100 µM. By addition to 25 mM Ca^{2+}, the target cleavage reactions were initiated.

Results and Discussions

The immobilized deoxyribozyme distinguishes the difference in the RNA foldings. To obtain information about the effect of target foldings, we first investigated whether the immobilized deoxyribozyme cleaves the site within the duplex (Figure 2). The immobilized deoxyribozyme (dCGCTGGCAggctacaacgaGTCTTC; small letters indicate its catalytic loop) was able to cleave the site only in the single–stranded RNA (rGAAGACA↓UGCCAGCG (r15); ↓ indicates its cleavage site), although the single–stranded DNA, pseudo single–stranded RNA [rGAAGAC(dA)UGCCAGCG], RNA/DNA hybrid, and RNA/RNA duplex were not cleaved.

The single–stranded RNA regions are basically unpaired, such as in the hairpin loops and internal loops within the RNA foldings. To investigate the effect of the RNA hairpin loop size, the cleavage reactions of eight RNA hairpin loops by the immobilized deoxyribozyme were carried out. The sequence of the hairpin loop with 15 nts is the same to that of r15 described above. All these RNA sequences melted with biphasic behavior and the T_m (melting tempreature) values were independent of the concentration of the RNAs (data not shown), indicating that the folded structure of these RNAs would be the only stable intramolecular hairpin loop, but not an intermolecular loop structure.[24] For RNA with 15 nts loop, the cleavage reaction was little observed. On the other hand, for ≥17 nts as a hairpin loop, the cleavage was clearly observed at only one site. These results suggest that at least one nucleotide spacer at both ends of the hairpin loop are required for efficient cleavage by the immobilized deoxyribozyme. The catalytic activity was saturated over ≥23 nts as the hairpin loop size. To investigate the effect of the cleavage position, the RNA hairpin loop cleavage reactions by the immobilized deoxyribozyme were carried. The RNA hairpin loop domain was systematically slid from the 5'–end to the 3'–end in the recognition site of the immobilized deoxyribozyme. The amount of cleavage product for these RNA by the immobilized deoxyribozyme were approximately equal, indicating that any target position in the RNA hairpin loop is cleaved only at one site.

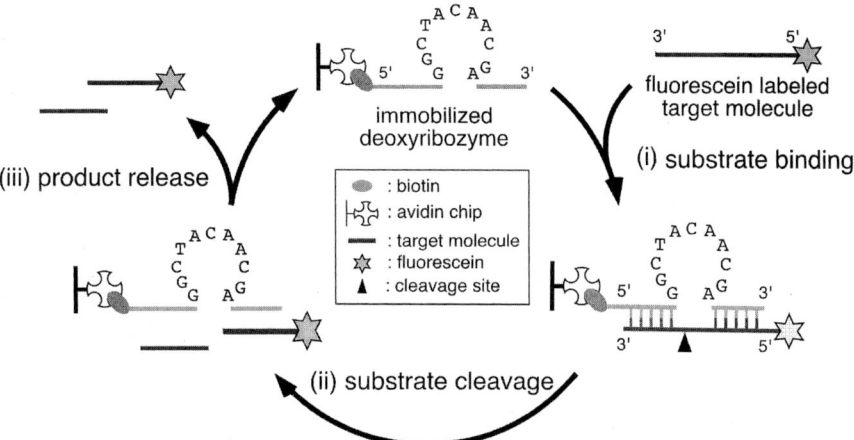

Fig. 2. Reaction scheme of the novel functional DNA chip with the immobilized deoxyribozyme.

Development of the deoxyribozyme SPR sensor chip. The construction and application of the immobilized deoxyribozyme on an SPR sensor chip to create a deoxyribozyme array has not been described. We have already demonstrated that the novel and useful biomaterial of the deoxyribozyme SPR sensor chip is possible, which may enable the rapid invention of numerous DNA biosensor elements with satisfactory performance characteristics.[16] The same method should apply to the use of DNA and many nucleic acids analogs that can provide greater versatility for analysis recognition and increased stability in harsh environments. In this regard, the catalytic nucleic acid arrays could have long storage lives, perhaps greater than biosensor components made from natural protein. To clarify the properties of the immobilized deoxyribozyme as a sensor chip, the cleavage reaction were directly measured by an SPR apparatus.

Figure 3 shows the typical SPR sensorgrams of the binding and cleavage steps between the target (r15 and pseudo RNA substrate [rGAAGAC(dA)UGCCAGCG]) and the deoxyribozyme SPR sensor chip. When its target binds to the immobilized deoxyribozyme on the SPR sensor chip, the observed response unit is increased by changes in the refractive index as shown by the RU (resonance units) values in real time.[25] When the pseudo RNA substrate was flowed over the sensor chip coated with the immobilized deoxyribozyme, about 150 RU value was retained by the addition to Ca^{2+}. However, in the case of r15, the RU value was slowly decreased by the addition to Ca^{2+}. The differences in the RU value between its RNA substrate and pesudo RNA substrate after 30 min running were 69 %. After the recovery of these samples, the cleavage products and targets were able to be separated by electrophoresis on denaturing gel, indicating that no cleavage band was observed for the pseudo RNA substrate, while only one cleavage product was observed for r15. The cleavage site was at only one site of rAp↓U in the asymmetric internal loop. These results indicate that the decrease in the RU value shows the process of the product release after the RNA cleavage reaction.

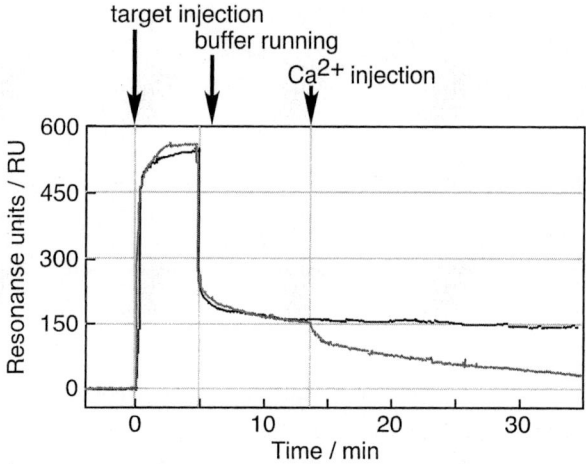

Fig. 3. The typical SPR sensorgrams of the binding and cleavage steps between the target molecule and the immobilized deoxyribozyme on the SPR sensor chip in a buffer containing 50 mM Tris-HCl (pH 8.0) at 37 °C. The flow samples were r15 (black line) or the pseudo RNA substrate (gray line).

NAPzyme has its RNA cleavage function in the absence of divalent metal ion. The position 10 in the catalytic loop of the deoxyribozyme has a larger influence on the RNA cleavage step than the binding step between the deoxyribozyme and its RNA substrate.[26] Thus, we designed mutant deoxyribozymes with an abasic nucleotide or a histidine NAP at position 10 as shown in Figure 1b. The original deoxyribozyme was able to cleave its RNA substrate at only one site in the presence of Ca^{2+}. Surprisingly, only one cleavage band was observed the mutant **3** in the absence of Ca^{2+}. The cleavage site by mutant **3** was identical to the cleavage site by the original deoxyribozyme. However, mutants **1**, **2** and **4** were not able to cleave its RNA substrate, even after a 120 min incubation.

Perspectives

In this study, we developed novel biomaterials as the immobilized small Ca^{2+}–dependent deoxyribozyme. These targets were most suitable for only single–stranded RNAs containing a hairpin loop with ≥17 nts loop size. Thus, the immobilized deoxyribozyme can distinguish the higher–ordered structures of RNAs. Based on our results, we propose a search system for RNA higher–ordered structures using the immobilized deoxyribozyme and will then be able to construct a database of higher–ordered structures of genome and other nucleic acids. Furthermore, we have developed a novel NAPzyme that consists of a small deoxyribozyme and a histidine NAP with a propylene linker at position 10 of its catalytic loop. The NAPzyme was shown to cleave its RNA substrate at only one site in the absence of divalent metal ions. These studies might serve as a starting point for the further development of deoxyribozymes (ribozymes) that more closely mimic the function of RNase A.

Acknowledgment

This work was supported in part by Grants-in-Aid for Scientific Research from the Ministy of Education, Science, Sports, and Culture, Japan to N.S.

[1]   C. Debouck, P. N. Goodfellow, *Nat. Genet.* **1999**, *21*, 48.
[2]   D. W. Selinger, K. J. Cheung, R. Mei, E. M. Johansson, C. S. Richmond, F. R. Blattner, D. J. Lockhart, G. M. Church, *Nat. Biotechnol.* **2000**, *18*, 1262.
[3]   I. Barrette, G. Poisson, P. Gendron, F. Major, *Nucleic Acids Res.* **2001**, *29*, 753.
[4]   T. Ohmichi, S. Nakano, D. Miyoshi, N. Sugimoto, *J. Am. Chem. Soc.* **2002**, *124*, 10367.
[5]   J. Li, Y. Lu, *J. Am. Chem. Soc.* **2000**, *122*, 10466.
[6]   S. W. Santoro, G. F. Joyce, *Biochemistry* **1998**, *37*, 13330.
[7]   T. R. Cech, B. L. Golden, in: *"The RNA World"*, *Second ed.*, R. F. Gesteland, T. R. Cech, J. F. Atokins, Eds., Cold Spring Harbor Laboratory Press, Plainview, NY **1999**, p.265.
[8]   T. Ohmichi, Y. Okumoto, N. Sugimoto, *Nucleic Acids Res.* **1998**, *26*, 5655.
[9]   E. A. Curtis, D. P. Bartel, *RNA* **2001**, *7*, 546.
[10]   J L. O'Rear, S. Wang, A. L. Feig, L. Beigelman, O. C. Uhlenbeck, D. Herschlag, *RNA* **2001**, *7*, 537.
[11]   S. W. Santoro, G. F. Joyce, K. Sakthivel, S. Gramatikova, C. F. Barbas, III, *J. Am. Chem. Soc.* **2000**, *122*, 2433.
[12]   C. Park, L. W. Schultz, R. T. Raines, *Biochemistry* **2001**, *40*, 4949.
[13]   R. T. Raines, *Chem. Rev.* **1998**, *98*, 1045.

[14]   N. Sugimoto, Y. Okumoto, T. Ohmichi, *J. Chem. Soc., Perkin Trans. 2* **1999**, 1382.
[15]   Y. Okumoto, N. Sugimoto, *J. Inorg. Biochem.* **2000**, *82*, 189.
[16]   Y. Okumoto, T. Ohmichi, N. Sugimoto, *Biochemistry* **2002**, 41, 2769.
[17]   E. G. Richard, in: *"Handbook of Biochemistry and Bilogy: Nucleic Acids"*, 3rd ed., G. D. Fasman, Ed.,
       CRC Press, Cleaveland, OH, **1975**, *vol.1*, p.597.
[18]   J. Kawakami, M. Yoneyama, D. Miyoshi, N. Sugimoto, *Chem Lett.* **2001**, 258.
[19]   T. H. Smith, M. A. Kent, S. Muthini, S. J. Boone, P. S. Nelson, *Nucleosides & Nucleotides* **1996**, *15*,
       1581.
[20]   J. Kawakami, H. Fujiki, S. Izumi, N. Sugimoto, unpublished results.
[21]   N. Sugimoto, R. Kierzek, D. H. Turner, *Biochemistry* **1988**, *27*, 6384.
[22]   T. Ohmichi, N. Sugimoto, *Biochemistry* **1997**, *36*, 3514.
[23]   T. Ohmichi, H. Nakamuta, K. Yasuda, N. Sugimoto, *J. Am. Chem. Soc.* **2000**, *122*, 11286.
[24]   T. Dale, R. Smith, M. J. Serra, *RNA 6*, 608 (2000).
[25]   P. C. Bevilacqua, R. Kierzek, K. A. Johnoson, D. H. Turner, *Science* **1992**, *258*, 1355.
[26]   Y. Okumoto, Y. Tanabe, N. Sugimoto, *Biochemistry* **2003**, *42*, *in press*.

Macromol. Symp. **2003**, *201*, 253—260

Assessing the Economic and Ecological Impact of Plastic Products

Karl-Heinz Feuerherd

Kobe Yamate University, 5-2 Nakayamatedori 6-chome, Chuo-ku, Kobe 650-0004, Japan
karl@kobe-yamate.ac.jp

Summary: The framework of the International Standard Organization's (ISO) 14040 series provides the practitioner with rules to deal with different aspects of analysis and assessment of the life cycle of products. The standard emphasizes the importance of accurate data to perform a valid analysis. This premise lets LCA (life cycle assessment) studies become time-consuming and expensive. But, sound decision-making by management in enterprises can be done even in cases where data are incomplete or vary, when the eco-efficiency analysis methodology is used that has been developed and applied by BASF in more than 150 projects during the last five years.

Keywords: calculations, eco-efficiency, life cycle assessment, plastics, Ultramid®

Introduction

Among worldwide operating chemical companies the environmental activities of BASF have again focused attention during the last years. With the dawn of the new century the company has anchored Sustainable Development even more firmly in its management system. In 1992, representatives of 172 governments adopted Agenda 21 in Rio de Janeiro. At that time, very few of the signatories envisaged the enormous power that the idea of Sustainable Development would attain within only a few years. With the creation of a Sustainability Council at the top corporate level, BASF is one of the first large concerns in the world to anchor this principle in its organization. In addition, the company established a Competence Center for Responsible Care to advance environment, safety and health worldwide throughout the BASF Group. And Fortune Magazine ranks BASF number one among world chemical companies and among German businesses in its annual listing of Global Most Admired Companies. Sustainability is not just window dressing as far as BASF is concerned. Instead, *sustainable success* has become a precondition for all activities. How closely economic success and ecological progress are related, is impressively highlighted by the *BASF type eco-efficiency analysis*.

          DOI: 10.1002/masy.200351128

At the heart of this analysis lies the question "What must BASF's products of the future look like?" The goal is to offer competitive, environmentally friendlier products with optimum benefits for the customers. Eco-efficient products like these also give the company a competitive advantage. BASF's analysis already includes financial and environmental criteria and the specialists are working to incorporate social indicators, too. BASF intends to then have a unique strategic instrument to optimize its product portfolio according to sustainability criteria.

Indeed, different concepts already exist to evaluate the environmental performance of products and services. But, from the viewpoint of BASF these concepts are not sufficient to help the management makes sound decisions on how to optimize the company's portfolio by selection of prosperous products.

BASF's Methodology

The principal concept of the eco-efficiency analysis method of BASF has been developed in 1996 by teaming up with management consultants Roland Berger + Partner. Since then the methodology has been refined continuously based on BASF's own growing expertise. The method is combining ecological and economic thinking. The ecological part is based on lifecycle thinking according to the International Standard Organization's (ISO) 14040 series. The functional unit in particular is defined from the viewpoint of the final customer, who is the driving force in industrial societies to manufacture goods and provide services. The economical part of the methodology assumes that the final customer has to pay for all activities directly in the form of prices for products and services or indirectly by taxes, charges, fees etc. whether he is aware of or not. At first glance the methodology looks simple, but it is also capable to perform detailed cost analysis, which is beyond the scope of this paper.

Thus, the BASF type of eco-efficiency analysis is a two-dimensional analysis that requires two or more products to be compared. Something like a "score figure" for a single product cannot be calculated. In addition, it is not allowed to calculate single figures like the "ratio of environmental and economical impact" to compare the performance of products and services. Therefore, the result must be presented graphically as so-called "eco-efficiency chart", which is similar to portfolio charting used in economics.

The lifecycle analysis is done by using the following set of selected categories.

Material: water, coal, oil, gas, lignite, limestone, bauxite, sulfur, sodium chloride, potassium chloride, feldspar and sand. This listing can be amended according to the necessities of a specific analysis.

Energy (Fuel): coal, oil, gas, hydro power, nuclear power, lignite, biomass and others.

Air emissions: carbon dioxide, sulfur oxides, nitrogen oxides, methane, non-methane organic volatile substances, halogen compounds, ammonia, laughing gas and hydrogen chloride.

Water emissions: BOD (biological oxygen demand), COD (chemical oxygen demand), nitrogen compounds, ammonia ions, phosphate ions, AOX (absorbable organic halogenated compounds), heavy metals, hydrocarbons, sulfate ions and chloride ions.

Waste: domestic waste, special waste and overburden (including building rubble).

The selection of these emission categories has been done from the viewpoint of existing "limit values" like scarcity factors, maximum allowable release concentrations, effects that harm the environment etc. These limits are fixed by laws and ordinances, which often differ country by country and region by region.

Furthermore the two categories "toxicity potential" and "risk potential" have been introduced, which are especially important for judgment in chemical companies.

The toxicity potential of products and services is determined by using figures from the German MAK (Maximale Arbeitsplatzkonzentration) list and/or R-phrases that are common in the EU (European Union) for characterization of chemicals. The calculation is done according to a special procedure[1] that has been developed by BASF and takes hazard and/or chemical risk considerations into account.

The concept of "risk potential" covers the idea of risk of accidents and injuries, which are based for example on statistics published by professional and trade associations. The calculation is done semi-quantitatively.

Figure 1 shows an example of the so-called "ecological fingerprint", which is a radar chart with five axes representing the main categories "material consumption", "energy consumption", "emissions" (air, water, solid waste), "toxicity potential" and "risk potential". The environmental impact caused by waste and emissions into air and water are treated as one total "emissions" category. In case renewable resources have to be investigated an additional axis is introduced that represents the use of land.

In the first step potential effects like GWP (global warming potential), ODP (ozone depletion potential), POCP (photochemical ozone creation potential) and AP (acidification potential) as well as water pollution and waste are quantified similar to a known procedure.[2] The aggregation is then done by using the "societal perception" scheme of Figure 2 to get one final figure representing the environmental load of a product's life cycle.

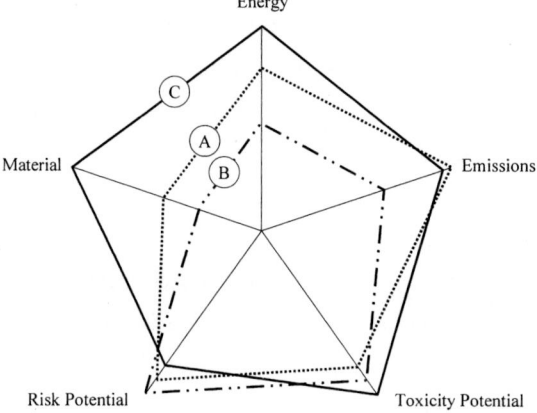

Fig. 1. Example of an ecological fingerprint for a representative scenario analyzing three products.

In addition a second weighting scheme is applied that is based on specific environmental categories occurring in the life cycle of a product. The absolute value of these individual resource and emission categories like carbon dioxide, nitrogen oxides etc. is related to the respective total value of these categories observed for example in Germany during one year. This weighting leads to so-called "relevance factors" which are then amalgamated on equal basis with the mentioned weighting scheme of societal perception to give one ultimate figure representing the environmental impact of a product based on "social and relevance weighting". It is important to recognize that the indicated percentage figures for social perception must not be regarded as fixed. The initial values are only used to calculate the so-called basic scenario, which then serves as starting point to perform variation calculations.

In parallel the total monetary burden imposed on the final customer is determined by taking into account the annual turnover this product (group) contributes to the GNP (gross national product).

Based on the respective figures for environmental load and monetary burden of two or more products to be compared the relative positions of these products are calculated to construct the so-called eco-efficiency chart. According to the labeling concept, which is different to Cartesian xy-plotting, products in the upper right quadrant are eco-efficient, because they have low environmental load and cause less monetary burden. The contrary is true for those products located in the opposite lower left quadrant. In addition the diagonal running from the upper left to the lower right corner is a boundary to separate less eco-efficient products located on its left side from more eco-efficient ones on its right side.

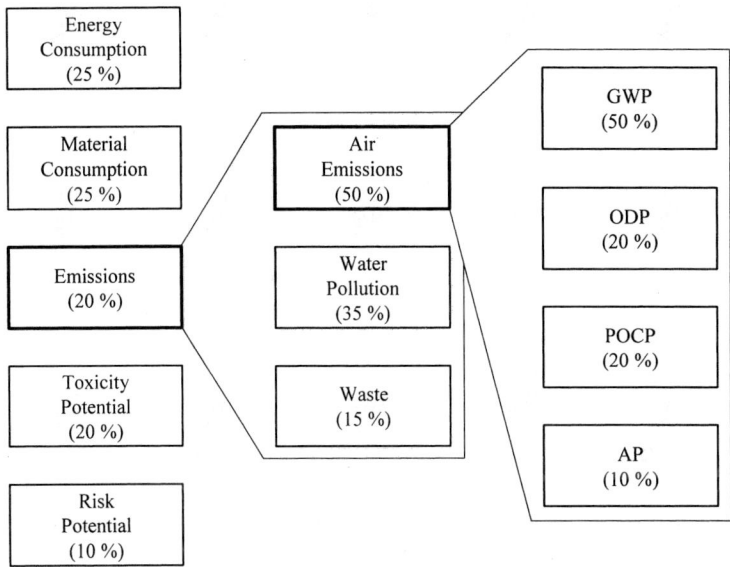

Fig. 2. Aggregation scheme based on "societal perception" for creating the ecological fingerprint and eco-efficiency chart.

There is always much debate that the selection of the weighting factors of Figure 2 will remarkably influence the result of a study. Indeed, this can be true in cases where the selection is over-emphasizing one or two categories like toxicity and risk potential.

As an example the initial values of the five weighting factors on the left side of Figure 2 are permutated. It becomes obvious that the product positions will shift as expected, but there is no dramatic change. Therefore, the principal decision based on the initial factors will not be affected as long as the monetary burden keeps constant as Figure 3 shows. For reference, the (x) marks show the result when all environmental categories are weighted equally. Even in this case the principal decision is not affected.

Case Study[3]

In vehicle construction, plastics have successfully replaced many conventional materials. Initially, most of the replacement occurred with interior parts. Use in the engine compartment represents a particular challenge for design and material properties. Intake manifolds made of Ultramid®, a BASF polyamide, are a classic example.

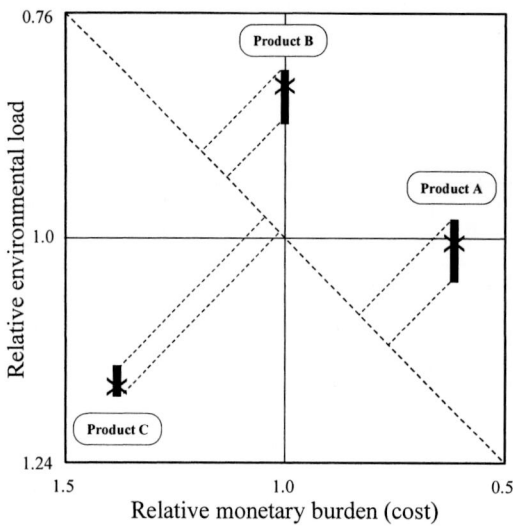

Fig. 3. Eco-efficiency chart that shows the effect when the weighting factors of Figure 2 are permutated.

These intake systems are geometrically complex components that in the past were sand cast from cast iron or aluminum. In the early 1980s, BASF began initial attempts to use plastics to manufacture these components. Even then, at a time when eco-efficiency or life cycle analyses were not widely known concepts, BASF engineers had set ambitious economic and environmental goals: The purpose of replacing metallic materials by polymers was to reduce costs and weight.

In order to be able to achieve these goals, it was first necessary to perfect the so-called lost core technology. In this process, a tin/bismuth core generally corresponding to the inner cavity of the intake system is first cast. This core is placed in a mold and plastic is injected around it. Then, the core is melted again, leaving behind the plastic intake system, which requires no further processing. The tin/bismuth alloy can be reused.

BASF has made key contributions to advancing this process to the large-scale production level and thereby helping plastic intake systems achieve a break-through. BMW was the first automobile manufacturer to use a system of this type in standard production: Beginning in early 1990, all of its 6-cylinder 4-valve engines were equipped with this manifold. Porsche followed in the same year with two intake manifolds for its 3.6-liter boxer engine used on the "911".

However, the lost core technology is not the only way to make intake manifolds from Ultramid®. In the meantime a half-shell technology has been developed, in particular for applications in which the shape of the components is not extremely complex. In this technology, two injection-molded parts are manufactured separately and then vibration-welded together. The advantages over the lost core technology are substantially lower capital spending and manufacturing costs.

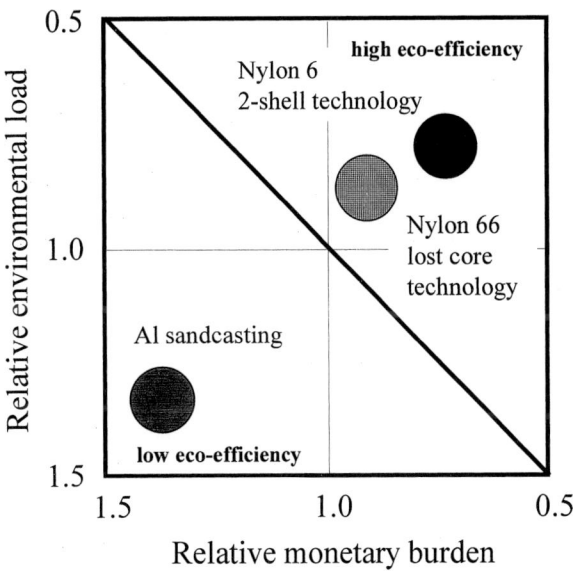

Fig. 4. Eco-efficiency chart showing the base case result when intake manifolds made of nylon 6, 66 and aluminum are compared.

Both processes have proven their effectiveness, and plastic intake systems are now the state of the art. About 70% of all passenger vehicles manufactured in Europe are equipped with such units. About 40% of these units are made of the BASF plastic Ultramid®. But, this development is not limited to Europe. Plastic intake systems have also gained a substantial market share in Asia and the USA.

There are economical and ecological reasons for this. First, in terms of economy: even though the lost core technology seems to be very complex and expensive at first glance, intake systems made of plastic are up to 50% less expensive than the equivalent metal design.

The life-cycle analysis is equally convincing. As Figure 4 shows, three different intake system designs were studied: one, manufactured from aluminum using the sand casting process, another manufactured from polyamide 6.6 using the lost core technology, and finally, a nylon 6 version fabricated from two individual shells. The analysis began by looking at the full life cycle of these components, from manufacturing, to use, to disposal.

Most important result: Compared to the metal version, both plastic versions have obvious environmental advantages. For one, less energy is used to manufacture them. For another, they are lighter and therefore reduce fuel consumption when the vehicle is being driven. However, there are differences between the plastic types. The complex lost core technology is somewhat more energy intensive than the two-shell technology. Added to this is the fact that nylon 6 has a somewhat better LCI (life-cycle inventory) sheet when manufactured than does nylon 66. Therefore, the intake system produced by joining two shells together comes out on top.

As for disposal of old parts, in the current case it has been assumed that the aluminum would be recycled while the plastic would end up in a landfill.

Conclusion

More than 150 eco-efficiency projects performed by BASF have proven that sound decision-making based on eco-efficiency charts can be done without meticulously create overmuch detailed LCI data, a procedure that is cost and time consuming. Existing data or wisely created data sets that are based on reasonable expertise are sufficient for decision-making in enterprises when a set of basic principles is observed consistently.

[1] P. Saling, A. Kicherer, B. Dittrich-Krämer, R. Wittlinger, W. Zombik, I. Schmidt, W. Schrott, S. Schmidt, *Int J LCA* **2002**, *7*, 203.
[2] W. Klöpffer, I. Renner in: *"Methodik der produktbezogenen Ökobilanzen"*, Texte 23/95, Umweltbundesamt, Berlin 1996, p.1ff.
[3] BASF Aktiengesellschaft, *"Ultramid® intake manifolds for internal combustion engines"*, http://www.basf.de /en/corporate/sustainability/oekoeffizienz/oea_projekte/saugrohre.htm, Ludwigshafen, 8-Nov-2002.

Macromol. Symp. **2003**, *201*, 261—269

Living Polymerization of Cyclic Esters - a Route to (Bio)degradable Polymers. Influence of Chain Transfer to Polymer on Livingness

Stanislaw Penczek, * *Ryszard Szymanski, Andrzej Duda, Jolanta Baran*

Centre of Molecular and Macromolecular Studies, Polish Academy of Sciences, Sienkiewicza 112, 90-363 Poland. E-mail: spenczek@bilbo.cbmm.lodz.pl

Summary: Polymerization of cyclic esters leads to (bio)degradable polymers of the increasing industrial importance. These polymerizations are of the living nature, although chain transfer to polymer with chain scission may cause deviations from the livingness and introduce structural differences (e.g. in end-groups), important for physical properties. Two different systems are discussed. In the first one two living macromolecules react one with another and reproduce two living macromolecules, retaining the same reactivities and the same end-groups. Polymerizations of ε-caprolactone and lactide belong to this category. On the other hand, polymerization of cyclic carbonates proceeds with chain transfer, in which disproportionation of the living chains takes place: from two living macromolecules one "dead" and one "doubly active" can be formed. Conditions of retaining the livingness in terms of the ratios of the rate constants of transfer, reinitiation, and propagation are discussed.

Keywords: chain-transfer, L,L-dilactide, living polymerization, MALDI, polyesters

Introduction

Polymerization of cyclic esters is not any more an area of academic curiosity but has become an important field in industrial polymers.[1] This is because on top of the previously technically important polymers like polyolefins, polymers of vinyl monomers or polymers made by ring-opening polymerization, a novel family of polymers appears, namely (bio)degradable polymers. Some are derived from the renewable resources. The major polymer based on the renewable resources is polylactide. Moreover, polylactide can be made from various wastes, as it has recently been demonstrated, and is perfectly biodegradable. Thus, polylactide is on the crossroads of three important tendencies: renewable raw materials, the use of the municipal waste, and biodegradability. Another important class of polymers is based on the polymerization of cyclic carbonates, degrading (e.g. hydrolytically) without giving accumulated acidic products.

Polymerization of cyclic esters has already been studied in detail.[2] Although some motivation to these studies may come from the above mentioned technological reasons, for a

DOI: 10.1002/masy.200351129

polymer chemist these processes open also a way to study novel living processes, in which there are phenomena non existing in vinyl polymerizations. One of the sources of this difference stems from the nature of the repeating units in the chains. In vinyl polymerization chains are built on the carbon-carbon linkages whereas in polymerization of cyclic esters the ester bonds are formed, the same ones as attacked in monomer and opened in the propagation step. Therefore, these repeating units are prone to similar attack and bond breaking in the chain. The understanding of these processes, apart from their fundamental importance, are also vital for the practical reasons, because depending on their nature and contribution to the overall polymerization process, polymers differing substantially in the molar mass distribution are formed. Different distributions influence the final physical properties of the involved polymers. Even more dramatic are these processes in highly branched, star-like polymers.[3]

Results and Discussion

(a) Chain Transfer in Polymerization of Cyclic Esters

There are two types of chain transfer to polymer with chain scission: namely the intra- and intermolecular transfers, as shown schematically below:

- intramolecular chain transfer- back biting (leading to cyclics formation):

$$(1)$$

- intermolecular chain transfer:

    (a) either conserving the activities of both chains

$$(2)$$

    (b) or disproportionating living chains into "dead" and "doubly active" chains

$$(3)$$

The first kind (intramolecular) of chain transfer has previously been studied quantitatively in our laboratory in the polymerization of ε-caprolactone:[4-6]

$$
...\text{-}\overset{O}{\overset{||}{C}}CH_2CH_2CH_2CH_2CH_2O\overset{O}{\overset{||}{C}}CH_2CH_2CH_2CH_2CH_2O...\overset{O}{\overset{||}{C}}CH_2CH_2CH_2CH_2CH_2O^* \xrightarrow{k_{tr(1)}}
$$

$$
\longrightarrow ...\text{-}\overset{O}{\overset{||}{C}}CH_2CH_2CH_2CH_2CH_2O^* + \overset{O}{\overset{||}{C}}CH_2CH_2CH_2CH_2CH_2O...\overset{O}{\overset{||}{C}}CH_2CH_2CH_2CH_2CH_2O
$$

(4)

(where ...O* denotes active species, either anions (e.g. alcoholate anion with omitted cation) or ...O-Mt, i.e. covalent alcoholate bond , e.g. ...-O-AlR$_2$, ...-OSnOC(O)R etc., on which propagation proceeds by concerted addition- insertion mechanism).

It has been established that $k_p/k_{tr(1)}$ (i.e. selectivity of polymerization) depends on both reactivity of active centers (given by k_p) and the size (i.e. the occupied volume) of these species.[5, 6]

For the intermolecular chain transfer there are, as indicated above, two possibilities: in the first one two living macromolecules (i.e. both fitted with active species) react one with another and give back two living macromolecules. Actually, in polymerization of lactones this is the only possibility, as shown below for ε-caprolactone (CL) (independently whether the active species are carboxylate anions, like in polymerization of β-propiolactone or alcoholate anions, like in polymerization of CL):

$$
...\text{-}\overset{O}{\overset{||}{C}}CH_2CH_2CH_2CH_2CH_2O\overset{O}{\overset{||}{C}}CH_2CH_2CH_2CH_2CH_2O^* +
$$
$$
...\text{-}\overset{O}{\overset{||}{C}}CH_2CH_2CH_2CH_2CH_2O\overset{O}{\overset{||}{C}}CH_2CH_2CH_2CH_2CH_2O^* \underset{k_{tr(2)}}{\overset{k_{tr(2)}}{\rightleftharpoons}}
$$

$$
\rightleftharpoons ...\text{-}\overset{O}{\overset{||}{C}}CH_2CH_2CH_2CH_2CH_2O\overset{O}{\overset{||}{C}}CH_2CH_2CH_2CH_2CH_2O\overset{O}{\overset{||}{C}}CH_2CH_2CH_2CH_2CH_2O^* +
$$
$$
+ ...\text{-}\overset{O}{\overset{||}{C}}CH_2CH_2CH_2CH_2CH_2O^*
$$

(5)

Thus, these processes do not change the number of living macromolecules: the total number of growing macromolecules remains constant, although a given living macromolecule is

264

changing its size a number of times during propagation due to the chain transfer reactions. Thus, propagation is not the only process responsible for the polymerization degree.

For the intramolecular chain transfer we were able to find the relationship between the selectivity and structure of active centers. Therefore, we have also been looking for the same dependence in the intermolecular transfer. The kinetic scheme that has to be considered is given below. Schematically:

$$P_i^* + M \xrightarrow{k_p} P_{i+1}^* \quad \text{(propagation)}$$

$$P_i^* + P_j^* \underset{k_{tr(2)}}{\overset{k_{tr(2)}}{\rightleftharpoons}} P_{i+n}^* + P_{j-n}^* \tag{6}$$

We were not able to find an analytical solution of this scheme although the corresponding differential equations could easily be formulated.[6-9] Nevertheless, we used another approach, namely we constructed nomograms, allowing to find the dependence of $k_{tr(2)}/k_p$ from the plot of M_w/M_n on monomer conversion (α), determined experimentally.[6, 7, 9, 10]

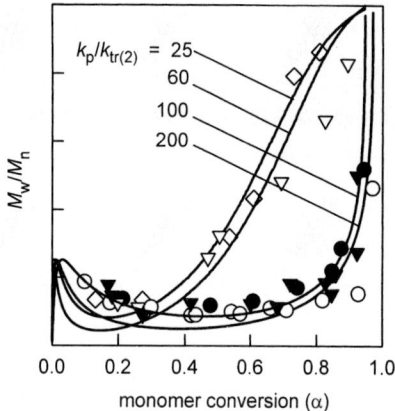

Fig. 1. Dependencies of M_w/M_n on monomer conversion determined for L,L-dilactide (LA) polymerization initiated with (in brackets are given $[M]_0/[I]_0$ ratios in which $[I]_0$ denotes starting concentration of the alkoxide group): ($\Diamond$, 10^2)Bu$_3$SnOEt, (∇, 2.4×10^2) Fe(OEt)$_3$, ($\blacktriangledown$, 3.3×10^1) Al(O^{i}Pr)$_3$, ($\bullet$, 5.6×10^1) Sn(Oct)$_2$, (O, 5.6×10^1) Sn(OBu)$_2$, THF solvent, 80 °C or 20 °C for Sn(OBu)$_2$. Points experimental, lines computed assuming $k_p/k_{tr(2)}$ = 25, 60, 100, and 200, respectively (ref.[11]).

The course of these lines depends also on the starting $[M]_0/[I]_0$ ratio (where $[M]_0$ and $[I]_0$ stand respectively for starting concentrations of monomer and initiator). This is clearly indicated in Figure 1, where two sets of lines (and corresponding experimental data) are related to different ratios [M]₀/[I]₀.

Another way to observe the rate of transfer, particularly in polymerization of lactide (LA), is to monitor the MALDI TOF mass spectrometry. When there is no transfer only macromolecules with even number of repeating units appear (LA is a dimer of lactic acid). Transfer leads to macromolecules with uneven number of units. This is illustrated in Figure 2a and b.

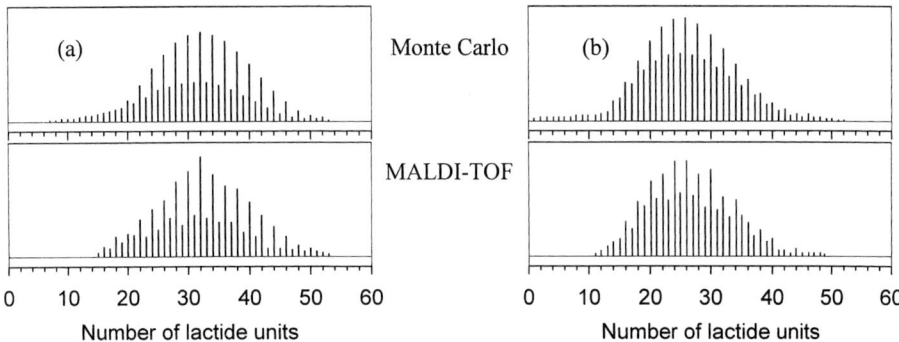

Fig. 2. Relative concentrations of oligomers in L,L-dilactide (LA) polymerization as observed by MALDI TOF and obtained by Monte Carlo computations. $[LA]_0$ = 1.2 mol/L, [Sn(II) octanoate]$_0$ = 0.05 mol·L^{-1}, [Butanol]$_0$ = 0.01 (a), 0.03 (b) mol·L^{-1}, solvent THF. Conversions, DP_n, and DP_w/DP_n, respectively: 76%, 30.1, 1.108 (a); 86%, 25.4, 1.116 (b). The estimated kinetic parameters: $k_p/k_{tr(2)}$ = 83.3, $k_{tr(1)}/k_{tr(2)}$ = 1.07×10^{-3} mol·L^{-1}.[12]

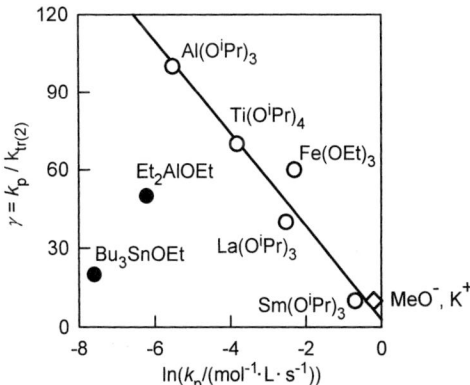

Fig. 3. Dependence of $\gamma = k_p/k_{tr(2)}$ on ln k_p determined in polymerizations of L,L-dilactide initiated by metal alkoxides. Covalent alkoxides: (O, ●); ionic alkoxide - MeOK: (◊). Conditions: $[LA]_0$ = 1.0 mol·L^{-1}, THF solvent, 80°C (MeOK initiated polymerization - at 20°C) (ref.[6]).

Then in Figure 3 (taken from ref.[6]), the dependence of $k_{tr(2)}/k_p$ on k_p is given. It follows that, this dependence holds only for a series of active species relatively similar in "size". Particularly low $k_{tr(2)}/k_p$ has been observed for aluminum *tris*-acetylacetonate (already mentioned in ref.[14] but not quantitatively studied). This preliminary result indicates, that the actual dependence of $k_{tr(2)}/k_p$ on the structure of active centers should include simultaneously reactivity and steric features. Similar phenomena have been observed in the polymerization of CL, where $k_{tr(1)}$ (for back-biting) depended on both k_p and the size of active centers.[4-6] The most striking example could be given by comparing ...-O-Al(C_2H_5)$_2$ and ...-OAlO^{i}Bu$_2$ active centers, that propagate CL with identical k_p but $k_{tr(2)}/k_p$ is for the former higher almost two times than for the latter.

(b) Chain Transfer with Disproportionation of Active Centers

In the previous paragraph chain transfer to polymer with chain scission and retention of activities in both interacting living macromolecules was discussed. However, as it has already been mentioned, this is only characteristic for some polyesters. In polycarbonates an additional reaction takes place as illustrated by Equation 3. Polycarbonates are "symmetrical esters" and in these symmetrical structures chain transfer with chain scission may proceed in an additional way: two living macromolecules, with one active center each, may react one with another, giving (cf. Equation 3) one "dead" macromolecule and one macromolecule with two active centers. For example, for trimethylene carbonate:

$$\ldots\text{-COCH}_2\text{CH}_2\text{CH}_2\text{OCOCH}_2\text{CH}_2\text{CH}_2\text{O}^* \ + \ \ldots\text{-COCH}_2\text{CH}_2\text{CH}_2\text{OCOCH}_2\text{CH}_2\text{CH}_2\text{O}^* \ \underset{k_{tr(2)}/2}{\overset{k_{tr(2)}/2}{\rightleftharpoons}}$$

$$\ldots\text{-COCH}_2\text{CH}_2\text{CH}_2\text{OCOCH}_2\text{CH}_2\text{CH}_2\text{OCOCH}_2\text{CH}_2\text{CH}_2\text{O}^* \ + \ \ldots\text{-CCH}_2\text{CH}_2\text{CH}_2\text{O}^* \quad (7a)$$

$$\ldots\text{-COCH}_2\text{CH}_2\text{CH}_2\text{OCOCH}_2\text{CH}_2\text{CH}_2\text{OCOCH}_2\text{CH}_2\text{CH}_2\text{OC-}\ldots \ + \ ^*\text{OCH}_2\text{CH}_2\text{CH}_2\text{O}^* \quad (7b)$$

"dead"   "doubly active"

For the first time the evidence of the existence of the process according to Equation 7b in the polymerization of cyclic esters has recently been given on the basis of the analysis of the

MALDI TOF spectra of poly(trimethylene carbonate) obtained with $(n\text{-}C_4H_9)_2SnOC_2H_5$ as initiator.[13]

According to the present knowledge, if the polymerization process were without any chain transfer with reshuffling of structures, then the end-groups should exclusively be **HO-** and **-OC$_2$H$_5$** as shown below:

$$R_3SnOC_2H_5 + n \quad \text{(cyclic carbonate)} \longrightarrow R_3SnOCH_2CH_2CH_2O\overset{O}{\overset{||}{C}}(OCH_2CH_2CH_2O\overset{O}{\overset{||}{C}})_{n-1}OC_2H_5 \longrightarrow$$

$$\xrightarrow[\text{termination}]{H_2O + "H^{\oplus}"} HOCH_2CH_2CH_2O\overset{O}{\overset{||}{C}}(OCH_2CH_2CH_2O\overset{O}{\overset{||}{C}})_{n-1}OC_2H_5 \tag{8}$$

Chain transfer according to Equation 7a would provide exclusively end-groups formed in initiation and termination. However, chain transfer according to Equation 7b (with disproportionation) would introduce two more sets of macromolecules, derived from "dead" and "doubly active" macromolecules containing either two $-OC_2H_5$ or two $-OH$ end-groups. The existence of macromolecules with these end-groups is clearly seen in the MALDI TOF spectrum given in Figure 4, where for every polymerization degree there are three different kinds of macromolecules, namely HO∼∼∼OC$_2$H$_5$ (1), C$_2$H$_5$O∼∼∼OC$_2$H$_5$ (2), and HO∼∼∼OH (3). This particular variety of the end-groups has never been observed in polymerization of either CL or LA, where there is no way for disproportionation during chain transfer to occur. Because of equal probabilities of breaking of two ester bonds of the carbonate moieties the equilibrium constant of the reaction 7b is equal to ¼. Consequently the concentration of monofunctional chains in equilibrium is twice as high as the concentration of "dead" or "doubly reactive" chains.

Comparison of the Monte-Carlo computations of the chain transfer with and without disproportionation revealed surprisingly small difference between these two mechanisms in their influence on the dependence of M_w/M_n on monomer conversion (α). This small difference can be explained by formation in the chain transfer reactions of new chains with the same molar mass distribution (MMD) for the processes with or without disproportionation (the only difference is in the structure of chain ends). The existing differences in total MMD of polymers stem from the two times faster propagation on the "doubly active" chains then on the "normal " chains (the dead chains, of course, do not propagate at all). However, the MMD of the sum of dead and "doubly reactive" chains is not much different from the MMD of the monofunctional chains. The difference between these MMDs decreases with increasing the rate of reshuffling of the polymer segments.

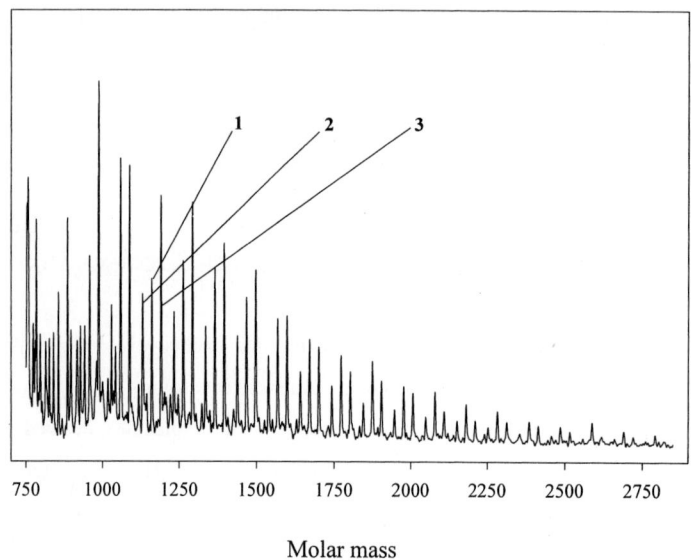

Fig. 4. MALDI-TOF-ms spectrum of poly(trimethylene carbonate) obtained in polymerization of trimethylene carbonate (TMC). Conditions: $[TMC]_0 = 1$ mol·L^{-1}, initiated with Bu$_3$SnOEt ($[Bu_3SnOEt]_0 = 0.01$ mol·L^{-1}) at 80°C in THF; conversion 12%; $M_n = 1200$, $M_w/M_n = 1.09$. Numbers: 1, 2, and 3 indicate signals of polymer chains differing in end-groups: HO~~OEt, EtO~~OEt, and HO~~OH, respectively.

(c) Chain Transfer and "Livingness" in Polymerization of Cyclic Esters.

According to the provisional IUPAC definition, living polymerization takes place when irreversible chain transfer or termination is absent. Thus, the reversible chain transfer is allowed; it means, that active center is transferred from a living macromolecule to another molecule, which starts sooner or later the growth of another chain (reinitiation with rate constant k_{ri}). If the rate constant of reinitiation is high enough (i.e. $k_{ri} \geq k_p$), than chain transfer is "ideal" and no retardation is observed. If $k_{ri} < k_p$ then the chain transfer is of degradative character. It looks that in the polymerization of cyclic esters, at least without disproportionation, chain transfer is ideal, no chain carriers die: all macromolecules are living all the time and all of the end-groups are the same. Therefore, this process belongs to living polymerizations. When disproportionation takes place then polymerization is on the borderline of livingness and whether all the macromolecules retain or not the ability to grow depends on the k_p/k_{tr} ratio. If $k_{tr} \ll k_p$ then transfer is ineffective but the "dead" macromolecules also ineffectively come back to propagation, making in this way polymerization nonliving. If, however, $k_{tr} \gg k_p$ then the dead macromolecules are quickly transformed into the living ones, and then the polymerization belongs to the living category.

Conclusion

In polymerization of cyclic esters, like ε-caprolactone (CL), lactides (LA), or cyclic carbonates (e.g. trimethylene carbonate (TMC)), taken as examples, chain transfer to polymer molecules with their scission takes place. This transfer, independently of the actual mechanism, broadens the molar mass distribution. When two living macromolecules react one with another in polymerization of CL or LA two other living macromolecules emerge and their ability to grow is identical to this of the parent ones. This behavior stems from the "unsymmetrical" structure of the ester bonds in contrast to the "symmetrical" (-OC(O)O-) in the case of polycarbonates. This symmetrical structure is responsible for two ways of breaking of the attacked chain. One way is identical to the way of breaking in "unsymmetrical" esters and the other one leads to disproportionation. In the latter reaction two living macromolecules give one "dead" (at least temporarily) and one "doubly active", i.e. with two active centers. This mechanism was for the first time established for the polymerization of cyclic carbonates by applying MALDI TOF analysis. Three different populations of macromolecules were observed with end-groups clearly indicating the presence of structural disproportionation of macromolecules undergoing chain transfer.

Acknowledgement

This work was supported by the Polish State Committee for Scientific Research (KBN), grant 7 T09A 144 21.

[1]   *Biopolymers*, Vol. 4: "Polyesters III - Applications and Commercial Products", A. Steinbüchel, Y. Doi, Eds., Wiley-VCH, Weinheim 2002.
[2]   *Biopolymers*, Vol. 3b: "Polyesters II – Properties and Chemical Synthesis", A. Steinbüchel, Y. Doi, Eds., Wiley-VCH, Weinheim 2002.
[3]   R. Szymanski, *Macromolecules* **2002**, *35*, 8239.
[4]   A. Hofman, S. Slomkowski, S. Penczek, *Makromol.Chem., Rapid Commun.* **1987**, *8*, 387.
[5]   S. Penczek, A. Duda, S. Slomkowski, *Makromol.Chem., Macromol.Symp.* **1992**, *54/55*, 31.
[6]   J. Baran, A. Duda, A. Kowalski, R. Szymanski, S. Penczek, *Macromol.Symp.* **1997**, *128*, 241.
[7]   S. Penczek, A. Duda, R. Szymanski, *Macromol.Symp.* **1998**, *132*, 441.
[8]   R. Szymanski, *Macromol. Theory Simul.* **1998**, *7*, 27.
[9]   R. Szymanski, J. Baran, *Macromol. Theory Simul.* **2002**, *11*, 836.
[10] J. Baran, A. Duda, A. Kowalski, R. Szymanski, S. Penczek, *Macromol.Rapid Commun.* **1997**, *18*, 325.
[11] S. Penczek, T. Biela, A. Duda, *Macromol.Rapid Commun.* **2000**, *21*, 950.
[12] J. Baran, R. Szymanski, *Polimery(Warsaw)* **2003**, in press.
[13] J. Baran, R. Szymanski, S. Penczek, in preparation.

Macromol. Symp. **2003**, *201*, 271—281

Bioprocessing – No Longer a Field of Dreams

Daniel J. Sawyer

Cargill Dow LLC, 15305 Minnetonka Blvd., Minnetonka, MN 55345, USA
E-mail: dan_sawyer@cargilldow.com

Summary: A Field of Dreams was a movie about baseball. The theme was "Build it and They Will Come." Bioprocessing as an industry has been waiting for someone to step up to the plate and build a large business, based on the technology. The large business was necessary to bring about a revolution to move the Chemical and Plastics Industries towards a new business model, based on more sustainable practices. By providing a unique combination of properties in fiber and packaging applications, PLA was the obvious first choice for companies looking to make bioprocessing a commercial reality.

Keywords: biopolymers, fibers, films, polylactide (PLA), recycling, renewable resources, sustainability

What is PLA?

PLA is the first product to be commercialized in the plastics industry on such a scale, based on renewable resources. Starches are extracted from common agricultural products such as corn or sugar beets. They are then converted by hydrolysis to sugars such as dextrose. Dextrose is utilized by microorganisms that ferment them to lactic acid, the starting material for the PLA production process.

PLA can be produced from lactic acid by direct coupling, using isocyanates. Typically, isocyanates are hazardous chemicals that aren't accepted as safe to use, or for the environment. Although direct condensation allows for easy use of a wide range of potential comonomers, producing polymers of useful molecular weights requires production in a solvent system, in order to handle the increasing viscosity as molecular weight increases.[1] This then means the extra burden of removing the residual solvent to purify the polymer. So far, this has limited polymer produced in this fashion to applications that don't involve food contact. Alternatively, using the lactide intermediate route, PLA with high molecular weight can be produced without the use of solvents (see Figure 1).[2]

          DOI: 10.1002/masy.200351130

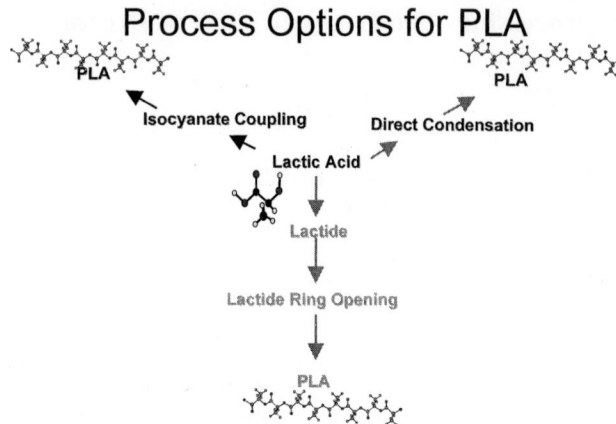

Fig. 1. Process options for PLA production.

LA Makes Multiple Lactide monomers which give rise to a family of polymers

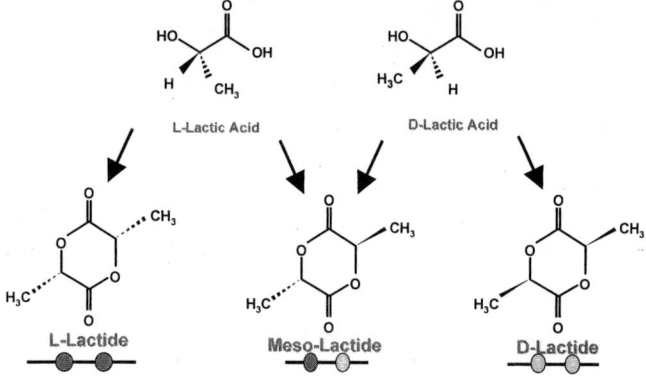

Fig. 2. Multiple lactide monomers arising from combining optically active lactic acid units.

The lactide route also allows for better control of the polymer properties by controlling the optical sequence of the polymer chains. For these reasons, the first commercial PLA process has utilized the lactide route. Since lactic acid is an optically active molecule, forming the cyclic dimer, lactide, allows for the production of three different stereoisomers. D-lactide and L-lactide maintain the stereoactivity of the two lactic acid units they are comprised of. Meso-lactide, made up of one L and one D-unit, is optically inactive (see Figure 2).

Reducing the optical purity from nearly perfect to about 10-12% D allows a family of polymers to be produced with a range of crystalline melting points from about 130°-172°C. Lower melting components are useful for binding or sealing in fibers or films.[3]

Whatever the production route, PLA is used in typical conversion processes to produce fiber, sheet or films (see Figure 3).

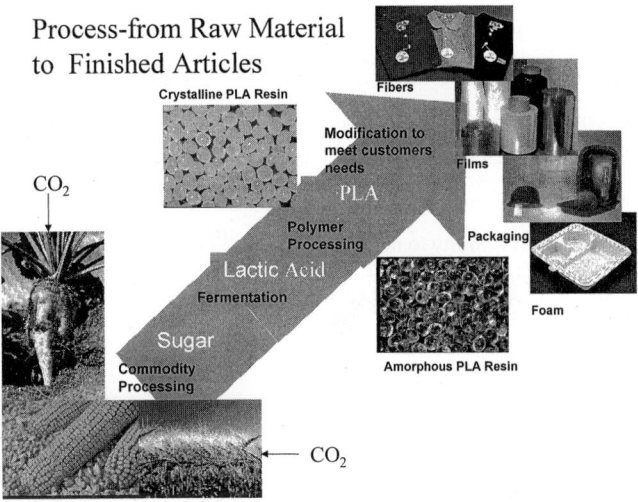

Fig. 3. Producing fibers, packaging and films from annually renewable resources.

Renewable Resource Base

When people consider sustainable materials, one of the most important aspects is the renewable resource base. In it's first commercial generation, PLA is being produced from corn. Although lactic acid is produced globally from sugar beets and could also be produced from wheat, casava, or other starch sources, corn is the most economically sensible option, initially. However, technology is being developed which will allow biomass to be used as the feed source. By-products like corn stover, which are not fully utilized today could conceivably be used as the feed material for lactic acid production. Although petrochemical resources could be used as the feedstock for PLA, there are important benefits which only the natural route provides. The precise control of the stereochemistry of the polymer chains is only possible using the natural route, since the microorganisms which produce lactic acid produce primarily one type of optically active form. Petrochemical-based lactic acid is of the racemic variety. Racemic mixtures will only allow for production of amorphous polymers, not the crystalline ones necessary to be used in heat-stable fiber and biaxially-oriented films.

Need for a More Sustainable Alternative

While population growth and increases in GDP are creating an increased demand for plastic, petrochemical resources and in some cases natural materials are not meeting the demand. For example, cotton production is not expected to meet the increased fiber demand as more people use more clothing. This increased demand will only be met through other fibers like polyester or new alternatives (see Figure 4). Continued depletion of the earth's fossil resources will eventually limit the earth's ability to meet the demands.[4]

The World Fiber Market is Expected to Grow

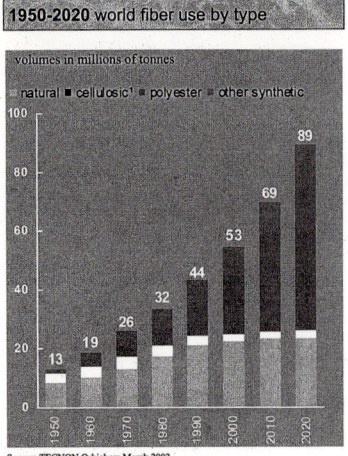

- world consumption of all fibers expected to continue upward trend
- polyester is key driver for market growth
- natural fibers have peaked
- cellulosic share of total consumption continues to be replaced by synthetic
- future consumer needs will be met through innovative fiber developments

Fig. 4. Cotton alone cannot meet the world's increasing demand for fibers.

Sustainability

Sustainability, in a single word explains the reason there is a significant market developing for PLA. Although Sustainability is somewhat complex to define, it obviously involves reducing the impact on the environment of the production of what people need and want. However, true sustainability also involves other aspects, which allow for a longer-term improvement. People often use the phrase "Triple Bottom Line," to include economic viability and social impact with the obvious environmental piece.[5]

Economic Viability

Clearly, companies trying to manufacture more sustainable products will not stay in business long-term, if they can't make money at the same time. It is only when products are commercially produced that the impact can be felt. Actually, economic viability can be extended to include price and performance. Consumers want products that perform at least as well as incumbents, but offer a sustainability advantage. However, they are not typically willing to pay more unless a product performs better than what they are used to.

Environmental Impact

The environmental impact portion of sustainability is the obvious part that most people discuss when looking at plastics in their applications. It involves not depleting the earth's ability to provide the same products for future generations. It not only involves using renewable resources, but also includes recycling processes, energy use, by-products of manufacturing and effects along the whole chain from resource to end-user, along with disposal methods. Traditional polymer recycling brings articles back to the melt state after their use. However, recycled polymer typically is not as pure and useful as "virgin" polymer. Thermal degradation occurs with all plastics to some extent, each time the material is melt processed. With some polymers, it is thermal oxidative degradation or crosslinking. With others, it is hydrolysis. In any case, traditional recycling is simply delaying the point at which the polymer reaches the end of it's use and needs to be disposed of. Traditional disposal is then done, either landfilling or incineration. In more complete, non-traditional recycling, work has been done with nylon, trying to return carpet fibers to caprolactam, but costs of separation and recovering the monomer have been prohibitive. With PLA, a more complete recycling loop can be performed. PLA can be brought back to the lactide intermediate. But more importantly, it can be hydrolyzed completely back to lactic acid. This means the recycling will allow articles to be produced in successive recycling steps that are as useful as in their first use. A key difference with PLA is that when the useful life is over, PLA adds the disposal option of composting, which is not an option for other synthetic polymers. Under very specific composting conditions-high temperatures and humidities, typically around $60^{\circ}C$, with 90% relative humidity, the PLA articles will degrade first by hydrolysis, then by microbial action, eventually degrading simply to carbon dioxide and water.[6]

Social Impact

Not only do sustainable products have an environmental and economic impact, but there is also necessarily a positive social impact. Products need to improve the quality of life for the consumers, the producers and the people affected by the production facility. Human resources must be viewed in the same way as natural resources in that abusing them will result in depletion of the resource. In thinking about sustainability, it is important to consider all three components of "The Triple Bottom Line."

Limited Supply of Petroleum Resources

As previously stated, sustainability is the driver that explains the need for PLA. Although petrochemicals provide the resource for traditional synthetic packaging and fiber articles, there is evidence that the world's oil reserves are inadequate to keep pace with the growth in demand. Even considering all projected reserves of oil, production is expected to peak in the next 20 to 75 years. This is assuming zero growth in demand! Assuming a more realistic demand increase of 1-3%, production will peak sooner.[7] As resources become scarce and reserves are used up, it becomes increasingly difficult to recover the oil and production will decrease. Furthermore, it is important to consider also where the oil reserves are located. Obviously, the concentration of oil reserves is in the middle east, with smaller reserves around the Pacific Rim and in the U.S.. Hubbart has predicted that peak production will be very close in time to the point at which 50% of the world's supply is depleted.[8] Considering this, as the reserves are depleted, very little will be available in some of the areas of highest demand (see Figures 5 and 6). Clearly, the resource for packaging and fibers must change![9]

Billions of Barrels of Known Oil Reserves Today

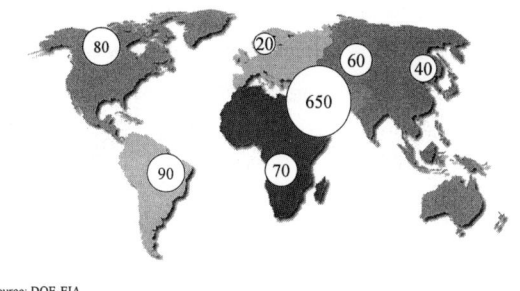

Source: DOE-EIA

Fig. 5. Known oil reserves today.

Future Look: 50% Evenly Depleted

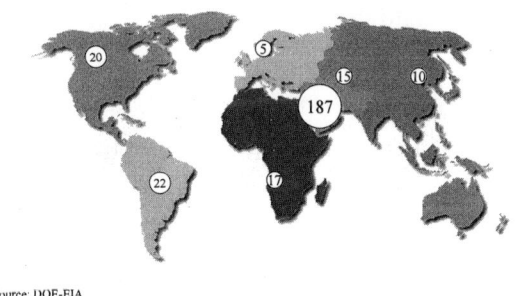

Source: DOE-EIA

Fig. 6. Oil reserves when 50% Depleted.

Performance of PLA

PLA polymer provides performance advantages in fiber and packaging applications. PLA film has FDA approval for food contact. PLA films do not contain any hormone mimics. PLA can be converted into film on standard processing equipment. The films have excellent gloss and clarity. The unique modulus of the films gives good twist and dead-fold properties. While PLA films do not provide the moisture vapor and oxygen barrier necessary for some types of food packaging, they do provide good barrier for certain flavors and aromas. They also provide good oil and grease resistance. With a range of melting temperatures possible through control of the stereochemistry, a broad range of 100% PLA heat-seal performance packages can be produced.

In fiber applications, end users take advantage of the loft and resilience of PLA fibers. PLA fibers wick moisture well, without absorbing large amounts of water. PLA burns with very low energy and under conditions of complete combustion, very little smoke is produced and no toxic chemicals are released. PLA fibers do not cause allergic reactions. Like other synthetic fibers, PLA fibers do not support bacterial growth. Due to a unique absorption spectrum, PLA fibers are naturally resistant to degradation by UV light. PLA fabric retains its strength in a similar way to acrylic, but does not yellow like acrylic or polyester. Lastly, PLA fibers give similar or better protection against stains, when compared with nylon.

PLA is Being Commercialized

Part of making PLA economically viable is producing it at low cost. To do so, the fermentation that supplies the lactic acid for polymer production and the polymer production must be done at large scale, to minimize manufacturing cost. This is why the first commercial PLA plant was built to produce 140,000 metric tons per year, while the lactic acid plant is scaled to 180,000 Metric Tons per year. The PLA plant has been operating since November, 2001 and the lactic acid plant is starting in the 1st quarter of 2003. As these plants move to full production capacity, the cost to produce PLA will come in line with more traditional performance polymers. Several packaging applications have been commercialized using PLA in Europe and Japan, including Dunlap Golf Ball sleeves in Japan and deli-food packages in grocery stores in Europe. In fiber applications, the largest launch to date has been Pacific

Coast Feather's launch of their natural balance^{TM} line of sleep products. Including pillows, comforters, mattress pads and fiber beds, consumer interest in the line of products has far exceeded expectations. End-users seem very positive about the performance of high-end synthetic fibers derived from annually renewable resources. Equally receptive have been grocery and electronics purchasers in Europe and Japan who have welcomed deli-trays and mini-disk packages, even blister packs for the popular Walkman^{TM} radios.[10]

Renewable Resources can be Used to Make a Range of Chemicals and Polymers

The fermentation process used to produce lactic acid actually is more meaningful than just for PLA. Successful large-scale production of lactic acid and lactide makes it possible to produce a whole range of monomers for polymers other than PLA and a wide range of chemicals, all from renewable resources. For example, propylene glycol, propylene oxide, acrylic acid, all are possible from lactide. Lactate ester solvents have also been produced and are actually moving to the market, being produced from lactide. Interestingly, the purity of ethyl lactate produced from lactide, rather than from petrochemicals, is significantly more pure and is thus better suited for demanding electronics cleaning. See Figure 7 for the range of products that can be produced from lactide.

What Role Does Cargill Dow Play?

Cargill Dow, a joint venture formed in 1997 has as it's mission to "Produce products which meet the needs of the wold today, without compromising the earth's ability to meet the needs of tomorrow." This means taking a constantly reflective approach on sustainability and taking continued steps to reduce the impact of producing polymers and chemicals on the earth.

Conclusions

PLA polymer has moved bioprocessing forward, from the "Field of Dreams" stage to one where polymers for packaging and fibers, along with a wide range of chemicals can be produced from annually renewable resources. Since PLA is now being produced at commercial-scale, it can become an increasingly viable alternative to petrochemical incumbents. The unique chemistry of PLA provides a set of performance characteristics that fiber and packaging applications will benefit from. As Cargill Dow, converting customers and end-users make PLA fibers and packaging a success, the bio-industrial revolution will boom. Companies will commonly produce products that make people's lives better, while reducing the impact making those products have on the earth.

Many Chemicals Can Be Made

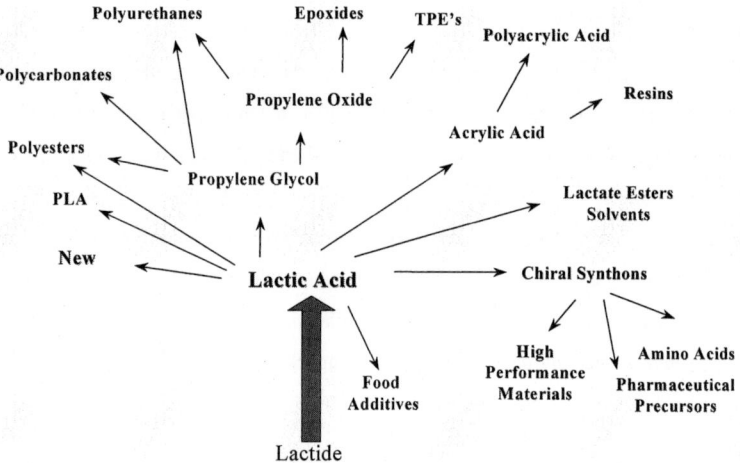

Fig. 7. Range of chemicals that can be produced from lactide.

Acknowledgements

I would like to acknowledge Dr. Patrick Gruber for providing many of the figures used in this paper. I would also like to acknowledge Dr. Jim Lunt for many figures used in this paper. Lastly, I would like to acknowledge the converters, brand-owners, retailers and especially consumers for beginning to change the way they think about plastics and sustainable practices.

[1] Mitsui Chemical, US Patent #571925

[2] Cargill, Inc., US Patent #5142023

[3] Sawyer, Daniel J., *PLA Technology and Applications*, Nonwovens World, Vol. 10 No. 2, April/May, 2001

[4] Technon Orbichem, March, 2002

[5] J. Elkington, *SustainAbility Conference*, 1996

[6] Press Release, Cargill Dow LLC, *Biocorp N.A. earns first certification for "compostable" food service items from the Biodegradable Products Institute*, September 17, 2002

[7] J. Wood and G. Long, *Long Term World Oil Supply*, Energy Information Agency, DOE, July 7, 2000, slide 19

[8] Dr. M. King Hubbert, *Energy from Fossil Fuels*, American Association for the Advancement of Science Centennial, pp. 171-177, 1950.

[9] Department of Energy, EIA

[10] Cargill Dow LLC internet Web Site, Various Press Releases, 2002

Microbial Degradation of Aliphatic Polyesters

Yutaka Tokiwa,[1] *Amnat Jarerat*[2]

[1]National Institute of Advanced Industrial Science and Technology (AIST),
Tsukuba Central 6, 1-1-1 Higashi, Tsukuba, Ibaraki 305-8566, Japan
Email: y.tokiwa@aist.go.jp
[2]C.P.R Co., Ltd., 4-102, Higashinobusue, Himeji, Hyogo 670-0965, Japan

Summary: Polyester-degrading ability of actinomycetes obtained from culture collections was investigated by the formation of clear zones on polyester-emulsified agar plates. Using 41 genera (43 strains) of actinomycetes with phylogenetic affiliations based on 16S rRNA sequences, poly(L-lactide) (PLA)-degraders were found to be limited to members of family *Pseudonocardiaceae* and related genera. On the other hand, poly(β-hydroxybutyrate) (PHB)-, polycaprolactone (PCL)-, and poly(butylene succinate) (PBS)-degraders were widely distributed in many families.

Keywords: actinomycetes, biodegradation, lipase, polyesters, *Pseudonocardiaceae*

Introduction

Serious problems regarding the treatment of plastic wastes have stimulated the development of material that can be decomposed after disposal to the environment by the activity of microorganisms to carbon oxide and water. The development of biodegradable plastics is considered to be one of the most desirable product innovations in order to resolve the problems on plastic waste.

We have reported that a fungus of *Penicillium* sp. strain 14-3, isolated from soil, almost completely degraded a kind of aliphatic polyester, poly(ethylene adipate) (PEA)[1] and its purified PEA-degrading enzyme was similar to lipase.[2] In 1977, we discovered that lipase and esterase from various microorganisms were able to degrade aliphatic polyesters such as PEA, PCL, etc.[3] According to this finding, aliphatic polyesters now are generally known to be susceptible to biological attack.

We also noted that the melting point (Tm) of plastics had a great effect on enzyme degradation, e.g., enzymatic degradabilities for the same series of polyesters decreased with increasing Tm. In general, the Tm of the polymer is determined by the values of ΔS (the change of entropy in melting) and ΔH (the change of enthalpy in melting). The high Tm of aromatic polyesters is caused by small ΔS value with increase in the rigidity of the polymer molecule based on an aromatic ring. Enzymatic degradation of copolyester (CPE) composed of aliphatic and aromatic polyester, by *Rhizopus delemar* lipase decreased with increasing rigidity (increase in aromatic polyester content) of the CPE molecule.[4] On the other hand, the high Tm of nylon (aliphatic polyamide) is caused by the large ΔH value based on the hydrogen bonds

© 2003 WILEY-VCH Verlag GmbH & KGaA, Weinheim

DOI: 10.1002/masy.200351131

among polymer chains. Enzymatic degradation of copolyamide-ester (CPAE), composed of aliphatic polyester and aliphatic polyamide, by *Rhizopus delemar* lipase decreased with an increase in the aliphatic polyamide content of the CPAE molecule.[5] Hence it was found that rigidity of CPE molecular chains, which is related to the ΔS value, and the ΔH value in the CPAE had a great effect, in addition to the chemical structure, on enzymatic degradation. The above described CPE and CPAE are now commercially available as biodegradable plastics under the names of Ecostar (Eastman Chemical Co., USA), Ecoflex (BASF Co., Germany) and BAK (Bayer Co., Germany), respectively. Furthermore, using various polymers such as PCL,[6,7] poly(β-hydroxybutyrate) (PHB),[8,9] poly(β-propiolactone),[10] poly(butylene succinate) (PBS),[11,12] poly(ethylene succinate),[13] polylactide (PLA),[14-16] poly(p-dioxanone),[17] polycarbonate,[18] etc, we made clear about their microbial degradation mechanisms.

Various kinds of biodegradable plastics are presently manufactured. Among the biodegradable plastics, aliphatic polyesters, such as PHB, PCL, PBS and PLA, have been widely explored. PHB is a natural aliphatic polyester that is produced by a wide variety of bacteria as an intracellular storage of carbon and energy. On the other hand, PCL and PBS are synthetic aliphatic polyesters that are currently available. PLA is polymerized from lactic acid, which can be prepared effectively by fermentation with renewable resources such as corn and tapioca starches.

Along with the development of biodegradable plastics, it is also important to gain much knowledge on microorganisms capable of degrading these polymers. We reported our study on the distribution of microorganisms able to degrade four representative aliphatic polyesters, e.g. PHB, PCL, PBS, and PLA. We used the plate count and clear zone methods to evaluate the distribution of polyester-degrading microorganisms in different soil environments and found the degrading microbes decreased in the order of PHB=PCL>PBS>PLA.[11,14,19,20] However, little is known about what family level or the group of microorganisms are responsible for the degradation of polyesters.[21]

A previous study among 25 strains of *Amycolatopsis* showed that they were able to form clear zone on more than one kind of emulsified-polyester plates, e.g. PCL, PHB, PBS and PLA. In addition to polyester-degrading activity, silk fibroin-degrading ability was also observed in these actinomycetes.[15] A similarity in chemical structures of a PLA monomer that is the L-lactic acid unit and L-alanine unit of silk fibroin were noted in our reports.[15,16,22,23]

To obtain further information on the microbial degradation of these commercially available degradable materials, here we attempt to investigate the polyester-degrading activity among actinomycetes obtained from culture collections with regard to their phylogenetic affiliation. The study was also extended to examine the silk fibroin-degrading activity of these actinomycetes.

Experimental Part

Materials

PLA, LACTY 1012 (number-average molecular weight, $M_n = 3.4 \times 10^5$) was obtained from Shimadzu Co. Ltd. Other biodegradable polyesters were obtained as follows: PHB, ($M_n = 1.4 \times 10^5$) from Mitsubishi Gas Chemicals; PCL, TONE P-767 ($M_n = 6.7 \times 10^4$) from Union Carbide Corporation, and PBS, BIONOLLE 1020 ($M_n = 3.5 \times 10^4$) from Showa High Polymer Co. Silk fibroin powder was obtained from Tosco Co. and washed with hot water (50°C) for 8 h under vigorous stirring to remove the water soluble components (repeated four times). Amino acid analysis showed that the silk powder was composed of glycine, 44.6 (mol-%); alanine, 29.5; and serine 11.7.

Microorganisms

All actinomycetes used were obtained from two culture collections, Japan Collection of Microorganisms (JCM) and Institute for Fermentation, Osaka (IFO). For the study of the phylogenetic positions of the actinomycetes, the published 16S rRNA gene sequences of the tested actinomycete genera in the DDBJ/EMBL/Genebank DNA sequence data libraries were used to construct a phylogenetic tree with the GENETYX 11.0 (Software Development Co., Japan) and CLUSTAL programs. The nucleotide sequence accession numbers used are shown in Figure 2.

Culture Media

Basal medium contained (per liter) 100 mg yeast extract, 200 mg $MgSO_4 \cdot 7H_2O$, 100 mg NaCl, 20 mg $CaCl_2 \cdot 2H_2O$, 10 mg $FeSO_4 \cdot H_2O$, 0.5 mg $Na_2MoO_4 \cdot 2H_2O$, 0.5 mg $Na_2WO_4 \cdot 2H_2O$, 0.5 mg $MnSO_4$, 1600 mg K_2HPO_4, 200 mg KH_2PO_4 and 1000 mg $(NH_4)_2 SO_4$ (pH 7.1). Polyester-emulsified culture plates were prepared by dissolving 1 g polymer pellets in 40 ml chloroform. Using an ultrasonic disruptor (Tomy UD-201, Tomy Seiko Co. Ltd.), the solution was emulsified into 1000 ml basal medium with 100 mg of surface active agent [Plysurf A210G Daiichi Kogyo Seiyaku, $RO(CH_2CH_2O)_nP(=O)(OH)OR'$; R, alkyl or alkylallyl group ; R',-H or $(CH_2CH_2O)_n$ R ; hydrophile-lipophile balance, 9.6]. After sonication, the emulsion was evaporated in vacuum at 40°C to remove the chloroform. Preparation of silk fibroin culture plates was carried out by dispersing the powder directly into basal medium with no addition of surfactant. Agar plates were prepared by the addition of 2% (wt./vol.) agar and pouring into petri dishes after autoclaving.

Degradation of Emulsified-polyester and Silk Fibroin on Agar Plates

The actinomycetes were streaked-out onto polyester-emulsified agar and silk fibroin agar plates. Degradability was determined by the clear zone formation around the colony on the opaque plates after 30 days of cultivation time at the optimum temperature of their growth.

Results and Discussion

The phylogenetic positions of the tested actinomycete genera with polyester-degrading ability were investigated using the clear zone method as previously described by Nishida and Tokiwa.[19] Figure 1 shows an example of clear zone formation on an agar plate emulsified with PLA after incubation at 30 °C for 2 weeks.

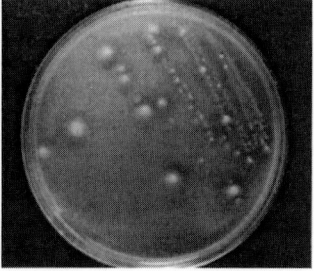

Fig. 1. Colonies of *Saccharothrix waywayandensis* JCM 9114 and clear zones formed on an agar plate with emulsified PLA.

The rationale for selecting the actinomycetes of this study was that (i) All of the strains used to study the relationship between phylogenetic positions and their polyester-degrading abilities are type strains, (ii) They have already been genetically classified and (iii) They are available from the culture collections. Since there were about 600 species in the genus of *Streptomyces*, three of the *Streptomyces* type strains were used to construct the phylogenetic tree. The phylogenetic positions of the tested genera based on 16S rRNA gene sequence data and their polyester-degrading abilities are shown in Figure 2. Interestingly, PLA-degraders were found to be limited to the family *Pseudonocardiaceae* and related genera. *Amycolatopsis, Saccharothrix,*

Lentzea, Kibdelosporangium and *Streptoalloteichus* listed in this family were found to have PLA-degrading abilities.

Several studies on microbial degradation of polyester have previously been studied by isolating the strains from the natural environment, and then analyzed their phylogenetic affiliations.[10,17,21,24] Suyama et al. studied polyester-degrading bacteria, i.e., PHB-, PCL- and PBS-degrading bacteria, using 16S rRNA gene sequence data and the polyester-degrading abilities of thirty-nine bacteria of classes *Firmicutes* and *Proteobacteria* isolated from soil samples. They reported the characteristic patterns of polyester substrate specificity of those strains but no PLA-degrading bacteria isolates were found.[21]

Out of the 41 genera tested in this study, only 5 genera showed clear zones around their colonies on PLA-emulsified plates. This approach revealed a restricted number of PLA-degraders among actinomycetes. The result in this study demonstrated that there is a relationship between the phylogenetic positions and PLA-degrading ability among the tested actinomycetes. On the other hand, PHB-, PCL-, and PBS-degraders were widely distributed in various families. From our result, it is considered that actinomycetes might play an important role in microbial degradation of polyesters. Additional strains will be investigated in order to substantiate our results, especially among a large number of *Streptomyces*.

Natural aliphatic polyesters are distributed widely in the environments, i.e. PHB (storage substance) in bacteria and cutin (the structural polyester of the plant cuticle) in plants. Since these polymers occur in nature, no doubts that enzymes degrading them are ubiquitous in living organisms. Nishida and Tokiwa found that some phytopathogenic fungi degraded a synthetic polyester PCL and proposed that their cutinase act on PCL as an analogue of cutin.[25] Further, Murphy et. al. confirmed that PCL-depolymerase of a fungal pathogen *Fusarium* is in fact the cutinase.[26] PBS is a kind of polyester in which component monomers are bonded via ester linkages. We found that lipase was able to degrade PBS.[27]

L-Alanine is a major amino acid constituent of silk fibroin within amino acid sequence: (Gly-Ala)$_2$-Gly-Ser-Gly-(Ala)$_2$-Gly-[Ser-Gly-(Ala-Gly)$_n$]$_8$-Try, where n is usually 2.[28] An isolated strain of *Amycolatopsis* used in silk degradation was confirmed able to form clear zones on PLA plate.[22] Surprisingly, many of the *Amycolatopsis* strains used in PLA degradation also formed clear zones on the silk plates.[15] Recently, we confirmed that the purified PLA-degrading enzyme produced by *Amycolatopsis* strain HT-41 has ability to degrade both PLA and silk fibroin powder.[23] Most strains forming clear zones on PLA-emulsified agar plates also formed clear zones on silk fibroin agar plates (Figure 2). It can be anticipated that the strains may regard the repeated L-lactic acid unit of PLA as L-alanine unit of silk fibroin.

288

As a hypothesis, it may be regarded that each chemically synthesized polymer is an analogue of a natural substrate. The research on biodegradable plastic is continuing with the aim of achieving harmony between human activities and the natural environment.

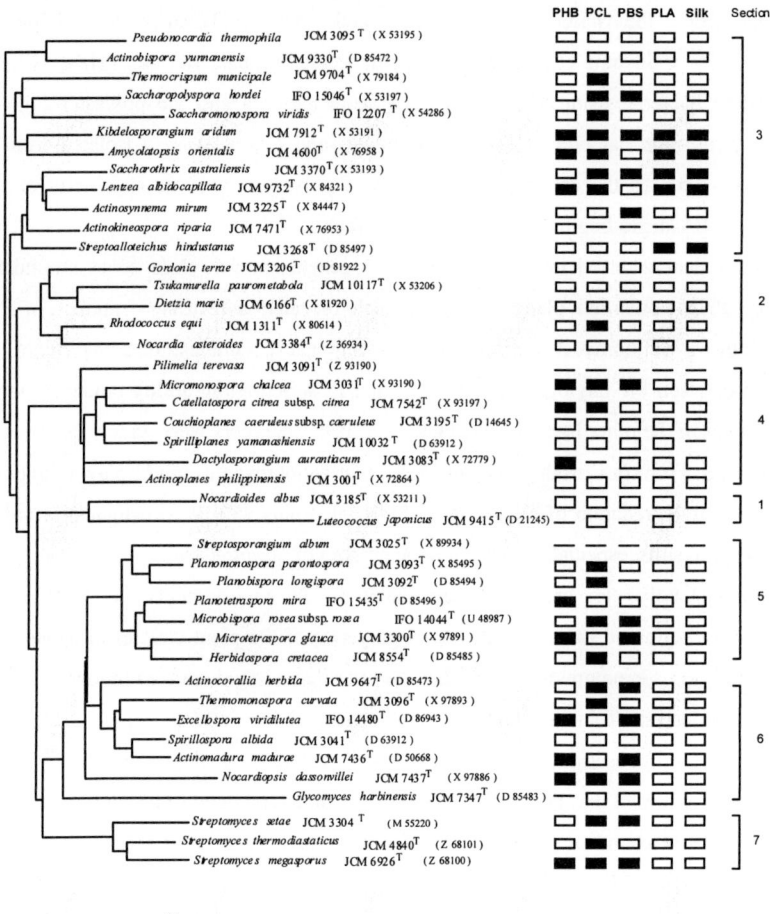

Fig. 2. Phylogenetic positions of actinomycetes and their polyesters, and silk fibroin degradabilities. The phylogenetic tree was constructed based on pairing of 16S rRNA sequences. The nucleotide sequence accession numbers are in the parentheses. The scale bar corresponds to a 10% difference in nucleotide sequence. T=type strain. Symbols: ■ degradation; □ no degradation and − no growth on the tested plates. Section 1: *Micrococcus, Microbacterium* and related genera. Section 2: *Mycobacterium, Nocardia* and related genera. Section 3: Family *Pseudonocardiaceae* and related genera. Section 4: Family *Micromonosporaceae*. Section 5: Family *Thermomonosporaceae*. Section 6: Family *Streptosporangiaceae*. Section 7: Family *Streptomycetaceae*.

[1] Y. Tokiwa, T. Suzuki, *J. Ferment. Technol.* **1974**, *52*, 393.
[2] Y. Tokiwa, T. Suzuki, *Agric. Biol. Chem.* **1977**, *41*, 265.
[3] Y. Tokiwa, T. Suzuki, *Nature* **1977**, *270*, 76.
[4] Y. Tokiwa, T. Suzuki, *J. Appl. Polym. Sci.* **1981**, *26*, 441.
[5] Y. Tokiwa, T. Suzuki, T. Ando, *J. Appl. Polym. Sci.* **1979**, *24*, 1701.
[6] Y. Tokiwa, T. Ando, T. Suzuki, *J. Ferment. Technol.* **1976**, *54*, 603.
[7] J. G.. Sanchez, A. Tsuchii, Y. Tokiwa, *Biotechnol. Lett.* **2000**, *22*, 849.
[8] H. Nishida, Y. Tokiwa, *J. Appl. Polym. Sci.* **1992**, *46*, 1467.
[9] H. Nishida, Y. Tokiwa, *J. Environ. Polym. Degrad.* **1993**, *1*, 65.
[10] H. Nishida, S. Suzuki, Y. Tokiwa, *J. Environ. Polym. Degrad.* **1998**, *6*, 43.
[11] H. Pranamuda, Y. Tokiwa, H. Tanaka, *Appl. Environ. Microbiol.* **1995**, *61*, 1828.
[12] A. Jarerat, Y. Tokiwa, *Biotechnol. Lett.* **2001**, *23*, 647.
[13] M. L. Tansengco, Y. Tokiwa, *World J. Microbiol. Biotechnol.* **1998**, *14*, 133.
[14] H. Pranamuda, Y. Tokiwa, H. Tanaka, *Appl. Environ. Microbiol.* **1997**, *63*, 1637.
[15] H. Pranamuda, Y. Tokiwa, *Biotechnol. Lett.* **1999**, *21*, 901.
[16] A. Jarerat, Y. Tokiwa, *Micromol. Biosci.* **2001**, *1*, 136.
[17] H. Nishida, M. Konno, A. Ikeda, Y. Tokiwa, *Polym. Degrad. Stab.* **2000**, *68*, 205.
[18] H. Pranamuda, R. Chollakup, Y. Tokiwa, *Appl. Environ. Microbiol.* **1999**, *65*, 4220.
[19] H. Nishida, Y. Tokiwa, *J. Environ. Polym. Degrad.* **1993**, *1*, 227.
[20] M. L. Tansengco, Y. Tokiwa, *Chem. Lett.* **1998**, 1043.
[21] T. Suyama, Y. Tokiwa, P. Ouichanpagdee, T. Kanagawa, Y. Kamagata, *Appl. Environ. Microbiol.* **1998**, *64*, 5008.
[22] Y. Tokiwa, M. Konno, H. Nishida, *Chem. Lett.* **1999**, 353.
[23] H. Pranamuda, A. Tsuchii, Y. Tokiwa, *Macromol. Biosci.* **2001**, *1*, 25.
[24] K. Nakamura, T. Tomita, N. Abe, Y. Kamio, *Appl. Environ. Microbiol.* **2001**, *67*, 345.
[25] H. Nishida, Y. Tokiwa, *Chem. Lett.* **1994**, 1547.
[26] C.A. Murphy, J. A. Cameron, S.J. Huang, R.T. Vinopal, *Appl. Environ. Microbiol.* **1996**, *62*, 456.
[27] Y. Tokiwa, T. Suzuki, K. Takeda, *Agric. Biol. Chem.* **1986**, *50*, 1323.
[28] D. J. Strydom, T. Haylett, R. H. Stead, *Biochem. Biophys. Res. Commun.* **1977**, *79*, 932.

Macromol. Symp. **2003**, *201*, 291—300

Trends in Industrial Polymer Research

Volker Warzelhan, * *Franz Brandstetter*

BASF Aktiengesellschaft, Polymer Research, D-67056 Ludwigshafen, Germany

Summary: In the past decades a shift in paradigm took place in industrial polymer research for structural materials. Only a few new polymers based on new monomeric building blocks were developed. The main focus is now on tailoring improved "old polymers" with well-defined structure and properties based on a set of low cost "old" monomers using controlled polymerization mechanisms.

Kewords: controlled polymer synthesis, industrial polymer research, modification and improvement of polymers, structural polymers

1 Overall Economic Background

Some one hundred years after they were first synthesized industrially, and eighty years after their molecular structure was solved, synthetic polymers continue to play an important role satisfying the material needs of today's society in the areas of diet and nutrition, health, accommodation, clothing, communication and mobility. The demand for polymers has rapidly grown over the last decades to generate a world market volume of 154 million metric tons in 2001. The secret of the success of structural polymers is to be found in their unique combination of hardness, lightness, resistance to corrosion, flame-retardant properties, weathering resistance, stiffness and toughness. In terms of production volume, structural synthetic polymers have overtaken steel in the last twenty years. The consumption per capita exhibited an extraordinary growth despite the two oil crises compared to other materials like wood or steel (Figure 1, example USA).

In defiance of this growth the margins for polymers have declined by 40 – 60% within the last decade due to over-capacities and severe competition (Figure 2). As a consequence a dramatic concentration process took place in industry. The capacities of world scale single strand plants have steadily been increased to reduce specific costs reaching 400 kt/a for polypropylene (PP) nowadays.

DOI: 10.1002/masy.200351132

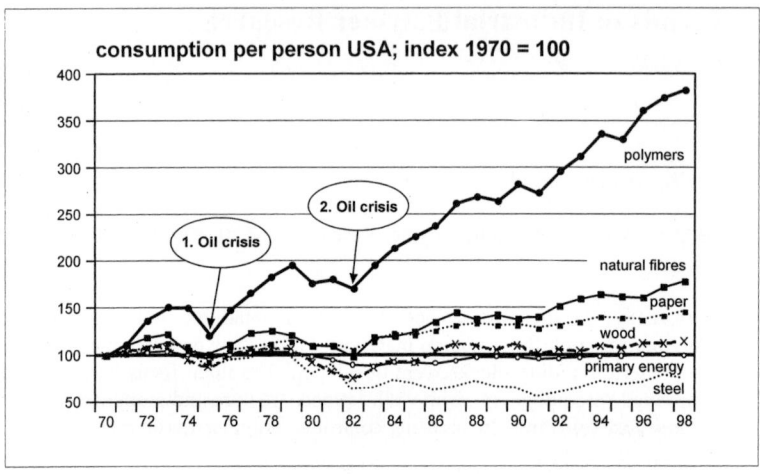

Fig. 1. The development of the demand per capita for different materials in the USA.

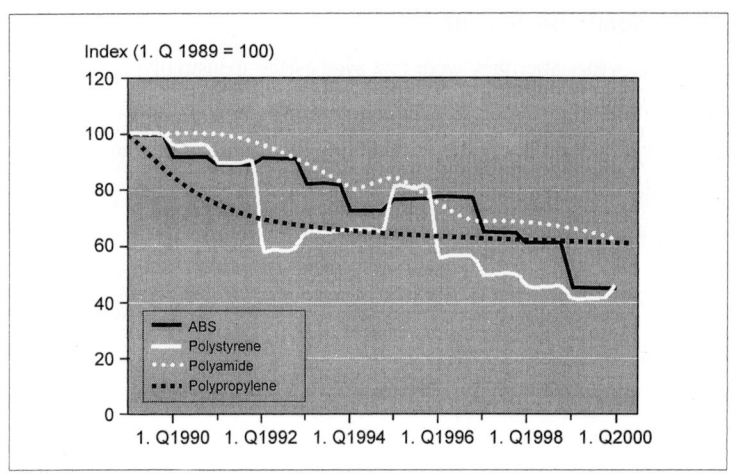

Fig. 2. Trends in margins for commodity polymers.

Many of the commercially most important classes of polymers were first manufactured industrially decades ago. Polystyrene (PS) and polyvinylchloride (PVC) both 70 years ago, polyethylene (PE) 60 years ago. Only a few commercially relevant polymer classes have been introduced within the last 40 years. In terms of volume, the four largest classes of synthetic polymers (PE, PP, PVC and PS) represent over 85% of the world market (Figure 3), a fact that is in sharp contrast to the predictions of the past.

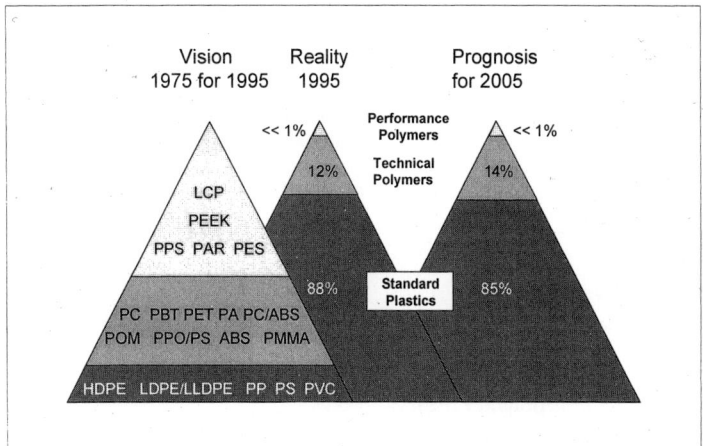

Fig. 3. Vision and reality of market development of thermoplastics within the last 25 years.

Seen the economic difficulties and the mature character of the involved polymers the question arises what the important drivers of the future will be in order to ensure a profitable business with polymers.

2 New Polymers from New Monomers

Do we really need new monomers to achieve new polymeric materials to fulfil customers' needs? The answer is quite clearly "no". The number of polymers based on new monomers decreased significantly in the last decade and such polymers only play a role in niche markets.

3 New Polymers from Old Monomers

In place of introducing new monomers it is more attractive to obtain new polymers by known "old" monomers. Due to the tremendous progress in catalyst design existing cheap monomers could be combined to new polymers with new structures achieving properties similar to engineering plastics. Several examples can be found during the last years (Table 1):

Are these new polymers from old monomers really successful in the market? After 20 years of intensive R&D efforts, and 5 years after the market launch, SHELL[1] has given up the development of polyketones.

DOW also announced to withdraw its interpolymers from the market.[2] SPS and COC are still being produced in semi commercial plants but growth rates are far smaller than forecasted.[3]

Table 1. New polymers from old monomers.

Polymer Trademark	competitive material	pros	cons
Polyketone Carilon® (Shell) Ketonex® (BP) $C=O + CH_2=CH_2 \rightarrow [C-CH_2-CH_2]_n$	Polyamide Polyacetal Polyethylene	toughness, chemical resistance, barrier properties	processing stability, property deterioration by fillers
Syndiotactic Polystyrene (sPS) Xarec® (Idemitsu) Questra® (Dow) $CH=CH_2$ Metallocene- cat.	Polybutylene terephthalate	low density, no shrinkage, no hydrolysis	toughness, long-term service temperature
Ethylene/Styrene Interpolymer (ESI) Index® (Dow) $CH_2=CH_2 +$ cat.	EVA S/B block copolymers flexible PVC	broad spectrum of properties, no outstanding single characteristic	
Cyclic Olefin Copolymer (COC) Topas® (Ticona) $n\,CH_2=CH_2 + m$ Metallocene- cat. $(CH_2-CH_2)_m$	Polycarbonate, Polymethyl- methacrylate	transparancy, moisture barrier, good electrical insulating property	costs, toughness- stiffness ratio

At first sight, new polymers from old monomers can be attractive since raw material costs are rather low. New polymerization technologies and new catalysts, however, have to be developed and implemented. Simultaneously, applications and processing at the customer side have to be worked out. These cost- and time-intensive procedures have often been underestimated.

A possible explanation can provide Figure 4, the so called "learning curve" that is simply based on an empiricism but holds for sewing machines as well as carbon fibers. The learning curve predicts that by doubling cumulative production a cost reduction of 20 – 30% is achievable simply by becoming more experienced with the product. Each new polymer, although having the potential to be lower in cost, is at the starting stage much more expensive than an existing polymer that is produced in millions of tons and therefore much further down on the learning curve.

One might come to the opinion that these new polymers simply have the fate to be too late on the marketplace to be successful.

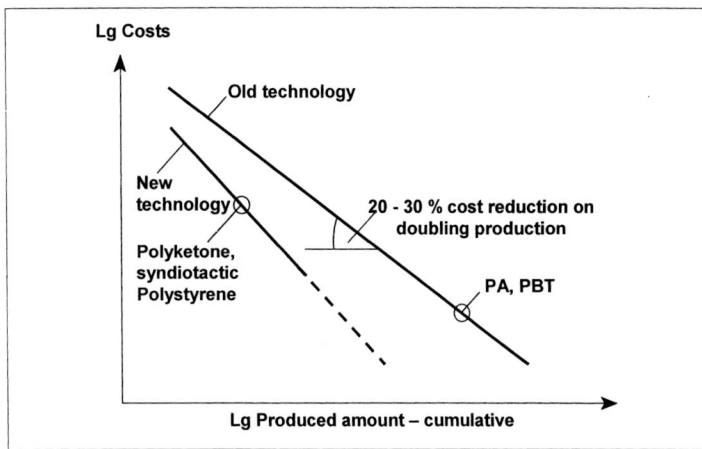

Fig. 4. The "learning curve" (explanation see text).

One example of the few new polymers, derived from a known monomer, that have been placed successfully on the marketplace is linear low density polyethylene (LLDPE). UCC licensed its technology aggressively from the beginning on, and a lot of licensees helped to ride down the learning curve.

4 Improved Old Polymers

A question that naturally arises from all these facts is whether we really need new polymers? We are convinced that the potential of the existing polymers is by far not yet exhausted considering the tremendous progress in polymer science. In particular the progress has been manifested by:

- An improved understanding of the relationship between primary and higher-order structures of synthetic macromolecules and the desired physical and applications-related material properties.
- The development of reaction conditions to achieve the controlled polymer synthesis for almost all different polymerization processes as well as for a large number of monomers to create macromolecules with specially tailored structures. Related to this is the development of catalysts of even higher selectivity and even greater efficiency.
- Exploiting the methods of high-throughput synthesis, high-throughput screening and data management to significantly accelerate research in these areas.

Based on this knowledge, each improvement of existing polymers, so called "drop in solutions", are by far easier to implement and commercialize. Many examples by BASF can be found:

1. Based on market studies BASF's researchers had the goal to develop a biodegradable film grade material with LDPE-like properties for packaging applications. As for this application price is decisive our strategy was to develop a biodegradable polymer based on existing monomers with existing technology in existing plants. In the end the problem solution was to modify PBT by adding the aliphatic adipic acid and to improve properties by branching and chain extension (Figure 5), so all in all quite conventional methods.[4]

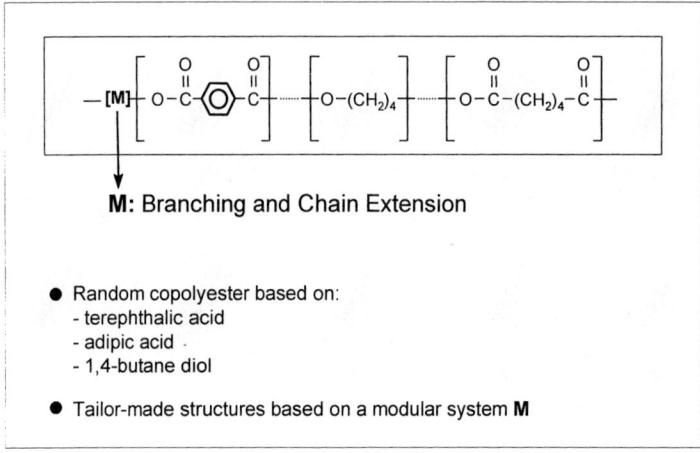

Fig. 5. Improved old polymers: Ecoflex® – a biodegradable "poly(butyleneterephthalate)" (PBT).

However, we obtained a 100% biodegradable polymer (Ecoflex®) with LDPE like properties that can be used as garbage bag, cling film or biodegradable cover sheet for lunch boxes out of starch.

2. Key for polyamide 6 fiber applications in textiles or carpets is the UV stability. We could show, that attaching the additive to the chain end rather than just simply blending it with the polymer is much more efficient because there is no loss of stabilizer during processing and life time of the product.[5] Most efficient are HALS- (hindered amine light stabilizers) type stabilizers that are incorporated into the chain during the polycondensation by chain end capping.

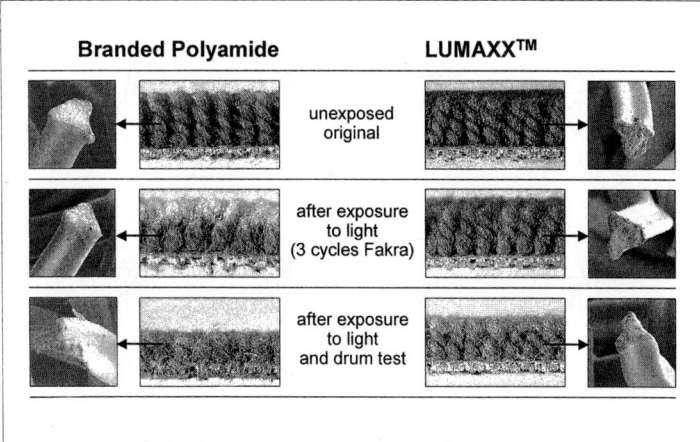

Fig. 6. Change of appearance of fibers after testing.

Figure 6 illustrates the effect of such attached additive in comparison to conventional polyamide. Our new material shows better color fastness and the pile height is much less reduced after usage because of the unique stabilization. Conventional carpet fibers get brittle and break leading to "walking streets".

3. The addition of magnesium or aluminum alkyls to butyllithium, forming an "ate"-complex, allows to slow down the anionic polymerization of styrene to same rates as the radical polymerization. It could be shown by BASF[6, 7] that up to 180 °C and in bulk this so called retarded anionic polymerization of styrene is still first order and living (Figure 7).

This enables us to produce high impact polystyrene (HIPS) anionically in existing radical HIPS plants under same conditions (drop in solution) with reduced level of residuals, with similar or even improved properties. The higher costs for initiator and solvent purification can be compensated by an integrated rubber production.

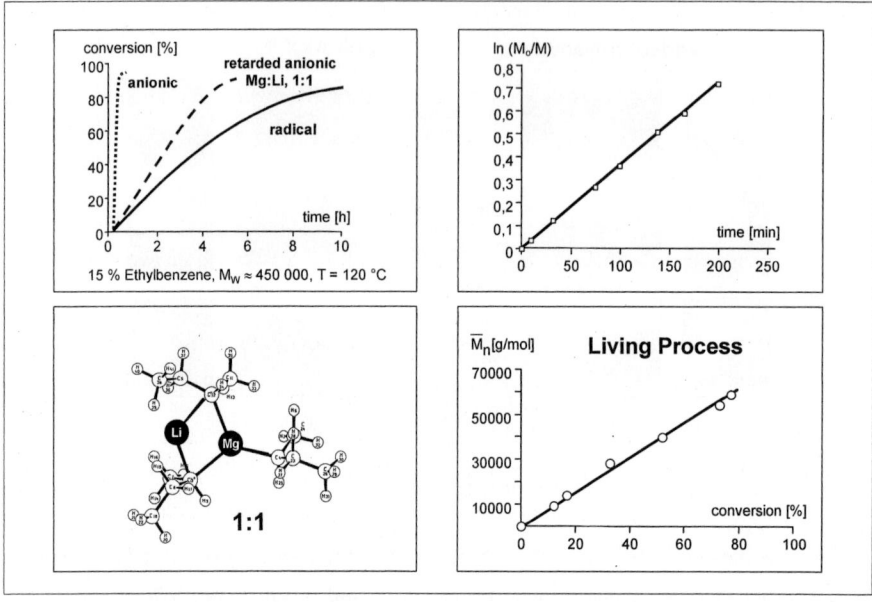

Fig. 7. Retarded anionic polymerization: kinetic data and proposed structure of the Li-Mg-"ate" complex.

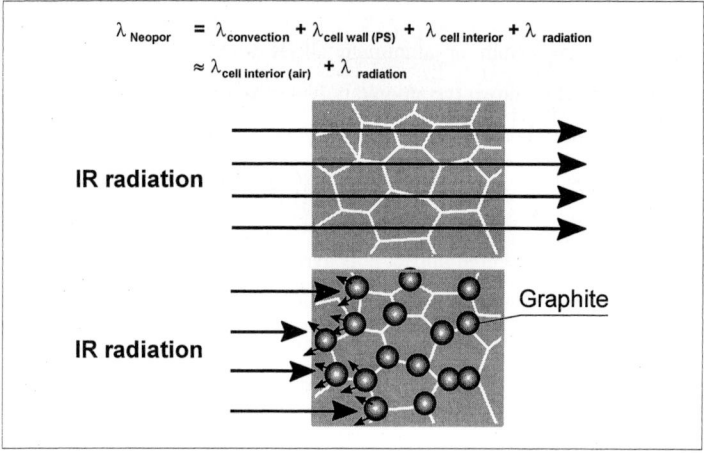

Fig. 8. How Neopor® cellular foam works.

4. A simple way of improving properties of existing polymers is by adding fillers from micro- down to the nano-scale. By adding graphite to expandable polystyrene it is possible either to reduce thickness of sheets by 30% at densities between 10 - 15 g/l to get

the same insulation properties compared to conventional Styropor® or to reduce the density of Neopor® by 50% compared to standard Styropor®.[9]

The IR radiation as a main contributor to the heat conductivity is reflected by the graphite particles that otherwise would penetrate the insulation foam almost without hindrance (Figure 8).

5. Integrated system solutions also allow new applications and growth potential for known polymers.

High off-line coating costs and/or expensive polymer blends for on-line coating have still limited the usage of plastic panels in automotive applications. BASF is developing a new concept of *Paintless Film Molding* (PFM; Figure 9) to significantly reduce costs:[9] A three layer coextruded and pigmented foil is backed by injection molding of glass reinforced ASA, ABS, PA, PBT or blends there of. By this new technique expensive coating costs are saved and cheap backing material can be used. The key is that the materials exhibit good compatibility to result in good adhesion to each other.

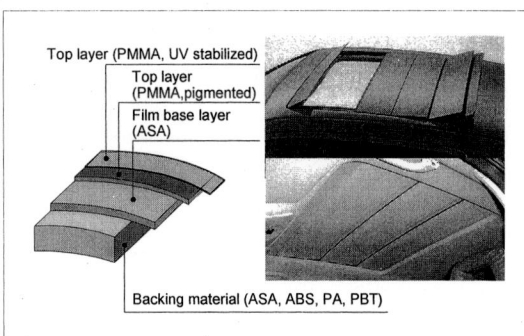

Fig. 9. Paintless Film Moulding (PFM).

These five examples demonstrated innovative solutions to further expand the potential of the old polymers. We are steadily improving properties while lowering the costs that is the key factor for further growth.

5 Outlook

The tremendous progress in catalyst research has led to new polymers with even lower potential manufacturing costs. Nonetheless, the effect of economy of scale, underestimated costs for processing and application development, the huge data base of the existing polymers to be developed for new polymers and also the progress of the old polymers will limit their success. New business models are necessary as the example of LLDPE has shown.

On the other hand, the progress in polymer science will bring us closer to the ultimate vision of "retro-synthesis":

- A deeper understanding of structure/property relationships permits to translate properties into morphologies and molecular structures.

- Such polymers are then tailored by more advanced methods of controlled synthesis (Figure 10).

As a consequence the conventional way of trial and error will become obsolete.

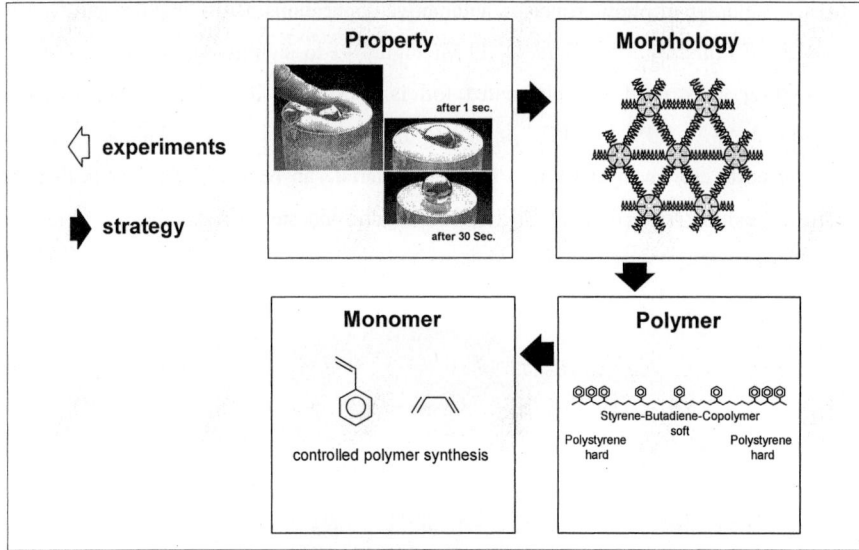

Fig. 10. The vision of polymer retro-synthesis.

[1]   Plastic News, **2002**, 14 (24), 11.
[2]   Plastic News, **2002**, 14 (25), 27.
[3]   Modern Plastic, **2001**, 78 (12), 49 (1); Chemical Week, **2002**, 164 (24), 20 (2).
[4]   U. Witt, M. Yamamoto, U. Seeliger, R.-J. Müller, V. Warzelhan, *Angew. Chem.*, **1999**, 111, 1540.
[5]   P.-M. Bever, U. Breiner, G. Conzelmann, B.-S. von Bernstorff, *Chemical Fibers International*, **2000**, 50, 176.
[6]   P. Desbois, M. Fontanille, A. Deffieux, V. Warzelhan, C. Schade, *Macromol. Symp.*, **2000**, 157, 151.
[7]   P. Desbois, M. Fontanille, A. Deffieux, V. Warzelhan, S. Lätsch, C. Schade, *Macromol. Chem. Phys.*, **1999**, 200, 621.
[8]   BASF-AG, product brochure: "Neopor® – Dämmstoff der Zukunft", **2002**.
[9]   A. Grefenstein, Intelligent material combinations by integration of filmbackmolded plastic body panels, VDI –Tagung Hamburg, May 7/8, **2002**

Efficient and Tailored Polymerization of Olefins and Styrene by Metallocene Catalysts

Walter Kaminsky,[*1] *Andreas Hopf,*[1] *Michael Arndt-Rosenau*[2]

[1]Institute for Technical and Macromolecular Chemistry, University of Hamburg, Bundesstr. 45, D-20146 Hamburg, Germany
[2]Bayer AG, F and E Elastomers, 41538 Dormagen, Germany

Summary: Different optimised new c_s-symmetric ansa zirconocenes were used for the homopolymerization of propene – activated by methylaluminoxane (MAO) with an Al : Zr molar ratio of 2000. In a series of experiments, the polymerisation temperature was varied in a range from 0 to 60 °C. The obtained syndiotactic polypropenes show highly isotacticities up to rrrr pentads of 98 % and melting points of 150 °C. Stereoselectivies are sensitive to the monomer concentration, showing decreasing syndiotacticities with decreasing propene concentration due to increasing amount of skipped insertion which is demonstrated for the zirconium catalyst.

Keywords: homopolymerization, methylaluminoxane (MAO), propene, syndiotactic polypropenes, zirconium catalyst

Introduction

Metallocenes are highly active catalysts for the production of precisely designed polyolefins and engineering plastics. Especially zirconocene methylalumoxane (MAO) catalysts and half-sandwich titanium complexes have opened a frontier in the area of new polymer synthesis and processing.[1-3] They show only one active site and produce polymers with a narrow molecular weight distribution and new properties. With the aid of metallocenes, plastics can be made for the first time with a property profile that is precisely controllable within wide limits, comprising temperature resistance, hardness, impact strength and transparency.

There is a variety of materials which can be produced with high activity by metallocene catalysts such as:

- Long chain branched polyethene;
- Polypropenes with low amounts of oligomers and different tacticities (atactic, isotactic, isoblock, stereoblock, syndiotactic);
- Copolymers with a high proportion (up to 60 mol%) of longer chain α-olefins such as LLDPE, POE, EP, and ethene/styrene copolymers;

DOI: 10.1002/masy.200351133

- Elastomers made of ethene, propene, and dienes (EPDM);
- Syndiotactic polystyrene with a high melting point;
- Homo- and copolymerisation of cycloolefins;
- Homo- and copolymers of butadiene with special microstructures;
- Polymerisation in emulsions or in the presence of fillers (nano particles).

Important are ethene-1-octene and ethene-styrene copolymers. These polymers show increased impact strength and toughness, better melt characteristics or elasticity, and improved clarity in films.[4,5]

Metallocenes are useful catalysts for the production of cycloolefin copolymers (COC) and α-olefin copolymers – new types of polymers with special properties and a high potential as engineering plastics[6-8]. Ethene/norbornene copolymers are most interesting for technical uses because of the easily available monomers. Due to different incorporation values of the cyclic olefin in the copolymer, the glass transition temperature can vary over a wide range independently from the used catalysts. A copolymer with 50 mol% of norbornene yields a material with a glass transition point of 145 °C. A Tg of 205 °C can be achieved at higher incorporation rates. The metallocene [Me$_2$C(tert-BuCp)(Flu)]ZrCl$_2$ shows not only high activities for the copolymerisation of ethene with propene or norbornene, and gives alternating structure, too.[9,10]

Fluorinated half-sandwich titanium complexes (R-Cp)TiF$_3$ are in combination with MAO very active catalysts for the polymerisation of styrene and butadiene. The obtained syndiotactic polystyrene shows a molecular weight of over 400 000 g/mol and a melting point of 275 °C. Metallocen-based polybutadiene has a special microstructure with 80 % cis 1,4-, 1 % trans 1,4- and 19 % 1,2-vinyl units. The activity is mainly influenced by the ligand structure while the microstructure is nearly untouched.[11]

C$_s$-symmetric group 4 ansa metallocenes were introduced first for the syndiotactic polymerisation of propene by Ewen and Razavi.[12] The mechanism for the synthesis of syndiotactic polypropene involves an alternating enchainment of monomer with inverse relative configuration and the formation of a syndiotactic chain. This so-called enantiomorphic-site control mechanism implies that the chiral catalytic centres are capable of discriminating between the two prochiral enantiofaces of a propene molecule.[13]

So far there are no studies dealing with a theoretical calculation of the influence of different substituents on catalysts of the type Ph$_2$C(Flu)(Cp)MCl$_2$ on stereoselectivity whereas many publications are known demonstrating the possibilities of variations on the ligand system.[14]

Experimental Part

The c_s-symmetric zirconocenes (Catalyst 1-4, Fig. 1) were synthesized analogue as described in the literature.[14,15] 5 g (18 mmol) 2,7-bis-t-buthylfluorene were dissolved in 80 ml ether and 11,25 ml of a solution of 1,6 molar butyllithium in hexane were slowly added. The solution was stirred until gas evolution was finished and then 4,14 g of the methyl substituted 6,6-diphenyl fulvene were added. After 10 – 12 hours, a small amount of water is added, the organic layer separated and dried over Na_2SO_4. The solution was reduced and crystallized at – 25 °C to yield colorless crystals (yield 65 – 70 %). This ligand was dissolved in ether, butyllithium added and after gas evolution the zirconiumtetrachlorid added. After filtering, the solution was reduced and crystallyzed. Orange-red crystals were obtained yield (40 – 50 %).

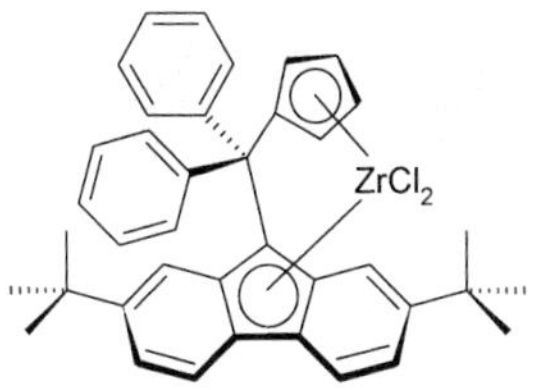

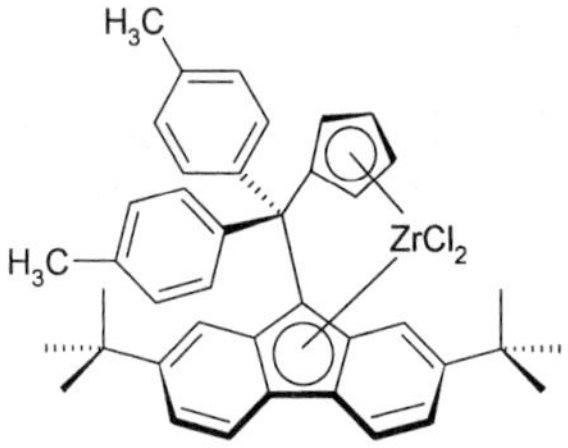

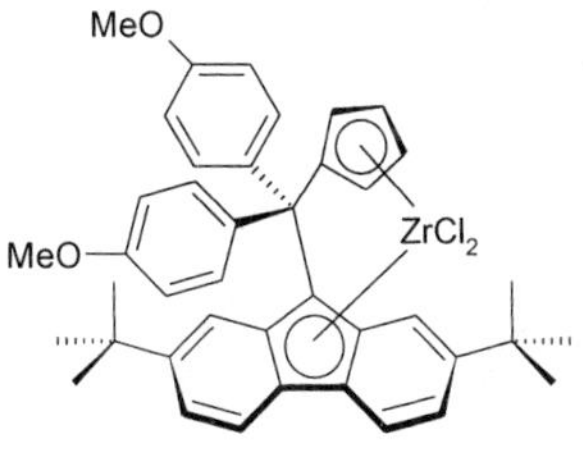

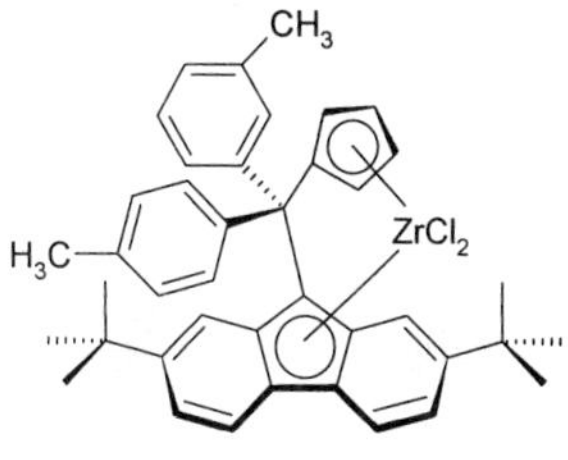

Fig. 1. New synthesized zirconocene complexes 2-4 and 1 for comparison.

Polymerizations were carried out in 200 ml toluene in a 1 l Büchi glass reactor equipped with a magnetical stirrer. The concentration of propene for every run was 1,38 mol/l and the pressure was kept constant during the polymerisation. The polymerisations were started through injection of the metallocene to the propene saturated toluene/MAO solution. After 1 h the reaction was stopped by addition of 2 ml methanol. The polymer solution was stirred overnight in a methanol/HCl solution, filtered followed by evaporation of the solvents and drying of the polymer under vacuo at 60 °C over night.

Toluene was purified through passing through two columns (molecular sieves and BASF copper catalyst), propene was passed through two columns packed with molecular sieves and Basf copper catalyst).

Polymer samples were recorded using a Bruker avance 400 spectrometer and referenced against $C_2D_2Cl_4$ at 100 °C, pulse angle 30 °, delay time 5 s, 1200 scans.

Results and Discussion

Syndiotactic Polypropene

By molecular modelling using a modified Biosym program analog to the calculation of the stereorigidity of ansa-bis-indenylzirconium complexes[16], the syndiotactic rrrr pentads in molar ratio were calculated. Starting from the unsubstituted $[Me^2C(Cp,Flu]ZrCl_2$ (Fig. 2), different substitutions at the fluorenyl ring and different bridges were investigated (Table 1 and 2 show the results of the calculations). The program makes it possible to calculate the rrrr pentads by different polymerisation temperatures starting from −30 °C up to 90 °C.

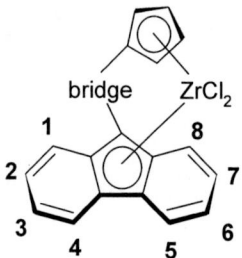

Fig. 2. Unsubstituted $[Me^2C(Cp,Flu]ZrCl_2$.

Table 1. Calculation of the syndiotactic rrrr pentads in molar ratio of polypropylenes produced by different fluorenyl substituted $[Me_2C(Cp,Flu)]ZrCl_2/MAO$ catalysts.

Polymerization temperature	-30°C	0 °C	30 °C	60 °C	90 °C
Substitution all H	0.9584	0.9451	0.9314	0.9178	0.9044
2,7-Me$_2$	0.9616	0.9488	0.9356	0.9223	0.9091
2,7-tBu$_2$	0.9644	0.9520	0.9391	0.9260	0.9130
4,5-Me$_2$	0.8242	0.8060	0.7904	0.7769	0.7648
4,5-Benzo	0.9083	0.8904	0.8738	0.8587	0.8447

It can be seen that from the calculated rrrr pentads, the zirconocene with the 2,7-t-Bu$_2$ substitution gives the highest values not only at –30 °C, but also at a polymerisation temperature of 90 °C.

Table 2. Calculation of the syndiotactic rrrr pentads in molar ratio of polypropylenes produced by different bridged c$_s$-symmetric [bridge(Cp,Flu)]ZrCl$_2$/MAO catalysts.

Polymerization Temperature	-30 °C	0°C	30 °C	60 °C	90 °C
Bridge					
Me$_2$C	0.9584	0.9451	0.9314	0.9178	0.9044
Me$_2$Si	0.9517	0.9370	0.9224	0.9082	0.8945
Ph$_2$C	0.9748	0.9643	0.9529	0.9410	0.9289
Ph$_2$Si	0.9402	0.9310	0.9252	0.9128	0.9007
C$_5$	0.9652	0.9527	0.9397	0.9265	0.9132
C$_6$	0.9561	0.9415	0.9265	0.9114	0.8965

If complexes with different bridges such as dimethylsilyl, diphenylcarbon, 1,1-cyclopentane, 1,1-cyclohexane, are calculated, the complex with a diphenylcarbon bridge gives the highest value of rrrr pentads of 97,48 % calculated for –30 °C and 92,89 % at 90 °C. Based on this theoretical studies, we optimized a zirconocene with a diphenylcarbon bridge and a 2,7-di-tert.butyl substitution at the fluorenyl ring. As the bridge shows the greatest effect, a fine tuning was carried out which was not possible to be calculated, the phenyl groups of the bridge were substituted by methyl groups (electron pressing) and by methoxy groups (electron drawing). The zirconocenes 1-4, Fig. 1 were synthesized and used for the polymerisation of propene together with MAO as cocatalyst over a wide range of polymerisation temperatures.

A maximum activity of 10 000 kg PP/mol Zr x h is given at 45 °C for the methyl substituted catalyst 2. Zirconocene 1 shows at 45 °C an activity of 6000 kg/mol Zr x h. Electron drawing substituents decrease the polymerisation activity (4500 kg PP/mol Zr x h, catalyst 4).

The picture is different when the melting points of the obtained syndiotactic polypropylenes are studied (Fig. 3). At low polymerisation temperatures, the melting points for complex 4 reaches 152 °C. This is a very high melting point for an unfractionated original sample of syndiotactic polypropylene. For all samples, the melting points decrease with increasing polymerisation temperature. Polymers produced at 60 °C show only melting points between 122 and 130 °C.

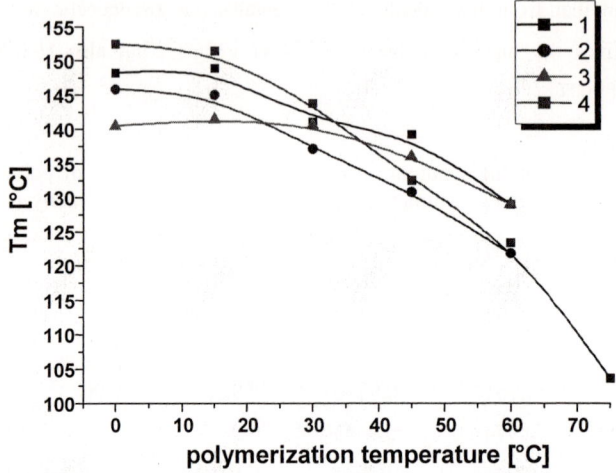

Fig. 3. Melting points of syndiotactic propene samples prepared by catalysts 1-4 and different polymerisation temperatures in toluene as solvent.

Unusual high are the molecular masses of the obtained polypropylenes. If catalyst 2 used, the synthesized polypropylene shows a molecular mass of 750 000 g/mol (0 °C), 600 000 (15 °C), 450 000 (30 °C), 300 000 (45 °C), and 170 000 (60 °C). The ^{13}C-NMR measured syndiotactic rrrr pentade reaches a molar ratio of 0,97 at 0 °C and 0,94 at 30 °C.

Even higher are the values if the polymerisation is carried out in liquid propene (bulk polymerisation). Table 3 shows for the complex 1 some data on activities and polymer properties.

Table 3. Bulk polymerisation of propene with [Ph$_2$C(Cp)(2,7-t-Bu$_2$Flu)]ZrCl$_2$/MAO (cata-lyst 1) at different temperatures.

T$_{pol}$ (°C)	Activity kg PP/mol Zr x h x Cp	Tm (°C)	Mη (kg/mol)	rrrr (%)
-20	350	146	1 200	93
-10	400	150	1 100	93
0	600	147	1 300	98
30	1 200	140	720	90
60	3 200	133	460	92

For the bulk polymerisation, the activities are always increasing with the polymerisation temperature. Molecular masses of more than 1 million can be reached. The value of the rrrr pentade is the highest for samples prepared at 0 °C and is in agreement with the calculation for the catalyst.

Conclusion

Molecular modelling leads to optimised c_s-symmetric zirconocenes which show in combination with MAO very high activities for the syndiotactic propene polymerisation. The produced polymers have extremely high molecular masses and melting points. The ^{13}C-NMR measured racemic pentads reaches values of more than 98 %.

[1] W. Kaminsky, Advances in Catalysis, 2001, Vol. 46, 89
[2] R. Blom, A. Follestad, E. Rytter, M. Tilsel, M. Ystenes (eds.), Organometallic Catalyst and Olefin Polymerization, Springer 2001, Berlin
[3] J. Scheirs, W. Kaminsky (eds.), Metallocene-Based Polyolefins, Vol. I + II, Wiley 2000, Chichester
[4] A. Torres, K. Swogger, C. Kao, S. Chum, in Ref. 3, p. 143
[5] I. Albers, W. Kaminsky, U. Weingarten, R. Werner, Catal. Communications 2002, 3, 105
[6] W. Kaminsky, A. Bark, M. Arndt, Makromol. Chem., Macromol. Sym. 1991, 47,8
[7] H. Cherdron, M.-J. Brekner, F. Osan, Angew. Makromol. Chem. 1994, 223 121
[8] D. Ruchatz, G. Fink, Macromolecules 1998, 31, 4669
[9] W. Kaminsky, M. Arndt, I. Beulich, Polym. Mater. Sci. Eng. 1997, 76, 18
[10] M. Arndt, I. Beulich, Macromol. Chem. Phys. 1998, 199, 1221
[11] W. Kaminsky, B. Hinrichs, D. Rehder, Polymer 2002, 43, 7225
[12] J.A. Ewen, R.L. Jones, A. Razavi, J.P. Ferrara, J. Am. Chem. Soc. 1988, 110, 6255
[13] A. Razavi, J.L. Atwood, J. Organomet. Chem. 1993, 459, 117
[14] H.G. Alt, R. Zenk, J. Organomet. Chem. 1996, 522, 39
[15] A. Hopf, W. Kaminsky, Catal. Communications 2002, 3, 459
[16] W. Kaminsky, O. Rabe, A.-M. Schauwienold, G.U. Schupfner, J. Hanss, J. Kopf, J. Organomet. Chem. 1995, 497, 181

Polyolefin: Changing Supply-Demand Framework and New Technology

Kissho Kitano,[*1] *Makoto Sugawara,*[2] *Mitsuyuki Matsuura*[1]

[1] Research and Development Division, Japan Polychem Corporation,
 10-1, Yurakucho 1-chome, Chiyoda-ku, Tokyo 100-0006, Japan
  Email: Kitano.Kissho@ma.pochem.co.jp
[2] Catalyst and Process Development Center, Japan Polychem Corporation,
 1, Toho-cho, Yokkaichi, Mie 510-0848, Japan

Summary: Polyolefin industry is now under a remarkable change of international supply-demand framework and its market is splitting into commodity and high performance products. It is getting more important for a material being harmless and comfortable, while the "life cycle cost", which includes the cost during use and the recycle cost after use, is regarded as more important to evaluate a material. Those changes are accelerating the inter-material penetration. Several examples of the material design and production technologies, which responded to the changing market needs and developed new applications of polyolefin, are discussed.

Keywords: automotive application, composites, environment impact, metallocene catalysts, poly(propylene)

Introduction

Polyolefin, combining polyethylene (PE) with polypropylene (PP), is the most broadly used plastic and its global production is estimated to have exceeded 80,000 kta in 2001. Recently, the supply and demand structure of polyolefin is remarkably changing. One of its major causes is the growing international competition of product supply. Gigantic global suppliers, which capacity is approaching 10,000 kta each, have emerged as the results of recent corporate consolidations. Meanwhile even the demand expansion in Asia Pacific area is not expected to absorb the ongoing capacity increase in that area including Middle East and China. Therefore, thorough cost reduction is essential especially for the polyolefin commodity products, to prepare the price competition anticipated in the near future. The improvement of polyolefin catalyst, which has been focused on the increase of catalyst efficiency and the development of efficient process, has largely contributed to reduce the catalyst cost and to eliminate the

© 2003 WILEY-VCH Verlag GmbH & KGaA, Weinheim   DOI: 10.1002/masy.200351134

310

process of catalyst decomposition and neutralization. Several simplified processes, such as gas phase process, which doesn't need the solvent recovery facilities, have been also developed to reduce the utility cost (Figure 1).[1] As the results of those improvements, the raw material cost has become to dominate the manufacturing cost of polyolefin and is almost determining the profitability especially of the commodity business. Japan doesn't have the major resource of low cost materials and is still highly depending on the old manufacturing process due to the long history of polyolefin industry. Therefore, the development of the high performance product market is getting more important to maintain the profitability in Japan. Those changes in the polyolefin industry are expected to accelerate the business to be separated into two major directions, that is, the rationalization in the commodity area and the development in the high performance area.

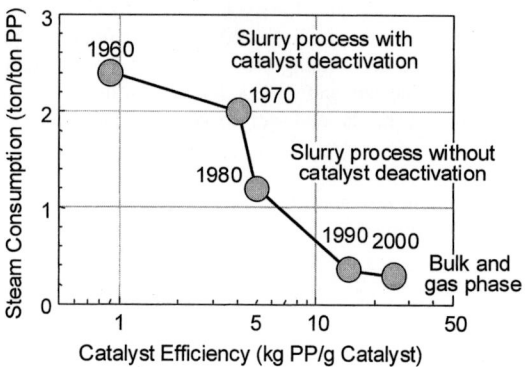

Fig. 1. Development trend of PP polymerization catalyst and process.

On the other hand, the market demand for polyolefin is also changing reflecting the recent transfiguration of social requirements, though there may be regional conditions. In Japan and several countries of North America and Europe, there is an increasing demand for the materials of environment friendliness, safety, convenience, comfort during use, and so on. The material cost is also being evaluated by "life cycle cost", which is the grand sum of all expenses including manufacturing cost, energy consumption during use, recycle cost after use, and environment impact. Those changes cause the recent inter-material penetration. Material design and production technologies are getting more important to provide the necessary products responding to the changing market demands. Several examples of new technologies, which expanded the applications of polyolefin, are discussed focusing on polypropylene.

PP Molecular Structure Control Technologies

The two major first order molecular structures of PP are molecular weight and crystallinity, and the latter depends on the stereo regularity at the polymerization. The catalyst efficiency and the stereo regularity of PP have been remarkably improved to give the commercial PP materials with very high crystallinity, high stiffness, and high heat resistance at the lower cost. Those materials are now widely used in the industrial applications.

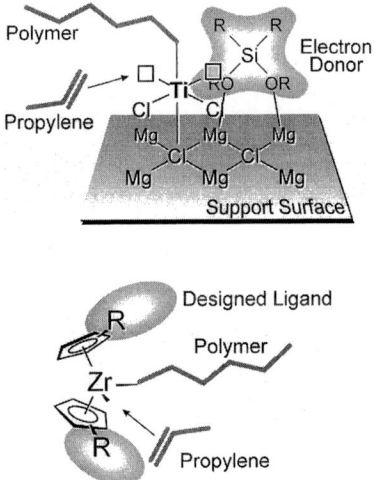

Fig. 2. Stereo regularity control models of propylene polymerization by Ziegler-Natta Catalyst (top) and metallocene catalyst (bottom).

Ziegler-Natta type titanium-based supported catalyst (Z-N catalyst) is most commonly used for the PP polymerization nowadays, but it has heterogeneity with the catalytic species and that provides by-product, or extractable fraction with the low molecular weight and low crystallinity. This fraction not only causes the lower strength and the lower heat resistance of PP material but also provides the several practical difficulties, for example, the fuming during processing and the odor of products. The largest factors to determine the stereo regularity of PP molecule are $MgCl_2$ support and the electron donor compounds, which are considered to restrict the geometry of monomer coordination and polymer propagation. Therefore, the selection of electron donor compound is important for Z-N catalyst to cope with both the high stereo regularity and the high catalyst efficiency.

On the other hand, metallocene type catalyst has become commercially used for PP recently. The stereo regularity of metallocene-based PP is mostly determined by the ligand structure of the transition metal complex instead of the electron donor (Figure 2). The ligand with the proper molecular design not only gives the high stereo regularity control capability and the high catalyst efficiency but also provides the uniform copolymerization structure of product. Japan Polychem Corporation has commercialized the metallocene-based propylene-ethylene random copolymer (RCP) WINTEC® using its proprietary catalyst technology.[2] Conventional Z-N catalyst tends to provide the larger amount of extractable fraction along with the lowering crystallinity of RCP for the lower melting point. Contrarily, metallocene catalyst dramatically reduces the amount of extractable fraction even at the very low melting point, and WINTEC® mRCP is highly evaluated as the clean material suitable for the food packaging and the medical use (Figure 3, 4).

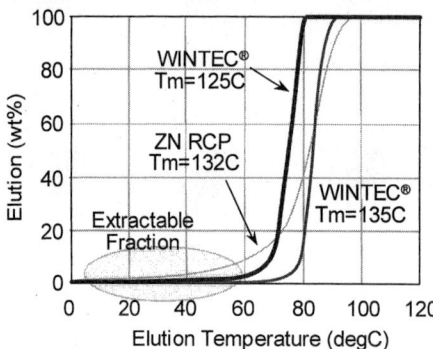 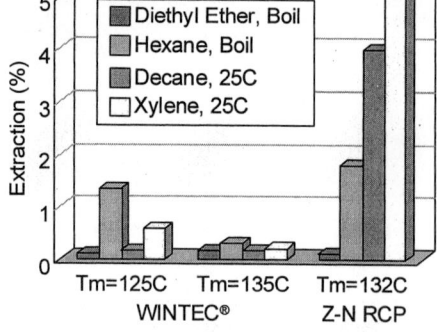

Fig. 3. Crystallinity distribution of PP random copolymers (RCP).

Fig. 4. Extractable fraction of RCP (WINTEC® is metallocene-based PP of Japan Polychem).

The crystalline morphology control is a commercial example of the higher order molecular structure control of PP. The size of spherulites in PP dominates scattering of light to lower the transparency, but the growth of spherulite can be controlled by addition of nucleation agent (Figure 5). This technology is driving the use of PP for the food-packaging sheet with the higher heat resistance compatible to the current A-PET and PS materials.

Asahi Denka

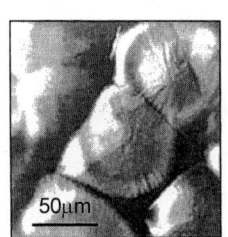 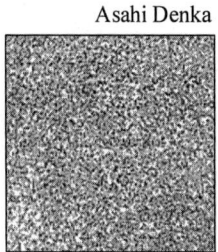

Fig. 5. TEM image of natural PP spherulite (left) and spherulite with phosphate salt nucleation agent (right).

Recent studies on metallocene-based PP have revealed several characteristic crystalline properties owing to its uniform molecular structure. Metallocene-based RCP, for example, shows high spherulite growth rate under the super-cooling condition, which results in the finely dispersed spherulite structure, and these are expected to provide the practical benefits.[3] On the other hand, PP block copolymer, or impact copolymer (ICP), which market is expanding around the automotive application, requires the precise molecular design and morphology control of rubber component. Metallocene catalyst technology is also expected to materialize the ideal molecular structure of ICP.

PP Composite and Other Technologies

Automotive materials are the most highly functionalized and differentiated application field of plastics. The weight reduction of auto parts using PP meets the objectives of recent automobile improvements, such as the better mileage, CO_2 reduction, and so on. Besides, the good balance of mechanical properties, practical performances, and cost of PP is expanding its application (Figure 6).[4]

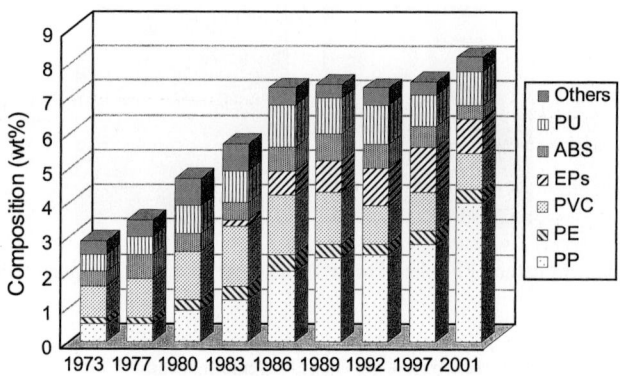

Fig. 6. Plastics materials and their composition used in Japan small cars (EPs=engineering plastics).

PP is widely used for various automotive parts, for example, large sized parts of bumpers and instrument panels, which require high safety and good appearance, other interior parts, and under-the-hood parts. The development of PP composite technology, which enabled the precise blend and distribution of rubber component and fillers, has expanded the viability of PP. PP composite meets the various parts design requirements owing to its wide range of blend composition and the versatility of compounding facilities with their size and functions. PP composite provides the mechanical properties compatible to engineering plastics and is replacing other materials taking advantage of its light weight and recycle suitability (Figure 7).

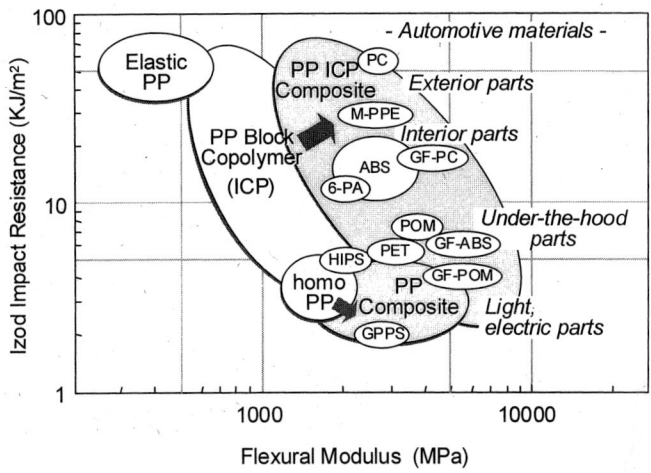

Fig. 7. Product range of PP composites for automotive applications.

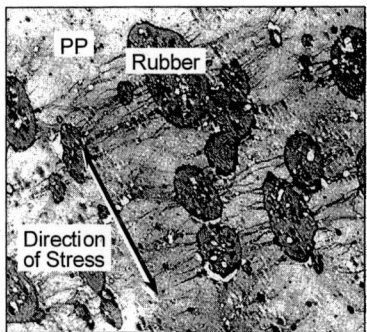

Fig. 8. TEM image of PP-rubber composite after impact absorption.

ICP is used directly or after compounding to form industrial parts. ICP is composed of homo PP and propylene-ethylene copolymer rubber components (Figure 8), which are commonly produced successively using cascade polymerization reactors. ICP should contain 30 to 40 percent of rubber to satisfy the high impact resistance required for the automotive exterior parts such as bumper. Z-N catalyst, however, has a limitation of the reactor-made rubber content due to the increased extractable fraction, which causes the operation difficulties and the inferior product performances. This makes the blend of additional rubber after polymerization necessary for ICP. There have been many efforts to optimize the manufacturing process, for example, molecule design improvement selecting polymerization catalyst and process and increased rubber content to reduce the additional rubber. Metallocene catalyst technology is expected to be suitable for this purpose.

Recently, the research for PP nano-composite (PPNC) is highlighted.[5] PPNC is composed of PP and finely dispersed filler such as clay, which commonly has the width or thickness of 1 nm scale and the length of more than 100 nm. PPNC is estimated to give the same stiffness of usual PP composite with c.a. 1/10 amount of filler and expected to contribute the weight reduction of auto parts. This high efficiency is assumed to be the combination of two factors, that is, the mechanical effect, which can be predicted from the filler aspect ratio, and the PP crystalline structure effect, which is induced by the finely dispersed filler (Figure 9, 10). Though PPNC is not widely commercialized yet due to the difficulties in the production technology, PPNC is expected to provide not only the mechanical strength but also other functions such as gas barrier and fire retardant capabilities.

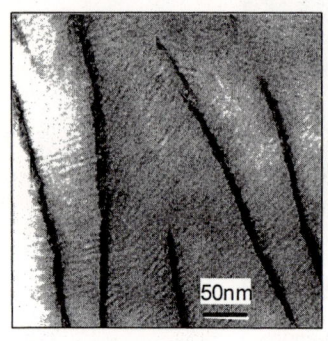

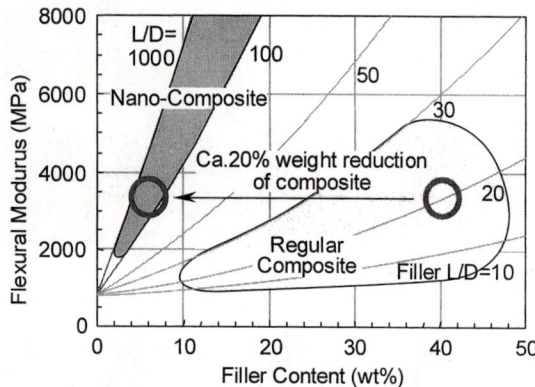

Fig. 9. TEM image of PP-clay nano-composite by melt extrusion method.

Fig. 10. Simulation results of filler aspect ratio effect on the filler content and stiffness correlation.

Foaming technology is rapidly spreading for the auto parts manufacturing in relation to the weight reduction, and PP foam has been used for impact absorption at the collision in the interior parts and bumpers. Recent molding technology is broadening the application of PP for the interior parts such as the door trims contributing to the further weight reduction. Foam material is required to be consistent with uniform bubble formation, which is enabled by the specific visco-elastic properties, or "strain hardening". Several methods have been developed and commercialized to control the visco-elastic properties of PP such as introduction of cross-linking or long chain branching. Besides the material technology, several new molding technologies have been developed, for example, the use of super critical fluid to control the melt flow and the combination of low pressure injection molding and expansion molding with the controlled mold opening ("Core back expansion").

Ikuyo, Japan Polychem

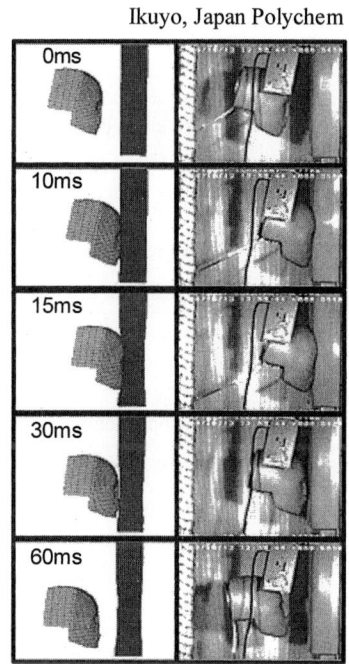

Fig. 11. Impact analysis for pillar design: results of simulation (left) and validation (dummy test: right).

As the polyolefin application spreads and the advanced material designs and new molding technologies are applied, the technologies of numeric simulation increase their significance.

Many kinds of software have been developed for computer-aided engineering (CAE), for example, flow analysis, mechanical analysis, impact analysis, and so on, and have become commonly used to optimize the parts design and processing conditions. Typical CAE applications are the calculation of polymer melt flow and pressure-temperature distribution in the mold during processing, the analysis of resin cooling process to predict the product qualities, and the estimation of mechanical properties of molded parts. It is important for the auto interior parts, such as pillar, to predict the potential damage to the human body in case of collision to secure the passengers' safety, and the simulation results are reflected in the parts design including the material properties (Figure 11).

Conclusions

Despite the long history beyond fifty years since commercialized, polyolefin industry is still expected to be spreading its applications. The ceaseless effort to create highly designed materials is important to keep up with the changing market demands using the various technologies as discussed above. The latest technologies, such as metallocene catalyst, nano-composite, and numeric simulation, are expected to support that effort.

Acknowledgements

The authors thank SPSJ for giving us this opportunity of presentation, and Japan Polychem Corporation for permitting it. We also thank Asahi Denka Co., Ltd. and Ikuyo Co., Ltd. for providing analyses data, and the researchers of Japan Polychem Corporation and Mitsubishi Chemical Corporation for their contributions.

[1] Research Report for Advanced Chemical Process Technology Development, JCII, **2000**
[2] T. Tayano, et al., Preprints, Metcon 2000 / Y. Nakamura, Preprints, FlexPO 2000 / T. Sugano, et at., Polymer Preprints, Japan, *51*, 59, **2002**
[3] G. Kanai, Preprints, Metcon 2002
[4] Annual Report, Japan Automobile Industry Association, **2002**
[5] A. Usuki, et al., Journal of Applied Polymer Science, *63*, 137, **1997** / K. Nakayama, et al, JP2002-167460, **2002**

New Methodology for Synthesizing Polyolefinic Graft Block Copolymers and their Morphological Features

Norio Kashiwa, Shin-ichi Kojoh,* Nobuo Kawahara, Shingo Matsuo, Hideyuki Kaneko, Tomoaki Matsugi

R&D Center, Mitsui Chemicals, Inc., 580-32 Nagaura, Sodegaura, 299-0265 Chiba, Japan E-mail: Shinichi.Kojoh@mitsui-chem.co.jp

Summary: This paper describes a new synthetic route for polyolefinic graft block copolymers by adopting coupling reaction between terminally hydroxylated polyolefins and maleic anhydride grafted polyolefins. Terminally hydroxylated polypropylene (PP-OH) was coupled with maleic anhydride modified polyethylene (PE-g-MAH) and such ethylene-propylene random copolymer (EPR-g-MAH) to give polyolefinic graft block copolymers (PE-g-PP and EPR-g-PP, respectively). The formation of PE-g-PP was confirmed by enhancement on molecular weight and it brought about distinctive decrease in size of dispersed domain in its phase separation morphology. Occurrence of coupling reaction to give EPR-g-PP was indicated by extreme decrease in its solubility to *n*-decane and it led to unique morphology demonstrating lamella microstructure that had never been reported for a comparable polyolefin composite.

Keywords: block copolymers, branched, compatibility, polymer coupling, polyolefins

Introduction

Since the birth of Ziegler-Natta catalysts, polyolefin industry has grown rapidly to the huge worldwide industry producing more than 80,000,000 tons of polymers per year owing to such innovations as discoveries and developments of $MgCl_2$-supported $TiCl_4$ and metallocene catalyst systems. Nowadays, a considerable amount of polyolefins is used for the molded products after blending two or more kinds of polyolefins, because polyolefin blends bring about unique and improved properties. Generally speaking, block copolymers play important roles in compatibilizing the blended polymers. Therefore, polyolefinic block copolymers are expected to compatibilize different kinds of polyolefins that are immiscible combinations in nature. Such effort will be able to create new class of plastic materials from polyolefins that are regarded as commodities so far. The polyolefinic block copolymers are classified into two types that are

© 2003 WILEY-VCH Verlag GmbH & KGaA, Weinheim   DOI: 10.1002/masy.200351135

linear and graft types. Until now, two and one methodologies have been reported for synthesizing the former and the latter, respectively.

To synthesize the former, living polymerization of olefins has been intensively studied and many useful catalysts have been discovered.[1-7] These catalysts enabled us to observe the well-defined morphologies of linear block copolymers with transmission electron microscopy (TEM).[8, 9] Alternative method for producing polyolefinic linear block copolymers is use of bis(2-arylindenyl) metallocenes to obtain stereoblock PP[10] and its morphological change during and after tensile extension was observed.[11] To synthesize the latter, copolymerization of macromonomers with olefins has been investigated.[12, 13]

Needless to say, the variety of the segments in the block copolymers is limited in the field of applicability of each method. Then, new methodology for synthesizing polyolefinic block copolymers is desired to diversify the combinations of the segments in polyolefinic block copolymers. Therefore, we investigated a new route to polyolefinic graft block copolymers with polymer coupling reaction as an application of our expertise in synthesizing terminally hydroxylated PP (PP-OH).[14] Suitable partners for PP-OH in coupling reaction would be maleic anhydride modified polyolefins that are conventional resins and expected to show high reactivity to hydroxy group of PP-OH.[15] In this paper, we introduce coupling reaction of PP-OH with maleic anhydride modified polyethylene (PE-g-MAH) and maleic anhydride modified ethylene-propylene random copolymer (EPR-g-MAH).

Experimental

Materials. Pyrolysis PP (py-PP) obtained in the conventional way[16] was used for hydroxylation to synthesize PP-OH. Its weight average molecular weight (Mw) was 8,000 and its molecular weight distribution (Mw/Mn) was 2.4. It was isotactic PP copolymerized with 2 mol% of ethylene. PE-g-MAH and EPR-g-MAH that were obtained in the conventional way[17] were used for coupling reaction with PP-OH. The values of Mw of PE-g-MAH and EPR-g-MAH were 45,000 and 130,000, respectively. The Mw/Mn values of PE-g-MAH and EPR-g-MAH were 2.4 and 2.0, respectively. PE was homopolyethylene and EPR consisted of 81 mol% of ethylene and 19 mol% of propylene. The contents of MAH in PE-g-MAH and EPR-g-MAH were 1.8 wt% and 1.0 wt%, respectively.

Hydroxylation. Into a nitrogen-purged 1 L glass reactor equipped with a mechanical stirrer, 26.6 g of py-PP was added with 34.6 mmol of *i*-Bu$_2$AlH and 800 mL of *n*-decane. It was heated to 100 °C and that temperature was maintained for 7 h with stirring. Then, dried air was fed into it at a rate of 200 L/h at that temperature for 6 h. The resulting solution was poured into a mixture of 2 L of methanol, 2 L of acetone and small amount of HCl, followed by strring with a magnetic stirrer chip for 2 h. Thus-obtained polymer (PP-OH) was recovered by filtration, washed with 1 L of methanol, and dried at 80 °C for 5 h.

Coupling Reaction. Into a nitrogen-purged 400 mL glass reactor equipped with a mechanical stirrer, 1.25 or 1.0 g of PP-OH was added with 150 mL of *n*-decane, catalyst amount of *p*-toluenesulfonic acid and 1.75 g of PE-g-MAH or 2.8 g of EPR-g-MAH, respectively. It was heated to 80 or 140 °C, respectively, and the temperature was maintained for 8 or 7 h with stirring. Then, it was poured into a mixture of 1.5 L of methanol and 1.5 L of acetone, followed by stirring with a magnetic stirrer chip for 5 min. The recovered polymer by filtration was stirred in 2 L of acetone with a magnetic stirrer chip for 2 h. Thus-obtained polymer was recovered by filtration, washed with 0.5 L of acetone, and vacuum-dried at 80 °C for 10 h.

Polymer Blend. For comparison with polyolefinic graft block copolymers, same procedures as described in Coupling Reaction except for using py-PP instead of PP-OH were carried out to prepare polymer blends.

GPC. Molecular weights of a polyolefinic graft block copolymer and a polymer blend were measured by a Millipore Waters 150C gel permeation chromatograph (GPC) equipped with a refractive index detector, using polyethylene calibration.

TEM. Morphologies of polyolefinic graft block copolymers and polymer blends were observed with TEM as following. Ultra-thin (ca. 100 nm) section of the polymer that had been pressed to give a sheet and dyed with RuO$_4$ was prepared with a Reica Ultracut microtome equipped with a diamond knife at −100 °C. The specimen was examined with a HITACHI H-810 transmission electron microscopy operated at 100 KV at 10,000 and 150,000 magnifications.

^{13}C NMR. The analysis with ^{13}C NMR was performed in the same manner as our previous paper.[14]

C10 Sol. Solubilities of a polyolefinic graft block copolymer and a polymer blend to *n*-decane at 23 °C (C10 Sol) were measured as following. Into a 1 L flask, 1 g of the polymer sample was added with 10 mg of 2,6-di-*t*-butyl-4-methylphenol and 500 mL of *n*-decane. The mixture was heated to 150 °C in order to dissolve the polymer sample. The obtained solution was cooled to 23 °C during 8 h and kept at that temperature for 8 h. The resulting slurry was filtered and the liquid phase portion was vacuum-dried until it reached constant weight. The percentage of thus-obtained constant weight in the weight of the initial polymer sample was C10 Sol.

Results and Discussion

Hydroxylation of py-PP

Chain-end structures of py-PP were investigated with ^{13}C NMR and the major group was vinylidene group as shown in Table 1, which is accordance with the literature on pyrolysis of PP.[18] It was used for preparing PP-OH through hydroalumination with *i*-Bu$_2$AlH, oxidation with dried air and methanolysis. The chain-end structures of the resulting polymer were analyzed with ^{13}C NMR and summarized in Table 1 in comparison with those of py-PP. The formation of 45 mol% of hydroxyl chain-end group from the vinylidene group accounting for 81 mol% in py-PP means comparable conversion with that observed in the hydroxylation of high molecular weight PP possessing alkylaluminum at its chain end.[14] Consequently, the obtained polymer possessed hydroxyl chain end in the content of 45 mol% of both ends of the polymer chain, namely, 0.9 hydroxyl group per chain on the average, although it would be a mixture of di-hydroxylated, mono-hydroxylated and non-hydroxylated polymers.

Table 1. The proportions of chain-end groups of pyrolysis and hydroxylated PP.

Sample	Chain-end group[a] / mol%				
	Vd	*n*Pr	*i*Pr	*i*Pr-OH	Others
py-PP	81	17	2	n.d.[b]	n.d.
PP-OH	6	15	34	45	n.d.

[a] Vd: vinylidene; *n*Pr: *n*-propyl; *i*Pr: *i*-propyl; *i*Pr-OH: hydroxy *i*-propyl.
[b] Not detected.

Coupling Reaction between PP-OH and PE-g-MAH

Thus-obtained PP-OH was reacted with PE-g-MAH at 80 °C for 8 h in *n*-decane with a molar ratio of 4 to 1 to synthesize PE-g-PP. This kind of coupling reaction between polyolefins has never been reported so far to the best of our knowledge, although coupling reaction between a maleic anhydride modified polyolefin and a polar polymer such as polyamide has been known well.[19] For its comparison, py-PP was blended with PE-g-MAH under the same conditions as the coupling reaction expect for the replacement of PP-OH by py-PP. Then, the both were compared with GPC analysis. As shown in Figure 1, a peak in low molecular weight region of (a) was obviously lower than that of (b) and, in its place, (a) gave the higher peak in high molecular weight region. They would clearly show that PP-OH forming the peak in low molecular weight region was bonded to PE-g-MAH by the reaction between –OH and –MAH.

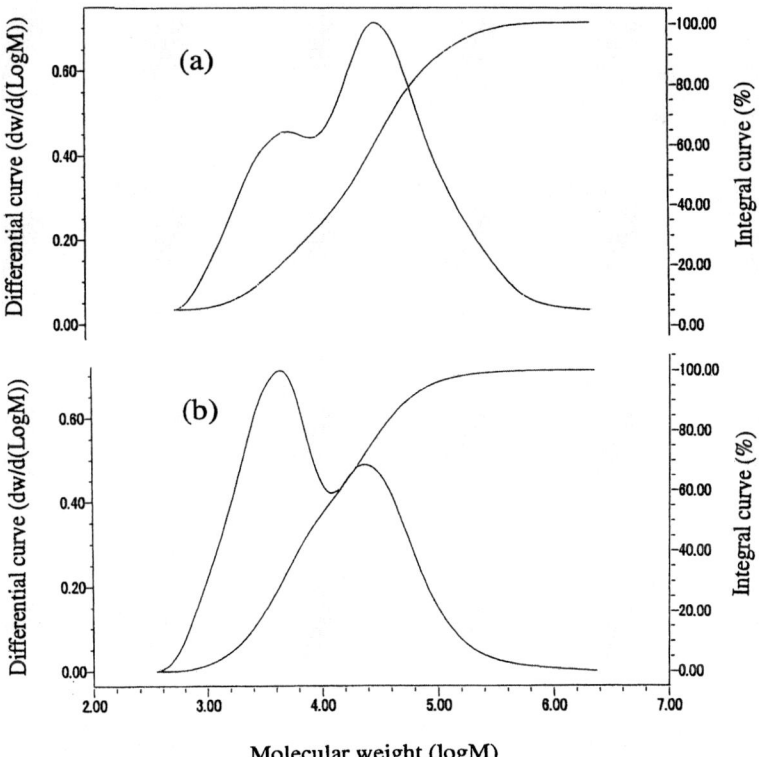

Fig. 1. GPC diagrams of (a) PE-g-PP produced by coupling reaction between PP-OH and PE-g-MAH and (b) polymer blend consisting of py-PP and PE-g-MAH. The molar ratio of PP segment to PE segment is 4 to 1 in any of the both.

Next, the both were compared in observation with TEM. The both showed phase separation morphology where the matrix was PE segment and the dispersed phase was PP segment as in Figure 2. Evidently, the size of the dispersed phase in PE-g-PP was much smaller than that in the polymer blend. Although the peak in low molecular weight region of PE-g-PP in Figure 1 means presence of PP-OH unreacted with PE-g-MAH, there were no coarse dispersed domains as shown in Figure 2 (a). It indicates that the formed PE-g-PP acted as a compatibilizer between unreacted PP-OH and PE segment.

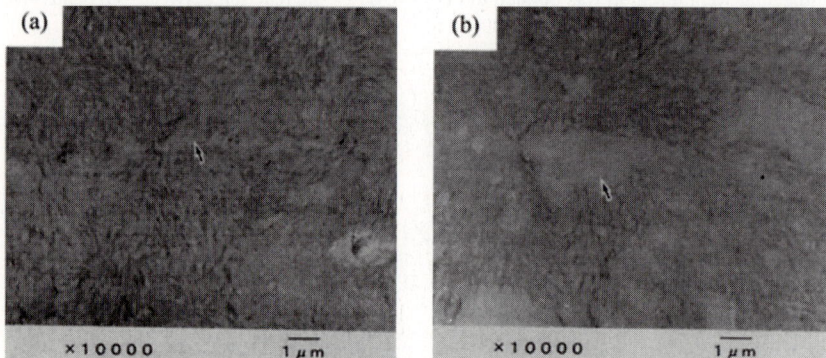

Fig. 2. TEM micrographs at 10,000 magnification from (a) PE-g-PP produced by coupling reaction between PP-OH and PE-g-MAH and (b) polymer blend consisting of py-PP and PE-g-MAH. The molar ratio of PP segment to PE segment is 4 to 1 in any of the both.

Coupling Reaction between PP-OH and EPR-g-MAH

Alternatively, studies on combinations of crystalline polyolefins and amorphous polyolefins are of importance to create new class of plastic materials. As an example, PP-OH was reacted with EPR-g-MAH at 140 °C for 7 h in n-decane with a molar ratio of 6 to 1 to synthesize EPR-g-PP. For its comparison, py-PP was blended with EPR-g-MAH under the same conditions as the coupling reaction expect for the replacement of PP-OH by py-PP. In the coupling reaction, its non-viscous initial solution changed to jelly-like product through highly viscous solution, although that in blending the polymers kept the state of non-viscous solution. It strongly suggests the proceeding of the aimed coupling reaction, although GPC analysis was not available for such jelly-like product due to its insolubility to solvents.

Each product was poured into a mixture of methanol and acetone, then the polymer recovered by filtration was washed with acetone followed by vacuum dry to be compared in C10 Sol. The

C10 Sol. of EPR-g-PP was 8.8 wt%, while that of the polymer blend was 75.3 wt% corresponding nearly to the weight proportion of EPR-g-MAH in it. It would indicate that PP segment bonded to EPR prevented the EPR segment from dissolving in *n*-decane.

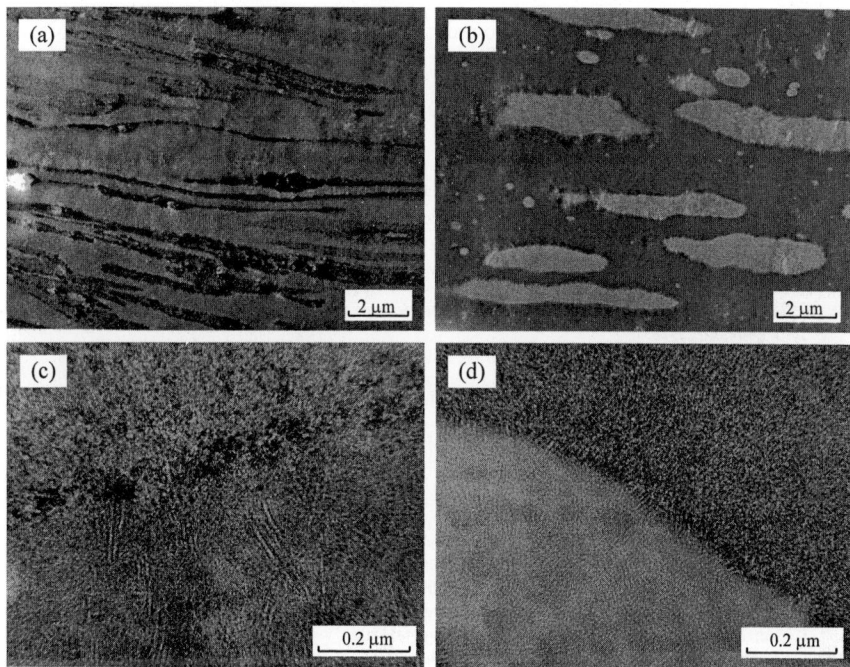

Fig. 3. TEM micrographs at 10,000 magnification from (a) EPR-g-PP produced by coupling reaction between PP-OH and EPR-g-MAH and (b) polymer blend consisting of py-PP and EPR-g-MAH and at 150,000 magnification of (c) the EPR-g-PP and (d) the polymer blend. The molar ratio of PP segment to EPR segment is 6 to 1 in any of the both samples.

Figure 3 shows morphologies observed with TEM for press sheets from the respective polymers. Phase separation morphology was observed in the polymer blend, where the matrix was EPR segment and the dispersed phase was PP segment (Figure 3 (b)). It was common morphology for polyolefins as done in Figure 2 and the dispersed phase was found to be considerably large and non-uniform. On the contrary, EPR-g-PP demonstrated unique lamella microstructure as shown in Figure 3 (a). Furthermore, its phase boundary was not distinct at high magnification as seen in Figure 3 (c), although the phase boundary was clear in the polymer blend even at high magnification (Figure 3 (d)).

Namely, the lamellar domains looked black or white at low magnification were turned out to include the other component on a nano-order scale by the observation at high magnification. Eventually it was discovered that PP could be compatibilized with EPR completely to give the novel polymer phase morphology by this coupling reaction.

[1]   Y. Doi, S. Ueki, T. Keii, *Macromolecules* **1979**, *12*, 814.
[2]   V. M. Mohring, G. Fink, *Angew. Chem. Int. Ed. Engl.* **1985**, *24*, 1001.
[3]   C. M. Killian, D. J. Tempel, L. K. Johnson, M. Brookhart, *J. Am. Chem. Soc.* **1996**, *118*, 11664.
[4]   J. D. Scollard, D. H. McConville, *J. Am. Chem. Soc.* **1996**, *118*, 10008.
[5]   R. Baumann, W. M. Davis, R. R. Schrock, *J. Am. Chem. Soc.* **1997**, *119*, 3830.
[6]   J. Saito, M. Mitani, Y. Yoshida, S. Matsui, J. Mohri, S. Ishii, S. Kojoh, N. Kashiwa, T. Fujita, *Angew. Chem., Int. Ed.* **2001**, *40*, 2918.
[7]   T. Matsugi, S. Matsui, S. Kojoh, Y. Takagi, Y. Inoue, T. Fujita, N. Kashiwa, *Chem. Lett.* **2001**, 566.
[8]   S. Kojoh, T. Matsugi, J. Saito, M. Mitani, T. Fujita, N. Kashiwa, *Chem. Lett.* **2001**, 822.
[9]   T. Matsugi, S. Matsui, S. Kojoh, Y. Takagi, Y. Inoue, T. Nakano, T. Fujita, N. Kashiwa, *Macromolecules* **2002**, *35*, 4880.
[10] G. W. Coates, R. M. Waymouth, *Science* **1995**, *267*, 217.
[11] R. L. Kravchenko, B. B. Sauer, R. S. McLean, M. Y. Keating, P. M. Cotts, Y. H. Kim, *Macromolecules* **2000**, *33*, 11.
[12] T. Shiono, S. M. Azad, T. Ikeda, *Macromolecules* **1999**, *32*, 5723.
[13] E. J. Markel, W. Weng, A. J. Peacok, A. H. Dekmezian, *Macromolecules* **2000**, *33*, 8541.
[14] S. Kojoh, T. Tsutsui, M. Kioka, N. Kashiwa, *Polym. J.* **1999**, *31*, 332.
[15] A. V. Machado, J. A. Covas, M. Duin, *Polymer* **2001**, *42*, 3649.
[16] M. Tanaka, M. Nakagawa, Japanese Laid Open Patent 2-6513.
[17] T. Yamanaka, E. Miura, Japanese Laid Open Patent 4-57808.
[18] T. Sawaguchi, T. Ikemura, M. Seno, *Macromolecules* **1995**, *28*, 7973.
[19] B. Lu, T. C. Chung, *Macromolecules* **1999**, *32*, 2525.